Springer Theses

Recognizing Outstanding Ph.D. Research

Aims and Scope

The series "Springer Theses" brings together a selection of the very best Ph.D. theses from around the world and across the physical sciences. Nominated and endorsed by two recognized specialists, each published volume has been selected for its scientific excellence and the high impact of its contents for the pertinent field of research. For greater accessibility to non-specialists, the published versions include an extended introduction, as well as a foreword by the student's supervisor explaining the special relevance of the work for the field. As a whole, the series will provide a valuable resource both for newcomers to the research fields described, and for other scientists seeking detailed background information on special questions. Finally, it provides an accredited documentation of the valuable contributions made by today's younger generation of scientists.

Theses are accepted into the series by invited nomination only and must fulfill all of the following criteria

- They must be written in good English.
- The topic should fall within the confines of Chemistry, Physics, Earth Sciences, Engineering and related interdisciplinary fields such as Materials, Nanoscience, Chemical Engineering, Complex Systems and Biophysics.
- The work reported in the thesis must represent a significant scientific advance.
- If the thesis includes previously published material, permission to reproduce this must be gained from the respective copyright holder.
- They must have been examined and passed during the 12 months prior to nomination.
- Each thesis should include a foreword by the supervisor outlining the significance of its content.
- The theses should have a clearly defined structure including an introduction accessible to scientists not expert in that particular field.

More information about this series at http://www.springer.com/series/8790

Marco Baity Jesi

Spin Glasses

Criticality and Energy Landscapes

Doctoral Thesis accepted by
the Complutense University of Madrid, Spain
and Sapienza University of Rome, Italy

Springer

Author
Dr. Marco Baity Jesi
Institut de Physique Théorique, DRF
CEA Saclay
Gif-sur-Yvette
France

Supervisors
Prof. Victor Martin Mayor
Departmento de Física Teórica I
Universidad Complutense de Madrid
Madrid
Spain

Prof. Giorgio Parisi
Dipartimento di Fisica
Sapienza Università di Roma
Rome
Italy

ISSN 2190-5053 ISSN 2190-5061 (electronic)
Springer Theses
ISBN 978-3-319-41230-6 ISBN 978-3-319-41231-3 (eBook)
DOI 10.1007/978-3-319-41231-3

Library of Congress Control Number: 2016943805

Printed on acid-free paper

This Springer imprint is published by Springer Nature
The registered company is Springer International Publishing AG Switzerland

Publications Related to This Thesis

- M. Baity-Jesi, V. Martín-Mayor, G. Parisi and S. Pérez-Gaviro, "Soft Modes, Localization and Two-Level Systems in Spin Glasses", Phys. Rev. Lett. **115**, 267205 (2015) [BJ15a].
- L. Yan, M. Baity-Jesi, M. Müller and M. Wyart, "Dynamics and Correlations among Soft Excitations in Marginally Stable Glasses", Phys. Rev. Lett. **114**, 247208 (2015) [Yan15].
- M. Baity-Jesi and G. Parisi, "Inherent structures in m-component spin glasses", Phys. Rev. B **91**, 134203 (2015) [BJ15b].
- M. Baity-Jesi *et al.*, "The three dimensional Ising spin glass in an external magnetic field: the role of the silent majority", J. Stat. Mech. (2014) P05014 [BJ14a].
- M. Baity-Jesi *et al.*, "Dynamical Transition in the $D = 3$ Edwards-Anderson spin glass in an external magnetic field", Phys. Rev. E **89**, 032140 (2014) [BJ14b].
- M. Baity-Jesi, L.A. Fernández, V. Martín-Mayor and J.M. Sanz, "Phase transition in three-dimensional Heisenberg spin glasses with strong random anisotropies through a multi-GPU parallelization", Phys. Rev. B **89**, 014202 (2014) [BJ14d].
- M. Baity-Jesi *et al.*, "Janus II: a new generation application-driven computer for spin-system simulations", Computer Physics Communications **185** (2014) 550–559 [BJ14c].
- M. Baity-Jesi *et al.*, "Critical parameters of the three-dimensional Ising spin glass", Phys. Rev. B **88**, 224416 (2013) [BJ13].
- M. Baity-Jesi *et al.*, "Reconfigurable computing for Monte Carlo simulations: Results and prospects of the Janus project", The European Physical Journal - Special Topics **210**, 33–51 (2012) [BJ12].

Supervisors' Foreword

This thesis combines theoretical physics with rigorous computational applications to achieve state-of-the art analyses of spin glasses. After a detailed historical introduction of the origin of spin glass theory, the author formulates and solves a series of open problems about the nature of these systems. Among the most noteworthy findings of the research exercise are the following: development of a new finite-size scaling ansatz for the study of the phase transition in a field, which accelerates convergence towards the thermodynamic limit; identification of the phase diagram for three-dimensional spin glasses with a random anisotropic exchange (this solution, obtained through a massive Monte Carlo campaign, likely represents the first open question on spin glasses to be solved through the use of GPUs); characterization of the effect of the number of spin components on the energy landscape of a spin glass; observation of correlations between soft spins that arise spontaneously during avalanches; and discovery that spin glasses are a non-trivial model exhibiting two-level systems.

These and other scientific results are presented in a clear and understandable format, in order for the reader to appreciate the latest research advancements in the field.

Madrid, Spain
Rome, Italy
December 2015

Prof. Victor Martin Mayor
Prof. Giorgio Parisi

Preface

The aim of the research discussed in this thesis is to make a small progress in the secular question on the nature of the glass transition. We focused, mainly with a numerical approach, on a paradigmatic glassy system, the spin glass, and we dealt with them by seeing their behavior at equilibrium as well as studying the features of their rugged energy landscape.

The equilibrium properties we were interested in concerned the universality in the glass transition and the fragility of the spin glass phase under an external magnetic field. On the side of the energy landscape, it is accepted that the energy landscape plays a major role in the slowing down of the glasses' dynamics. We tried to get a better insight by studying zero-temperature dynamics, by studying how the energy landscape becomes trivial when tuning certain parameters, and by analyzing the lowest modes of the density of states.

Scope and Organization of This Dissertation

The text is organized in four parts. In the following paragraphs we introduce briefly each of them.

Part I of this thesis is completely introductory on the systems we studied in this thesis, spin glasses. Section 1.1 aims to put the reader into context by introducing spin glasses in the frame of the glass transitions in general, by posing a historical basis about the birth of spin glasses, mentioning and explaining the development of some major theories. We get more technical in Sect. 1.2, where we detail the observables that will be analyzed throughout the rest of the text. In Sect. 1.3 we recall the reader some main concepts on scaling and renormalization group that will be useful to understand the analyses we performed.

Part II is dedicated to the study of critical properties of spin glasses through equilibrium simulations. We study the presence and the features of critical lines in the presence of perturbations on paradigmatic Hamiltonians.

In Chap. 2, which comes from Baity-Jesi et al. (J. Stat. Mech. **2014**, P05014 (2014)) and some unpublished results, we investigate, through Monte Carlo simulations with the dedicated computer Janus, whether the SG phase survives the imposition of a small external magnetic field, and thus whether there is a phase transition under the field. The two main theories on the SG phase have different predictions, so understanding whether there is or not a phase transition would be a strong factor for discrimination between the two. We find very large fluctuations in the observables we measure, and the average turns out to be a bad descriptor for our populations of measurements. Thus, we develop statistical methods and a new finite-size scaling ansatz that let us detect very different behaviors. Some of the measurements present strong signs of criticality, while others do not. It is not possible to determine which of the two behaviors will dominate in the thermodynamic limit, but we are able to set a temperature range where the would-be phase transition should be searched.

The material in Chap. 3 comes from Baity-Jesi et al. (Phys. Rev. **89**,014202 (2014)). To produce it I had the opportunity to work on large GPU clusters in Spain and in China. We did equilibrium Monte Carlo simulations on the Heisenberg spin glass with random exchange anisotropies. According to the Kawamura scenario, the chiral and the spin glass channels couple when anisotropies are introduced. We find a phase transition for each of the order parameters, and through a careful finite-size scaling analysis we conclude that the phase transition is unique. Moreover, the universal quantities we measure are compatible with the Ising universality class, instead of Heisenberg, indicating that the anisotropy is a relevant perturbation in the renormalization group sense.

Part III is on spin glasses in the absence of thermal vibration. The energy landscape appears to play a fundamental role in the sluggish dynamics that characterize a glass. It is a feature with a diverging number of dimensions, and still, it is most commonly described through a single number. This simplification is not always suitable and it is necessary to resort to different descriptors.

Chapter 4 that comes from Baity-Jesi, Parisi (Phys. Rev. B **91**, 134203 (April 2015)) is a study of the energy landscape of spin glasses as a function of the number of spin components m. When m is small the energy landscape is rugged and complex, with a large amount of local minima. An increase of m involves the gradual disappearance of most of those minima, along with a growth of the correlations and a slow down of the dynamics.

In Chap. 5, which is the result of my stay at the Center for Soft Matter Research of the New York University, we show how athermal dynamics in spin glasses is related to crackling noise, exposing the studies from Yan et al. and Baity-Jesi et al. (Phys. Rev. Lett. **114**, 247208 (2015), Range of the interactions in selforganized criticality (2015)) and unpublished material. We focus on the hysteresis of the SK model that describes spins in a fully connected graph. The dynamics along the hysteresis loop is in form of abrupt spin avalanches. We show that these avalanches cannot occur if the interactions are short-range, and that long-range interactions are a relevant perturbation to the short-range Hamiltonian. During the avalanches, furthermore, correlations between soft spins arise spontaneously, leading naturally the system to marginally stable states.

Chapter 6, based on the work of Baity-Jesi (Soft modes, localization and two-level systems in spin glasses (2015)), examines soft plastic modes of Heisenberg spin glasses in a RF, which we impose on the system in order to get rid of the soft modes due to the rotational symmetry. At low frequencies, the density of states has a non-Debye behavior, revealing the presence of a *boson peak*, a typical feature of structural glasses. These soft modes are localized, and they connect very near states, separated by very low energy barriers, which we identify as classical *two-level systems*. This helps to find a connection between the two main theories on the boson peak. On the one hand, the replica theory gives a mean field description that attributes the soft modes to a fractal energy landscape, and on the other there is the phenomenological picture of the two-level systems that attributes the excess of soft modes to a quantum tunneling between near states.

In Part IV we give our conclusions, resuming the main results chapter by chapter.

We also include several appendices. Appendix A is on Monte Carlo algorithms and on parallel computing for spin glass simulations. Appendix B is on the measurement of connected propagators in a field. Appendix C gives details on the creation of the *quantiles* defined in Chap. 2. In Appendix D we derive some identities that were crucial to make sure that our programs gave correct output. Appendix E is about error managing. Appendix F explains the energy minimization algorithms that were used in Chaps. 4 and 6.

High-Performance Computing in This Thesis

In this thesis, we present the results of several research projects on spin glasses, principally obtained through numerical simulations. Since this is a thesis in physics, we will mainly talk about the physical results, relegating to the background the numerical details.

Nevertheless, it is important to mention that extremely powerful numerical resources were necessary to arrive at some conclusions. Especially the work of Baity-Jesi et al. (J. Stat. Mech. **2014**, P05014 (2014), Phys. Rev. **89**, 014202 (2014)) would have been unthinkable with normal computing resources.

For the work of Baity-Jesi et al. (J. Stat. Mech. **2014**, P05014 (2014)), I enjoyed the chance of being part of the JANUS Collaboration, a partnership of physicists and engineers that work with the (FPGA)-based machine JANUS (Belletti et al., Computing in Science and Engineering **8**, 41 (2006), Yllanes Rugged Free-Energy Landscapes in Disordered Spin Systems (2011), Baños et al., Proc. Natl. Acad. Sci. USA **109**, 6452 (2012), and the recently launched *Janus II*, Baity-Jesi et al.,

Comp. Phys. Comm **185**, 550–559 (2014)),[1] devised expressly for Monte Carlo simulations of spin glasses. The JANUS computer has been able to thermalize much larger lattices than conventional computers, at lower temperatures, and it can reach times comparable with those of the experiments done by others (Belletti et al., Phys. Rev. Lett. **101**, 157201 (2008), Alvarez et al., J. Stat. Mech. **2010**, P06026 (2010), Alvarez Baños et al., Phys. Rev. Lett. **105**, 177202 (2010), Baños et al., Proc. Natl. Acad. Sci. USA **109**, 6452 (2012)).

In the case of the work of Baity-Jesi et al. (Phys. Rev. **89**, 014202 (2014)), I was part of SCC-Computing as a member of BIFI,[2] a FP7 project that aimed at developing connections between European and Chinese scientists by giving European groups the possibility to run simulations on the supercomputer *Tianhe-1A*, which had been the most powerful machine in the world, and at the time was ranked number two in *Top 500*.[3] Only thanks to these extraordinary resources, added to a careful tuning of our simulations in order to get the maximum performance, it has been possible to obtain the results shown in this dissertation.

In addition to the aforementioned facilities, I had the chance to use the small cluster of my group in Madrid, the *Minotauro* GPU cluster in the Barcelona Supercomputing Center, the *Memento* and *Terminus* CPU clusters and some GPU for benchmarking from BIFI, and the *Mercer* cluster of the New York University.

[1] http://www.janus-computer.com/.

[2] Strategic collaboration with China on super-computing based on Tianhe-1A, supported by the EU's Seventh Framework Programme (FP7) Programme under grant agreement n°287746. http://www.scc-computing.eu.

[3] Top 500 is the annual ranking of the 500 most powerful computers in the world, in terms of flops. http://www.top500.org.

Acknowledgments

This dissertation is the result of a Ph.D. thesis in co-tutorship between the two universities *Universidad Complutense de Madrid*, Madrid, Spain and *Sapienza, Università di Roma*, Rome, Italy. My supervisors were Víctor Martín Mayor (Spanish side) and Giorgio Parisi (Italian side). I am very grateful for the time spent with both, and for the unquantifiable amount of things I learned from them during these years. Also I must acknowledge Luis Antonio Fernández Pérez for he has been at my side helping me with programming and sharing with me his complicated codes, and David Yllanes and Beatriz Seoane, senior Ph.D. students in the group, were very nice receiving me in the group, and supported me when I needed one. I also wish to mention José Manuel Sanz González, who guided my steps through a large part of my first article. During my thesis I also had the opportunity to stay in the group of Matthieu Wyart, whom I desire to thank for giving me the privilege of working with him, as well as Le Yan, my extremely valid colleague during and after my months there. I also thank Andrea Liu for having me in her research group during the last few months of my Ph.D.

I also acknowledge the funding from the FPU program of the Ministerio de Educación, Spain, which funded me with a four-year fellowship, and with extra allowances for my research stays at NYU[4] and UPenn.[5] Further research costs such as materials and trips have been funded by MINECO, Spain, through the research contract No. FIS2012-35719-C02, by the European Research Council under the European Union's Seventh Framework Programme (FP7/2007-2013, ERC grant agreement no. 247328), by the Seventh Framework Programme (EU-FP7) through the research contract No. 287746, and by the GDRE 224 CNRS-INdAM GREFI-MEFI. I am also grateful to the BIFI[6] for letting me use their CPU and GPU resources, and to its very professional staff.

[4]New York University, New York, NY, USA. Stay from May 2 to July 31, 2014.

[5]University of Pennsylvania, Philadelphia, PA, USA. Stay from September 15 to December 14, 2015 (to be concluded).

[6]The Institute for Biocomputation and Physics of Complex Systems of the University of Zaragoza.

In aggiunta ai ringraziamenti anteriori, accademici e formali, non voglio rinunciare a sottolineare il mio amore verso mia madre e la mia famiglia, il *clan Jesi*, al cui supporto incondizionato fuori dal comune devo gran parte degli obiettivi che ho raggiunto finora. Ci regaliamo ogni giorno il privilegio della nostra unità.

Content

Acronyms

1-RSB	One-step replica symmetry breaking
A	Random (aleatory) dynamics
CG	Chiral glass
CPU	Central processing unit
Cu	Copper
CV	Conditioning variate
dAT	De Almeida-Thouless
DM	Dzyaloshinskii–Moriya
dof	Degrees of freedom
DOS	Density of states
EA	Edwards–Anderson
EMCS	Elementary Monte Carlo step
FP	Fixed point
FP7	Seventh Framework Programme
FPGA	Field programmable gate array
FSS	Finite-size scaling
G	Greedy dynamics
GPU	Graphics processing unit
HB	Heat bath
HPC	High-performance computing
IEA	Ising–Edwards–Anderson
i.i.d.	Independent identically distributed
IS	Inherent structure
JK	Jackknife
l.h.s.	Left-hand side
MC	Monte Carlo
MF	Mean field
Mn	Manganese
MPI	Message passing interface
MSC	Multi-spin coding

OR	Overrelaxation
pdf	Probability distribution function
PRNG	Pseudo-random number generator
PT	Parallel tempering
R	Reluctant dynamics
REM	Random energy model
RF	Random magnetic field
RG	Renormalization group
r.h.s.	Right-hand side
RKKY	Ruderman–Kittel–Kasuya–Yosida
RS	Replica symmetric
RSB	Replica symmetry breaking
RW	Random walk
SG	Spin glass
SK	Sherrington–Kirkpatrick
SOC	Self-organized criticality
SOR	Successive overrelaxation
TAP	Thouless–Anderson–Palmer

Notations

$\dots$	Subleading corrections
$[\dots]$	Estimator of the thermal average
$\langle\dots\rangle$	In Chap. 5: average over the avalanches
$\langle\dots\rangle$	In the other chapters: thermal average
$\langle\dots\rangle_t$	Time average
$\overline{(\dots)}$	Average over the disorder
$\lVert\dots\rVert_1$	1-norm
$\lVert\dots\rVert_2$	2-norm
$\langle s \mid s'\rangle$	Scalar product between configurations $\lvert s\rangle$ and $\lvert s'\rangle$
a_1, a_2	Nontrivial components of the pion
a	Parameter for the width of the bins of the $P(\hat{q})$
a, b, c, d	Replica indices
A	Area of the hysteresis loop
$A(\langle\vec{\pi}_0\rvert)$	Amplitude parameter for forcings along $\lvert\vec{\pi}_0\rangle$ [see also Eq. (6.15)]
$A(\lvert\vec{\pi}_{\mathrm{RAND}}\rangle)$	Amplitude parameter for forcings along $\lvert\vec{\pi}_{\mathrm{RAND}}\rangle$ [see also Eq. (6.15)]
A_{F}	Amplitude of the forcing
average, annealed	$\log(\overline{\mathcal{Z}_J})$
average, quenched	$\overline{\log(\mathcal{Z}_J)}$
α	In Sect. 1.3: critical exponent of the specific heat
α	In Chap. 6: $\rho(\lambda)\sim\lambda^{\alpha}$ for small λ
b	In Sect. 1.3: size of the patches in the block Hamiltonian
b	In Appendix B.3.3: bit
$b_j^{(\mathrm{JK})}$	jth jackknife block
$\mathcal{B}$	Local basis
$\mathcal{B}'$	Full set of measurements
β	In Sect. 1.2: inverse temperature
β	In Sect. 1.3: critical exponent of the order parameter
β	In Chap. 5: scaling exponent for ΔM

c_1, c_2	Terms of the variance
$C(\lambda)$	Correlation between the softest spin and the spins with local stability λ
$C_f(\lambda), C'_f(\lambda)$	Correlations between the spin s_0 and the spins with local stability λ before and after the flipping event
C_{h}	Specific heat
$C^{(\mathrm{plane})}_{\mathrm{link}}(r)$	Link plane correlation function
$C^{(\mathrm{point})}_{\mathrm{link}}(r)$	Link point correlation function
$C(\boldsymbol{r})$	Two-point correlation function (with translational invariance)
$C(r)$	Point-to-plane correlation function
$\hat{C}(\boldsymbol{k})$	Fourier transform of $C(r)$
$C^{(\mathrm{plane})}(r)$	Plane tensorial correlation function
$C^{(\mathrm{point})}(r)$	Point tensorial correlation function
$C(\boldsymbol{x}, \boldsymbol{y})$	Two-point correlation function
$\mathcal{C}(\boldsymbol{r})$	Green's correlation function
configuration	Position of all the spins in the system: $\lvert s\rangle$
χ	Susceptibility
χ_{CG}	Chiral glass susceptibility
χ_{SG}	Spin glass susceptibility
$\chi(\boldsymbol{k})$	Wave-vector dependent susceptibility
χ_L	Susceptibility of a lattice of size L
$\chi_{\mathrm{R},i}(\boldsymbol{k})$	Per-quantile $\chi_{\mathrm{R}}(\boldsymbol{k})$
$\chi_{\mathrm{R}}(\boldsymbol{k})$	Wave-vector dependent replicon susceptibility
d	Dimensions of space
$\hat{d}$	Scaling function
D	Amplitude of the anisotropy
$\mathrm{D}^{\mathrm{dip}}_{x,y}$	Dipolar anisotropy matrix
$\mathrm{D}^{\mathrm{DM}}_{x,y}$	Dzyaloshinskii–Moriya anisotropy matrix
D_{Ising}	Ising anisotropy
$D(\lambda, t)$	Diffusion constant
D_x	Per-site random single-ion anisotropy
D_{XY}	XY anisotropy
$\mathcal{D}(n)$	Avalanche size distribution
$\mathcal{D}^{\alpha\beta}_{xy}$	Diagonal part of $\mathcal{M}^{\alpha\beta}_{xy}$
δ	In Sect. 1.3: field critical exponent
δ	In Chap. 5: correlation exponent (number of spins)
δ	In Chap. 6: $g(\omega) \sim \omega^{\delta}$ for small ω
$\delta^{\alpha\beta}$	Kronecker delta
$\delta(\ldots)$	Dirac delta function
ΔE	In Appendix A: energy difference in the PT swap
ΔE	In Chap. 6: energy difference in terms of $\mathcal{H}_{\mathrm{RF}}$
$\Delta E(\mathcal{F})$	Energy cost of flipping all the spins belonging to $\mathcal{F}$

ΔE_{av}	Energy change in a single avalanche
ΔE_{flip}	Energy change in a single flip
ΔE^{*}	Reverse energy barrier
ΔE_{xy}	Energy cost of flipping s_x and s_y
ΔM	Magnetization jump
ΔM_{c}	Magnetization jump cutoff
$\Delta M^{(\text{tot})}$	Total magnetization change
ΔW	Variation of W after the forcing
$\hat{e}_{1,x}, \hat{e}_{2,x}$	Unitary vectors forming the basis $\mathcal{B}$
e_{IS}	Energy of the IS
E	Mean number of spins triggered by a flip (see also $E(n_{\text{unst}})$)
$\tilde{E}_j(\mathcal{O})$	Estimator of $\mathcal{O}$ for the jth block
$\tilde{E}_j^{(\text{JK})}(\mathcal{O})$	Estimator of $\mathcal{O}$ for the jth jackknife block
$E(n_{\text{unst}})$	Indicator of the random walk bias
$\tilde{E}(\mathcal{O})$	Estimator of the average
$E(\mathcal{O})$	Thermal and disorder average of observable $\mathcal{O}$
$\hat{E}(\mathcal{O})$	$E(\mathcal{O} \mid \hat{q})$
$\hat{E}(\ldots)$	Estimator of the average
$E^{(\text{tot})}$	Total energy change in a full hysteresis loop
ϵ	In Chap. 2: width of the interval
ϵ	In Chap. 6: amplitude of the perturbation
η	Anomalous dimension
η_{eff}	Effective anomalous dimension
η_x	Gaussian noise
f	Extension factor of the simulations
$f_{\mathcal{O}}, g_{\mathcal{O}}, h_{\mathcal{O}}$	Scaling functions for an observable $\mathcal{O}$
$f_\lambda(x)$	Distribution of kicks x given to spins of stability λ
$f_{\max}$	Maximum extension factor of the simulations
$f_n^{(\text{JK})}$	$f(\hat{E}(\mathcal{O}))$ (the last extra jackknife block stores the average)
$f(t, h)$	Intensive-free energy
$F(\lambda)$	Cumulative of $\rho(\lambda)$
$\mathcal{F}$	In Chap. 5: a set of soft spins
$\mathcal{F}$	In the other chapters: free energy
$\mathcal{F}_{\text{coarse}}$	Coarse-grained free energy
$\mathcal{F}_J$	Sample-dependent free energy
$g(\omega)$	Density of states
$G_2^{\text{c}}(r)$	2-replica connected plane correlation function
$G_2^{\text{nc}}(r)$	2-replica non-connected plane correlation function
G_{A}	Anomalous propagator
G_{L}	Longitudinal propagator
$\hat{G}_{\text{R}}(k)$	Fourier transform of $G_{\text{R}}(r)$
$G_{\text{R}}(r)$	Replicon propagator
$\mathcal{G}$	Green's function

GA	G inherent structure, A avalanche
GG	G inherent structure, G avalanche
GR	G inherent structure, R avalanche
γ	In Sect. 1.3: critical exponent of the order parameter
γ	In Chap. 5: correlation exponent (stability)
γ	In Chap. 6: $F(\lambda) \sim \lambda^{\gamma}$ for small λ
$\Gamma_1(x,y), \Gamma_2(x,y)$	Connected correlators
h	External magnetic field
h_{MIN}	Smallest local field
$h(\hat{q})$	Histogram of $\hat{q}$
$\tilde{h}$	Rescaled field
$\vec{h}_x$	In Chap. 6: random field of amplitude H_{amp} on site x
h_x	Local field in the SK model
H_{amp}	Amplitude of the random field
$\mathcal{H}$	Hamiltonian
$\mathcal{H}_{\mathrm{ANI}}$	Heisenberg spin glass with a random anisotropic exchange
$\mathcal{H}_{x,y}^{\mathrm{DM}}$	Dzyaloshinskii–Moriya anisotropy term
H_{EA}	Edwards–Anderson Hamiltonian
$\mathcal{H}_h$	Edwards–Anderson Hamiltonian in a field
$\mathcal{H}_{\mathrm{F}}$	Forced Hamiltonian
$\mathcal{H}_{\mathrm{IS}}$	Hamiltonian $\mathcal{H}_{\mathrm{RF}}$ calculated at the IS
$\mathcal{H}_{\mathrm{Ising}}$	Hamiltonian with Ising anisotropy
$\mathcal{H}_{\mathrm{r}_1}$	Hamiltonian with random single-ion anisotropy on an established axis
$\mathcal{H}_{\mathrm{r}_2}$	Hamiltonian with random single-ion anisotropy on a random axis
$\mathcal{H}_{\mathrm{RF}}$	Hamiltonian of the Heisenberg SG with a random field
$\mathcal{H}_{\mathrm{si}}$	Hamiltonian of a ferromagnet with single-ion anisotropy
$\mathcal{H}_{SL}$	Hamiltonian that mixes short- and long-range interactions
$\mathcal{H}_{\mathrm{SR}}$	Finite-connectivity Hamiltonian
$\mathcal{H}_{\mathrm{XY}}$	Hamiltonian with XY anisotropy
Heisenberg spin	Spin with $m = 3$ components
i_h	Index of the forcing
inherent structure	The local energy minimum reached by relaxing the system
IS	The subscripted quantity is referred to an inherent structure
$\lvert \mathrm{IS}(i_h)\rangle$	IS with the perturbed Hamiltonian
$\lvert \mathrm{IS}^*\rangle$	IS after the forcing
Ising spin	Spin with $m = 1$ components
J_{typ}	Typical coupling
J_{xy}	Coupling between sites x and y
$J_{xy}^{(\ell)}$	Amplitude of the long-range coupling
$J_{xy}^{(s)}$	Amplitude of the short-range coupling
k	Wave vector

k_{11}	Wave vector along a diagonal of the lattice
k_B	Boltzmann constant ($k_B = 1$)
k_{min}	Lowest wave number
k_n	Wave vector along an axis
K	Kick in the local stability of its neighbors
K	Kurtosis
$K_n(\mathcal{O})$	nth moment of $\mathcal{O}$
$\hat{\mathrm{K}}(q)$	$\hat{E}([q - \hat{E}(q)])^4$
L	Linear lattice size
$\lambda(m)$	The mth smallest stability
λ_{min}	Lowest eigenvalue
λ_{MIN}	Smallest local stability
λ'	Local stability after the flip
λ^*	Crossover λ
λ_x	Local stability in the SK model
Λ	Parameter of the SOR
m	In Sect. 1.3: magnetization
m	In Chap. 5: the number of chosen softest spins
m	In the other chapters: number of spin components
$\hat{m}$	Polarized magnetization along the pion
$(m/L)_1$	Ratio over which there is only one minimum of the energy
$(m/L)_{SG}$	Ratio under which the number of (ISs) is exponential
M	Magnetization
$\mathcal{M}_{xy}$	Matrix element of the Hessian matrix x and y stand for the position
$\mathcal{M}_{xy}^{\alpha\beta}$	Element of the Hessian matrix x, y indicate the position, $\alpha\beta$ indicate the component
master	The node that manages the instructions
μ	Generic principal axis
n	In Chap. 5: size of the avalanche
n	In Chap. 6: number of visited valleys
n_{av}	Number of avalanches
n_b	Number of bits
n_c	Avalanche size cutoff
n_λ	Number of calculated eigenvalues
$n(T)$	Number of coarse-graining steps
n_{unst}^*	$n_{unst} : E(n_{unst}) = 1, r(n_{unst}) = 0$
$n_{unst}(t)$	Number of unstable spins after t spin flips
$\hat{n}_x$	Random vector on the sphere of radius 1
N	Total number of spins
$\mathbb{N}$	Set of the natural numbers
N_A	Avogadro number
N_{EMCS}^{min}	Minimum number of EMCS
N_F	Number of forcings

N_{GPU}	Number of GPUs
N_{m}	Number of offline measurements
N_{MC}	Number of EMCS
N_{replicas}	Number of replicas
N_{samples}	Number of samples
N_{sweeps}	Number of full sweeps of the lattice
N_{t}	Number of quadruplets of configurations
N_T	Number of temperatures
N_τ^{min}	Minimum number of EMCS in units of τ
N_{threads}	Number of threads
$\mathcal{N}(X)$	Number density of having fluctuation $\mathcal{N}$
$\mathcal{N}$	Number of equivalent wave vectors
$\mathcal{N}_{\mathrm{m}}$	Total number of measurements
$\mathcal{N}_{xy}^{\alpha\beta}$	Non-diagonal part of $\mathcal{M}_{xy}^{\alpha\beta}$
ν	Critical exponent of the correlation length
ν_{CG}	Exponent of the chiral glass sector
ν_{SG}	Exponent of the spin glass sector
$\mathcal{O}$	Generic observable
$\mathcal{O}_i$	ith measurement of observable $\mathcal{O}$
ω	Largest irrelevant exponent
$\hat{p}$	Scaling function
P_{1d}	Return probability of a 1d RW
P_{2d}	Return probability of a 2d RW
$P_L^H(A_F, n)$	Probability of changing valley
$P_J(q)$	Sample-dependent distribution of the overlap
$P(\ell)$	X_1 averaged over the ℓth plane
$P(q)$	Probability distribution function of the overlap
$P(\hat{q})$	Probability distribution function of the CV
$P(q_{\mathrm{IS}})$	Overlap distribution of the inherent structures
$P^{(\mathrm{sym})}(q)$	Symmetrized probability distribution function
$\mathcal{P}(\Delta M)$	Magnetization jump distribution
$\hat{\Phi}_k^{(\mathrm{ab;cd})}$	Fourier transform of $\Phi_x^{(\mathrm{ab;cd})}$
$\Phi_x^{(\mathrm{ab;cd})}$	Connected 4-replica field
$\lvert\widetilde{\pi}_0\rangle$	Softest mode
$\lvert\widetilde{\pi}_{\mathrm{RAND}}\rangle$	Random mode
$\vec{\pi}_x$	Pion
$\tilde{\pi}_x$	Pion as a function of two components
ψ_n	Eigenvector
q	Overlap
q^{ab}	Overlap matrix
q_{av}	Average overlap
$q_{\mathrm{link,av}}$	Average link overlap
q_{b}	Overlap between initial and perturbed IS

q_{EA}	Edwards–Anderson parameter
$\hat{q}$	Conditioning variate
$\hat{q}_{CG}(k)$	Fourier transforms of the chiral glass overlap
$\hat{q}_i$	Value of the conditioning variate (CV) for measurement i
$\hat{q}_{SG}(k)$	Fourier transforms of the spin glass overlap
$\tilde{q}_i$	ith quantile
q_{if}	Overlap between configurations before and after the forcing
$\hat{q}_i$	Value of the CV at the ith measurement
q_{link}	Link overlap
q_{max}	Maximum overlap
$q_{link,max}$	Maximum link overlap
q_{med}	Median overlap
$q_{link,med}$	Median link overlap
q_{min}	Minimum overlap
$q_{link,min}$	Minimum link overlap
q_{CG}	Chiral glass overlap
q_{SG}	Spin glass overlap
q_x	Local overlap
$q_{x,t}$	Instant measurement
Q	Tensorial overlap
Q_{link}	Tensorial link overlap
quenched disorder	The disorder that appears in the Hamiltonian
$\boldsymbol{r}$	Euclidean distance between two points
r	Modulus of the distance
$\hat{r}$	Unit vector
$r(n_{unst})$	Indicator of the random walk bias
$\vec{r}_x$	Position of $\vec{s}_x$
$R_{12,i}$	Per-quantile R_{12}
RA	R inherent structure, A avalanche
replicas	Independent copies of the same sample
RG	R inherent structure, G avalanche
RR	R inherent structure, R avalanche
R_{12}	$\chi(k_1)/\chi(k_{11})$
ρ	Power law exponent for $\mathcal{D}(n)$
$\rho(\lambda)$	Distribution of local stabilities
ρ_{ss}	Steady-state local stability distribution
$\|\vec{s}\rangle$	Configuration
$\|\vec{s}^{(IS)}\rangle$	Inherent structure
$\vec{s}^{(IS)}$	Spin of the inherent structure
$\vec{s}_x^{(IS)*}$	Spin of the IS after the forcing
$\vec{s}_x$	Vector spin on site x
s_x	Ising spin on site x
$s_{x,z}$	z component of the spin in site x
$S_2(q)$	Two-replica skewness

$S_3(q)$	Three-replica skewness
$\hat{S}(\mathcal{O})$	$S(\mathcal{O}\vert\hat{q})$
σ_x	Block variable
sample	A single realization of the disorder
slave	Node that receives the instructions
σ	Scaling exponent for n
σ_f	Jackknife error
σ_x	Amplitude of the noise
$\sum_{x,y}$	Sum over all the x and y
$\sum_{x\neq y}$	Sum over all the x and y, except $x = y$
$\sum_{\vert x-y\vert=1}$	Sum over all the nearest neighbors x and y
$\sum_{y:\vert x-y\vert=1}$	Sum over all the y neighbors of x
t	In Sect. 1.3 and Chap. 2: the reduced temperature
t	In Chap. 4: time in the quench
t	In Chap. 5: number of elapsed spin flips in the avalanche
$\tilde{t}$	Rescaled reduced temperature
t_{w}	Waiting time
T	Temperature
T_0	Final temperature of a quench
T_{CG}	Chiral glass critical temperature
$T^{L,2L}$	Temperature where $\xi_L(T)/L$ crosses $\xi_{2L}(T)/(2L)$
$T_{\min}, T_{\max}$	Lowest and highest simulated temperatures
T_{SG}	Spin glass critical temperature
$T_{\mathrm{CG}}^{L,2L}$	Crossing temperature of χ_{CG} with sizes L and $2L$
$T_{\mathrm{SG}}^{L,2L}$	Crossing temperature of χ_{SG} with sizes L and $2L$
$T^{\mathrm{up}}(h)$	Upper bound for the critical temperature
τ	In Chap. 2: exponential autocorrelation time
τ	In Chap. 5: power law exponent for $\mathcal{P}(\Delta M)$
τ	In Appendix A: integrated autocorrelation time
T_{c}	Critical temperature
θ	Pseudogap exponent
u_x^{a}	Word
$v(\lambda, t)$	Drift term
$v_{\mathrm{ss}}(\lambda)$	Steady-state drift term
$\mathcal{V}(x)$	Set of sites that are connected to x
valley	Local minimum of the energy
$\mathrm{var}(\mathcal{O})$	Variance of $\mathcal{O}$
$\mathrm{var}(\mathcal{O}\mid\hat{q})$	Conditional variance
w_x	Local spin variation
W	Global spin variation
$\mathcal{W}$	Cumulant of the spin variation
Word	n_b-bit integer
x	Site

X	Fluctuation of $\Delta E(\mathcal{F})$
$\hat{X}_1(k)$	Fourier transform of $P(\ell)$
$X_1(x), X_2(x)$	Fields of differences
X_{MIN}	Most extreme fluctuation of $\Delta E(\mathcal{F})$
$\chi_c(\hat{q}_i)$	Characteristic function
XY spin	Spin with $m = 2$ components
ξ	Correlation length
ξ_L	Correlation length measured in a lattice of size L
$\xi_{L,i}$	Per-quantile ξ_L
$\xi_2^{(\mathrm{plane})}$	Second moment plane correlation length
$\xi_2^{(\mathrm{point})}$	Second moment point correlation length
y_h	Field scaling exponent
y_t	Temperature scaling exponent
Y_λ	Inverse participation ratio
$\mathcal{Z}$	Partition function
z	In Chap. 3: exponent that relates space and time
z	In the other chapters: connectivity
Z_2	Reflection symmetry group
$\mathcal{Z}_J$	Sample-dependent partition function

Part I
Introduction

Chapter 1
An Introduction to Spin Glasses: History, Simulations and Phase Transition

1.1 Background

After briefly introducing the glass phase in general terms, showing how it appears in many aspects of modern society, we make a historical presentation on birth and evolution of the SG theory. It is hard to propose oneself a historical approach on a research topic, since any quoted argument could need a whole treatise for itself, so we choose the starting point that looked mostly appropriate to us, and refer to an exhaustive bibliography the interested reader.[1] Moreover, since the SG theory has by now evolved over half a century under disparate aspects, and it has fused with many other domains of science, such as biology and computer science, it is unthinkable to use this introduction to mention all the aspects of this stimulating branch of physics. We will instead focus on the origin of SGs as they are known at present, and we will only touch on those aspects of SG theory that are useful to expose the results of this thesis.[2] Since its aim is to get into the topic and set the bases for further discussion, the introduction on SGs is left open, and recent developments are left to the introduction of each chapter.

1.1.1 The Glass Transition

If we cool a liquid quickly enough, it can happen that the sudden lack of thermal vibration arrest its dynamics before it is able to end in the lowest-entropy

[1] In particular, in [Mat81] there is an extended historical introduction on magnetism (but not on SGs). Historical comments on SGs appear in [Myd93]; a perspective is given in [She07].

[2] The references herein come from an intensive bibliographic research, and are in the author's opinion the most representative of a part of the history of SGs. It may occur to the reader that some notable publication or remark, that should appear in this thesis, has been not been cited. If this were the case, the author would thank such reader if he could inform him in order to add the missing work to further versions of this introduction.

M. Baity Jesi, *Spin Glasses*, Springer Theses, DOI 10.1007/978-3-319-41231-3_1

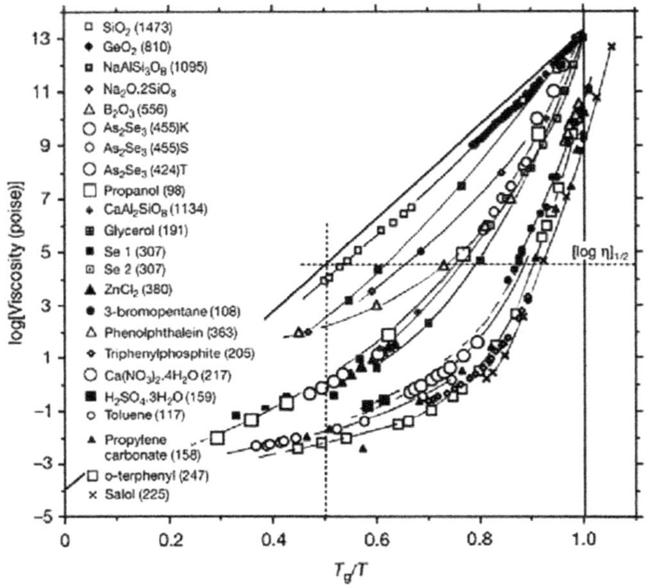

Fig. 1.1 Logarithm of the viscosity η against the inverse of the temperature T, normalized with a constant T_g. T_g is the temperature where the viscosity is 10^{13} poise (with the exception of some curves that do not meet at $T_g/T = 1$ because T_g was defined as the temperature where the enthalpy relaxation time is $\approx$200 s). It represents an experimental cutoff over which the relaxation times $t \sim \eta$ are too long to perform equilibrium experiments. On the other hand, $\eta = 10^{-4}$ poise is the roughly common high-temperature limit of the viscosity. Reprinted by permission from Macmillan Publishers Ltd: Nature 410, 663–667 (doi:10.1038/35070517), copyright (2001) [Mar01]

configurations and crystallize. Once this happens, a glass is formed, and the material behaves as a solid even though apparently no symmetry was broken and no phase transition took place. Simply, the viscosity and the relaxation times grow so fast in a very short range of temperatures, that the liquid stops flowing and appears solid. In Fig. 1.1, a famous plot by Angell shows this steep behavior in a set of glass formers. With a factor 2 change in temperature the viscosity grows 8–11 orders of magnitude. A so large growth of the relaxation times is hard to explain in the absence of a phase transition, and no completely satisfying theory has been found. So many scenarios have been proposed to explain this phenomenon, that is it often said that there are more glass theories than theorists.

Besides the natural interest in amorphous solid states, called structural glasses, the reason why much emphasis is put in the study of the glass transition is the huge amount of applications that glasses have, and the immense amount of disparate systems that exhibit a glassy state.

The most commonly known glasses are silica compounds. They are fused to a temperature where the viscosity is low and they are malleable, and the glassy phase is obtained by quickly taking them back to room temperature. For their properties of manufacturability, low dilatancy, uncorrosiveness and transparency they are present in many objects of our everyday life, such as windows, bottles, optical fibers, beakers and touchscreens.[3]

Still, the glassy phase presents itself in numerous different forms in technology and nature [Ang95]. Metallic glasses are used for high efficiency transformers for their magnetic properties, or as an alternative to silicon to make molds for nanocomponents [Gre95]. Automobile bodies and parts of boats are made of fiberglass, that is obtained by embedding extremely fine fibers of glass in an organic polymer plastic, trapping the air in order to make it a good thermal insulator [May93, Mar06]. Vitrification takes places in processes related to the stabilization of labile biochemicals for commercial use [Cro98], and in the preservation of insect life under extreme conditions of cold or dehydration [Cro98]. Protein folding exhibits glass-like behavior [Web13], many foods and their industrial production chain involve glassy states and dynamics [Bla93], and so do instances of optimization and combinatorial problems [Mez87].

Spin glasses are yet another instance of the glassy phase, characterized by an amorphous magnetic low-temperature state. Despite a very peculiar phenomenology [Nag79, Myd93, Vin97, Jon98, Her02], few or none industrial applications of SGs exist at the moment, and it would be reasonable to query why SGs are apparently overrepresented in theoretical physics.

The main reason is due to their simplicity. Very simple Hamiltonians defined on uncomplicated graphs capture highly non-trivial behaviors, making them probably the most understandable models that display a glassy phase. Their study is useful to get an insight on the study of the glass phase in a more general sense and on complexity, since

- experimental measurements are easier through the use of very sensitive magnetometers called SQUIDs (Superconducting QUantum Interference Devices). See e.g. [Dru07, Kum14].
- in the context of SGs it was possible to develop very advanced theoretical tools that can be reused in other contexts [Mez87, Bia12, Cha14].
- differently from structural glasses, the SG transition is well identified in finite dimensions [Bal00, Lee03].
- they are easier to simulate, because e.g. they are defined on graphs where the neighbors do not change with time, the degrees of freedom are binary or limited. It is possible to simulate far more degrees of freedom than on structural glass, making finite-sizes effects less overwhelming [Fer15].
- it is possible to construct dedicated hardware for more effective numerical studies [Bel06, Bel08a, BJ12, BJ14c].

Finally, SGs are often used as toy models to test the phenomenology of more complicated systems, and not seldom SG theory was of crucial importance for relevant advances in numerous fields. For example the Random First Order Transition theory for structural glasses is inspired on the p-spin SG model [Cav09] (Sect. 1.1.2);

[3]Devices such as tablets and smartphones require high-tech glasses. The recently-developed Gorilla Glass (http://www.corninggorillaglass.com/), for example, enjoys wide popularity.

neural networks are now a branch of SG theory, and for example the Hopfield model is known to display a SG phase and is studied with SG tools [Mez87]; protein folding codes can be successfully obtained with SG theory [Gol92], and many ideas from SGses were used to understand this phenomenon [Wol92]. In this dissertation we use SGs to understand marginal stability and two-level systems (Chaps. 6 and 5).

1.1.2 The Origins of Spin Glass Theory

During the beginning of the second half of the 20th century much attention has been devoted to the study of solutions of (Mn) in copper (Cu), that displayed peculiar properties that puzzled the condensed-matter community [Owe56, Nob59, Zim60]. A cusp in the susceptibility was observed at a temperature T_c roughly proportional to the concentration of Mn (with concentration of 0.1–10 % T_c ranged between 1 K and 100K), separating the paramagnetic phase from a peculiar phase in which no order was identified, though several features discriminated it from a paramagnetic phase. It lacked spontaneous magnetization, but after applying reasonably large fields one could observe remnant magnetization. Also, the susceptibility χ was practically constant instead of being inversely proportional to the temperature T, $\chi \propto 1/T$ as the Curie law suggests for a paramagnet, and the low-temperature specific heat was linear in T instead of being proportional to $1/T^2$.

This surprising low-temperature behavior was attributed to the $s - d$ interaction [Mar60], that couples electrons of unfilled inner shells and conduction electrons. Depending on the involved metal, this interaction can lead both to ferromagnetism and antiferromagnetism. In order to explain the atypical ordered phase the $s - d$ interaction was supposed to be the dominant one.

This interaction was first pointed out by Zener in 1951, with a phenomenological model that did not involve the possibility of antiferromagnetism [Zen51a, Zen51b, Zen51c]. Few years later the theory was further developed by Kasuya [Kas56], that found that the $s - d$ interaction can imply antiferromagnetism and spin waves, and Yosida [Yos57], that notices that the model from Ruderman and Kittel [Rud54], for the coupling between two magnetic moments through their hyperfine interaction with the conduction electrons, successfully describes the $s - d$ interaction.[4]

The resulting coupling $J_{xy}^{(\mathrm{RKKY})}$ between two Mn ions separated by $\boldsymbol{r}$ resulting from this description is called (RKKY). It has a sinusoidal form that to our purposes can be represented as a pairing

$$J_{xy}^{(\mathrm{RKKY})} \sim \cos\left(\frac{\boldsymbol{k}\cdot\boldsymbol{r}}{|\boldsymbol{r}|^3}\right), \tag{1.1}$$

[4]A Hamiltonian for the $s - d$ interaction is also derived in [Mit57]. More useful references on the subject: [Fro40, Blo55, VV62, Mat81].

between two spins $\vec{s}_x$ and $\vec{s}_y$ at distance $\boldsymbol{r}$ one from the other. The $\boldsymbol{k}$ is of the order of the Fermi vector, meaning that the oscillations of the cosine are very quick. So, expression (1.1) tells us that, besides decaying as $1/r^3$, depending on the distance between the ions the couplings can be ferromagnetic or antiferromagnetic.

The interactions of the Cu substrate were assumed negligible for the study of the magnetic properties of the examined CuMn alloys, and the cusp in the susceptibility was entirely attributed to the RKKY interaction between the Mn ions [Mar60]. Being the positions in the alloy of these ions random, both the module and the sign of the couplings had to be treated as a random variable, and random ferromagnets became popular [Bro59]. First modelizations involved systems of spins under independent effective random local fields [Mar60, Kle63], and later on disorder is assumed in the interactions [Mon70].

The birth of spin glass theory.

The term *spin glass* is first used in a paper by Anderson in 1970,[5] in analogy with structural glasses, to stress the presence of a low-temperature phase with unidentified order. He defines a formally simple model where the Hamiltonian has an explicit dependence on the disorder [And70]. He assumes that the dominant role is not assumed by the electrons, that have only the function of transmitting the interaction, but by the Mn ions and their exchange interactions. The interaction between the Mn spins is given by the RKKY interaction (1.1), whose sign depends on the distance $\boldsymbol{r}_{xy}$ between two spins $\vec{s}_x$ and $\vec{s}_y$ and that decreases in magnitude as $\boldsymbol{r}_{xy}$ increases. Since $\boldsymbol{r}_{xy}$ is random and depends on the single realization of the alloy and of its disorder, that we will call *sample*, also the coupling J_{xy} is a random variable. So, Anderson proposed the first SG Hamiltonian as a Heisenberg model

$$\mathcal{H} = \frac{1}{2} \sum_{x \neq y} J_{xy} \vec{s}_x \cdot \vec{s}_y, \qquad (1.2)$$

where the J_{xy} are random constants distributed through an unknown distribution that should reproduce roughly the RKKY interaction. The essential novelty is thus that the "experimental" couplings $J_{xy}^{(\mathrm{RKKY})}$ are replaced by the random variables J_{xy}. We call *quenched disorder* the randomness of the J_{xy}s, that appears directly in the Hamiltonian. Notice that being the couplings J_{xy} randomly negative and positive, it is impossible to satisfy simultaneously the energy along all the bonds (we will come back to this later on). This feature is called *frustration*. Hamiltonian (1.2) possesses both quenched disorder and frustration, that become the distinctive features of a SG model [You05, Kaw10]. Anderson tried a mean field approach without, yet, averaging over the disorder. He also assumed the possibility of purely nearest-neighbor interactions on a regular lattice, and treated the system as a set of independent clusters

[5] Under suggestion of B.R. Coles.

each with its critical temperature, bringing back the problem of localization that in his view had been disregarded. This cluster-based interpretation was well embraced by the scientific community. Experimental observations of the susceptibility cusp were done also in other types of alloy such as AuFe, with similar results. The dominant interpretation was an arisal of ferro- and antiferromagnetic clusters with short-range order that as the temperature is lowered interact at long range [Bec71, Can72, Smi74], or seeing the SG as a sort of macroscopic antiferromagnet [Adk74].[6]

The Edwards-Anderson model.

The milestone year for the definition of SGs as a branch of theoretical physics is 1975. A solid basis on SG theory was given in [Edw75, Edw76] by Edwards and Anderson through a very simple model that was able to describe qualitatively the experimental observations. Their starting idea is that in the low temperature *spin glass phase* there must be some local ordering of the spins along a random preferred direction. Even though this direction is unknown, one can see whether an alignment is taking place by examining if after a time t the single spins $s_x(t)$ have a tendency of pointing in the same direction. In quantitative terms, they define the *overlap*

$$q = \lim_{t\to\infty} \frac{1}{N} \sum_{x}^{N} \langle \vec{s}_x(0) \cdot \vec{s}_x(t) \rangle_t \,, \tag{1.3}$$

where $\langle \mathcal{O}(t) \rangle$ is the time average of a generic observable $\mathcal{O}$, $\langle \mathcal{O}(t) \rangle_t \equiv \frac{1}{t} \int_0^t dt' \mathcal{O}(t')$. Equation (1.3) is one of several ways to define the order parameter of a SG. Assuming that the equilibrium phase is ergodic, one can rewrite Eq. (1.3) by replacing the time average $\langle \ldots \rangle_t$ with an ensemble average $\langle \ldots \rangle$ to give an alternative expression for the overlap,

$$q = \frac{1}{N} \sum_{x}^{N} \langle \vec{s}_x \rangle^2 \,. \tag{1.4}$$

In the paramagnetic phase there is no favored direction, so $q = 0$. On the other side, in the SG phase each spin will align along a privileged direction and $q \neq 0$. In [Edw75] Hamiltonian (1.2) is taken into account and it is shown with a mean field approach that a phase transition occurs with q as order parameter, accompanied by a cusp in the susceptibility. Hamiltonian (1.2), with nearest neighbor interactions on a regular lattice, assumes the name of EA model. Assuming a unitary distance between nearest neighbors, the EA Hamiltonian is

$$\mathcal{H}_{\mathrm{EA}} = \frac{1}{2} \sum_{|x-y|=1} J_{xy} \vec{s}_x \cdot \vec{s}_y, \tag{1.5}$$

[6] This latter interpretation tried to explain the rounding of in the cusp of the susceptibility under an applied magnetic field. As we will discuss more thoroughly in Chap. 2, it is still an open issue whether a SG in a field undergoes a phase transition

where for[7] simplicity reasons the J_{xy} were assumed by Edwards and Anderson to come from a Gaussian pdf $P(J)$. Different samples of an EA spin glass will have a different realization of the coupling, but on average they must have the same behavior, and the larger the lattice more similar the behavior will be. This assumption, that gives sense to the free energy of the SG model, is called *self averageness*. So, calling $\mathcal{F}_J$ and $\mathcal{Z}_J$ the free energy and the partition function of a sample with a set J of couplings, one is interested in the average free energy

$$\mathcal{F} = \int \mathcal{F}_J P(J) dJ = -k_B T \int P(J) \log \mathcal{Z}_J dJ, \tag{1.6}$$

that by writing with an over bar $\overline{(\ldots)}$ the average of the disorder assumes the form $\mathcal{F} = -k_B T\, \overline{\log \mathcal{Z}_J}$. Equation (1.6) encloses a central difficulty in SG theory, that is taking the average of the logarithm of $\mathcal{Z}_J$. This is called a *quenched average*, in opposition with the easier approach, called *annealed average*, of taking the logarithm of the average of $\mathcal{Z}_J$, resulting in the annealed free energy $\mathcal{F}_{\mathrm{Ann}} = \log \overline{\mathcal{Z}_J}$, that results incorrect at low temperatures (see e.g. [Mez87]). To overcome the problem of this integration, Edwards and Anderson propose the *replica trick*, that consists in using the identity $\log(x) = \lim_{n\to 0} \frac{x^n - 1}{n}$ to transform the annoying logarithm in a power law,[8]

$$\mathcal{F} = -k_B T \overline{\log \mathcal{Z}_J} = -k_B T \lim_{n\to 0} \frac{\overline{\mathcal{Z}^n} - 1}{n}. \tag{1.7}$$

By artificially assuming that n is an integer, one could think about $\overline{\mathcal{Z}^n}$ as the partition function of n independent *replicas* of the same system, that share the same instance of the couplings but are independent one from the other. With the help of replicas the order parameter can be rewritten as [Par83]

$$q^{\mathrm{ab}} = \left\langle \vec{s}_x^{(\mathrm{a})} \cdot \vec{s}_x^{(\mathrm{b})} \right\rangle. \tag{1.8}$$

where (a) and (b) indicate different replicas. Treating $\overline{\mathcal{Z}^n}$ as a set of independent replicas simplifies the calculations, although it implies a few mathematical forcings such as taking the limit $n \to 0$ with $n \in \mathbb{N}$. Notwithstanding, although the EA model still nowadays lacks a full analytical understanding, the replica trick became a very popular tool for disordered systems.

---

[7]It is the case to make clarity on the notation for the summations. $\sum_{\boldsymbol{x},\boldsymbol{y}}$ is a sum over all the choices of $\boldsymbol{x}$ and $\boldsymbol{y}$. $\sum_{\boldsymbol{x}\neq\boldsymbol{y}}$ is a sum over all the choices of $\boldsymbol{x}$ and $\boldsymbol{y}$, except $\boldsymbol{x} = \boldsymbol{y}$ (in our models the positions $\boldsymbol{x}$ are discretized). $\sum_{|\boldsymbol{x}-\boldsymbol{y}|=1}$ is a sum over all the choices of $\boldsymbol{x}$ and $\boldsymbol{y}$ that are nearest neighbors. In all the previous cases each coupling is counted twice, so we put a factor 1/2 in front of the summation. $\sum_{\boldsymbol{y}:|\boldsymbol{x}-\boldsymbol{y}|=1}$ is a sum over all the choices of $\boldsymbol{y}$ that are neighbors of $\boldsymbol{x}$, so the summation runs over a number of terms equal to the connectivity z. Writing $\sum_{|\boldsymbol{x}-\boldsymbol{y}|=1}$ is equivalent to $\sum_{\boldsymbol{x}} \sum_{\boldsymbol{y}:|\boldsymbol{x}-\boldsymbol{y}|=1}$.

[8]The identity comes from a first order expansion of the exponential function: $x^n = e^{n \log(x)} = 1 + n \log(x) + o(n^2)$.

The EA model was promptly be extended to quantum spins [She75b, Fis75], but we will not treat quantum SGs in this thesis, so we will leave these models aside.

The Sherrington-Kirkpatrick model.

Also in 1975, with the aim of giving a model for which mean field theory be valid, Sherrington and Kirkpatrick propose to slightly modify Hamiltonian (1.5) by imposing fully-connected interactions and Ising spins $s_x = \pm 1$ [She75a]

$$\mathcal{H}_{\mathrm{SK}} = \frac{1}{2}\sum_{x,y} J_{xy}\vec{s}_x \cdot \vec{s}_y, \tag{1.9}$$

where the couplings J_{xy} are Gaussian distributed with $\overline{J_{xy}} = 0$, and their variance is such that the energy is extensive, $\overline{J^2} = 1/N$. This model, for which mean field theory is valid, will be called SK model. Their solution, yet, has unphysical features such a negative entropy at low temperatures. Sherrington and Kirkpatrick attributed this to an assumption they made, in their calculations, of commutativity between the limit $n \to 0$ and the thermodynamic limit $N \to \infty$ (N indicates the number of spins). Yet, it slowly became clear that the problem resided in the (yet reasonable) ansatz they made of *replica symmetry* [Alm78b, Bra78], that the overlap (1.8) is the same no matter what two replicas are chosen [the (RS) ansatz],

$$q^{\mathrm{ab}} = q(1 - \delta^{\mathrm{ab}}), \tag{1.10}$$

especially after it was shown that in the SK model the inversion of the limits is valid [Hem79].

It is worth to mention also another interesting model with disorder proposed in 1975, the Random Field Ising Model [Imr75],[9] that depicts an Ising ferromagnet in which each spin feels a random field that is not correlated with the rest of the sites. This is not a SG because there couplings are ferromagnetic, so there is no frustration. A way to define frustration quantitatively is through the Wilson loop. For each closed circuit in the lattice, we can take the ordered product of all the links that form it. If this product is negative it is not possible to find a configuration that minimizes simultaneously the local energy along each of the links, and the loop is said to be frustrated [Tou77, Bla78].[10,11]

The RS solution of the SK model given in [She75a] was shown to be stable only at high temperatures by de Almeida and Thouless [Alm78b] (this result was

---

[9] We will take inspiration from this model in Chap. 6 to work on a system with broken rotational symmetry.

[10] See the introduction of [Mez87] for a definition of frustration from every-day life examples, and [Par95] for an intuitive discussion on Wilson loops.

[11] In this text, when we will talk about the system being more or less frustrated we will be referring to the presence of a larger or smaller number of frustrated loops. When instead we say that two spins are mutually frustrated, we mean that the energy is not minimized along the bond(s) connecting the two spins.

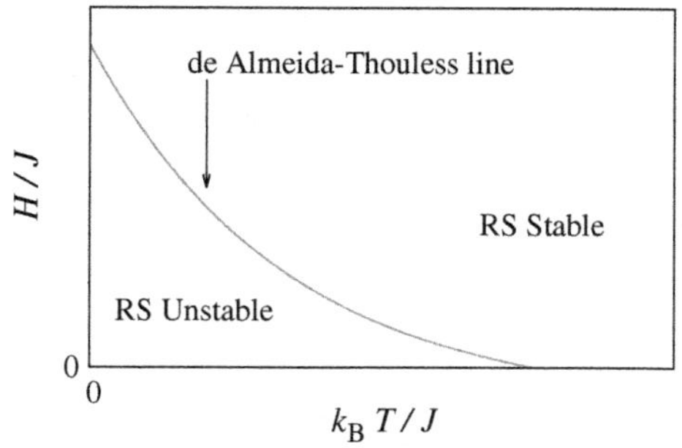

Fig. 1.2 Stability of the RS solution of the SK model in the paramagnetic phase. H is the applied magnetic field, T is the temperature, J is the typical coupling energy, and k_B is Boltzmann's constant (in this thesis, we will usually set $k_B = 1$). The RS solution is stable only at high temperatures or at high fields. The (dAT) line separates the zone of the phase diagram where the RS phase is stable from the one where magnetic ordering appears

promptly generalized to spins with any finite number m of components [Alm78a]). The paramagnetic phase is RS, but under a certain temperature massless modes in the overlap correlation functions (replicon modes) become unstable [Bra79]. Replica symmetry, thus, becomes unstable in favor of a yet undefined SG phase. Therefore all the results obtained under that temperature, including the critical temperature, are not very useful. Also in the presence of an externally applied magnetic field it was shown that for low temperatures and fields the RS phase is not stable, so at least in the SK modelization, there exists a SG phase in a field (Fig. 1.2). The critical line where the RS phase becomes unstable will be called the dAT line. Even though the reason of this instability was suspected to be replica symmetry [Bra78, Alm78b], it was not clear how to break the symmetry between replicas in order to obtain a physically reasonable solution.

Perhaps with the additional stimulation of these initial failures of the replica approach, different approaches have been tried, such as expansions in $6 - \epsilon$ dimensions of space [Har76, You76, Che77, Sou77] of alternative formulations of the mean field. In opposition with the replica method, that constructs a mean field theory after having averaged over the disorder with the replica trick, Thouless, Anderson and Palmer formed a mean field theory first, including in the free energy the rebound effect of each spin on itself (Onsager's reaction term [Ons36, Bar73]), and only after averaged over the disorder [Tho77]. Still, the (TAP) approach was shown to be useful only at high temperatures (see e.g. [Mez87]). Numerical simulations confirmed the validity of all the aforementioned analytical results only at high temperature [Kir78].

Apparently no theory was satisfactory describing the low-temperature phase of a SG, and no ansatz for RSB was fully satisfactory.

The Parisi solution.

In order to find the good solution of the SK model the replica symmetry needed to be broken, but q^{ab}, an $n \times n$ matrix (with $n \to 0!$) could be parametrized in infinite

$$
\begin{pmatrix}
0 & & & & & & & \\
& 0 & & & & & & \\
& & 0 & & & q_0 & & \\
& & & 0 & & & & \\
& & & & 0 & & & \\
& & & & & 0 & & \\
& q_0 & & & & & 0 & \\
& & & & & & & 0
\end{pmatrix}
\longrightarrow
\left(\begin{array}{cccc|cccc}
0 & & & q_1 & & & & \\
& 0 & & & & q_0 & & \\
& & 0 & & & & & \\
q_1 & & & 0 & & & & \\
\hline
& & & & 0 & & & q_1 \\
& q_0 & & & & 0 & & \\
& & & & & & 0 & \\
& & & & q_1 & & & 0
\end{array}\right)
\longrightarrow \tag{1.11}
$$

$$
\left(\begin{array}{cc|cc|cc|cc}
0 & q_2 & \multicolumn{2}{c|}{q_1} & & & & \\
q_2 & 0 & & & & q_0 & & \\
\cline{1-4}
\multicolumn{2}{c|}{q_1} & 0 & q_2 & & & & \\
& & q_2 & 0 & & & & \\
\hline
& & & & 0 & q_2 & \multicolumn{2}{c}{q_1} \\
& q_0 & & & q_2 & 0 & & \\
\cline{5-8}
& & & & \multicolumn{2}{c|}{q_1} & 0 & q_2 \\
& & & & & & q_2 & 0
\end{array}\right)
\longrightarrow \cdots \tag{1.12}
$$

Fig. 1.3 Sketch of the first two steps of replica symmetry breaking. The first $n \times n$ matrix represents the RS ansatz, where there is total symmetry with respect to replica exchange. The second matrix shows the first step of RSB, the matrix is divided in blocks, and the overlap q^{ab} can now assume two values. In the SK model the process needs to be iterated infinite times to obtain the exact solution. The iteration procedure is clear from the 2-step RSB: the inner blocks are subsequently divided in smaller blocks, up to having a continuum of solutions at the full RSB level. More details in the main text

ways, and the only *modus operandi* with new ansatz for a RSB overlap matrix was by trial and error [Bra78]. It appeared also that adding new order parameters to the model, that is giving q^{ab} the possibility to assume more than one value, shifted the negative zero-temperature entropy towards zero [Par79b]. Each new order parameter is equivalent to a new breaking of the replica symmetry, so an ansatz with 2 order parameters is called with 1RSB. It became quickly clear that the SG phase has intriguing unseen properties when finally the good ansatz was found by Parisi in 1979, with infinite steps of RSB, that we call full RSB [Par79a].

The Parisi ansatz for the matrix q^{ab} consisted in an iterative process starting from the RS ansatz $q^{ab} = q_0(1 - \delta^{ab})$ (Fig. 1.3) [Par80b, Par80a, Par80c]. The $n \times n$ matrix is then parted in n/m_1 blocks of size $m_1 \times m_1$. The off-diagonal blocks stay unchanged, but the off-diagonal terms of the diagonal blocks now assume the value q_1. This is the first step of RSB, and is called 1RSB. The second step of RSB is identical, and consists in iterating the symmetry breaking in each of the n/m_1 diagonal blocks. Each is subdivided in m_1/m_2 sub-blocks of size $m_2 \times m_2$. The off-diagonal sub-blocks stay the same, while the off-diagonal elements of the the diagonal sub-blocks assume the value q_2. The process can be iterated infinite times, up to the full RSB solution. An overlap matrix constructed this way has any two rows (or columns) identical up to permutations. This property is called replica equivalence, and both the RS and the RSB matrices benefit from this property.

In the RS phase $q^{\mathrm{ab}} = 0\ \forall \mathrm{a, b}$, so the pdf of the order parameter, $P(q)$, is a $\delta(0)$. The full RSB ansatz implies instead that in the SG phase the pdf of the order parameter is non-trivial. By simply counting the $n(n-1)$ non-diagonal values q^{ab} can assume, one has

$$\begin{aligned} P(q) &= \frac{1}{n(n-1)} \sum_{\mathrm{a} \neq \mathrm{b}} \delta\left(q - q^{\mathrm{ab}}\right) = \\ &= \frac{n}{n(n-1)} [(n - m_1)\delta(q - q_0) + (m_1 - m_2)\delta(q - q_1) + \\ &+ (m_2 - m_3)\delta(q - q_2) + \cdots]. \end{aligned} \tag{1.13}$$

Once the $n \to 0$ limit is taken,

$$P(q) = m_1\delta(q - q_0) + (m_2 - m_1)\delta(q - q_1) + (m_3 - m_2)\delta(q - q_2) + \cdots, \tag{1.14}$$

the $P(q)$ is positive definite only if $0 < m_1 < m_2 < \ldots < 1$. One can hypothesize, as also numerical simulations suggest, that the q_i constitute an increasing sequence, and since the sequence is infinite it is convenient to define a function $q(x)$ such that

$$q(x) = q_i \quad \text{if } m_i < x < m_{i+1}, \tag{1.15}$$

so after a k-step RSB $q(x)$ is a piecewise function that takes at most $k+1$ different values, and when k is sent to infinity it becomes a continuous function in the interval [0,1] [Par80b]. In this representation the free energy becomes a functional of $q(x)$, and has to be maximized with respect to it. It is also shown by Parisi that

$$q(x) = q_{\mathrm{m}} \quad \text{for } x \leq x_{\mathrm{m}}, \tag{1.16}$$

$$q(x) = q_{\mathrm{M}} \quad \text{for } x \geq x_{\mathrm{M}}. \tag{1.17}$$

This means that the pdf can be rewritten as the sum of two delta functions connected by a smooth function $\tilde{P}(q)$ which is non-zero only in the interval $x_{\mathrm{m}} < x < x_{\mathrm{M}}$

$$P(q) = x_{\mathrm{m}}\delta(q - q_{\mathrm{m}}) + \tilde{P}(q) + x_{\mathrm{M}}\delta(q - q_{\mathrm{M}}). \tag{1.18}$$

Practically, given two random states α and β (chosen from $P(q)$), with mutual overlap $q^{\alpha\beta}$, with probability x_{M} α and β will be the same state and they will have maximal overlap q_{M}, with probability x_{m} they will be as different as it is possible, with $q^{\alpha\beta} = q_{\mathrm{m}}$, and with probability $1 - x_{\mathrm{m}} - x_{\mathrm{M}}$ the situation will be something in between. The lower limits q_{m} and x_{m} depend on an external magnetic field as $h^{2/3}$. In the interval $x_{\mathrm{m}} < x < x_{\mathrm{M}}$ the function $q(x)$ depends weakly on the field, and so does x_{M}. When the critical field is approached from the SG phase the distance between the two peaks in the $P(q)$ decreases, $x_{\mathrm{m}} \to x_{\mathrm{M}}$ and $q_{\mathrm{m}} \to q_{\mathrm{M}}$, until the $P(q)$ becomes trivial (a $\delta(q - q_{\mathrm{EA}})$) at the de Almeida-Thouless (dAT) line. Figure 1.4 gives a better intuition on the $P(q)$.

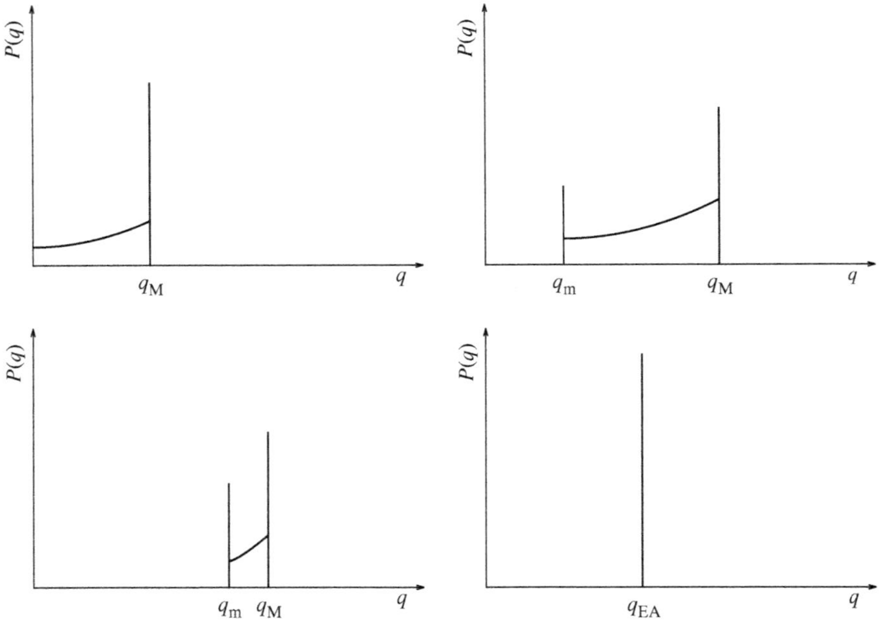

Fig. 1.4 Different instances of $P(q)$ in the SK model. *Top left* At zero field, close to T_c, q_M is proportional to $T - T_c$. *Top right* At small magnetic field h, q_m is proportional to $h^{2/3}$. *Bottom left* At large magnetic field h the dAT line is approached and the difference $q_M - q_m$ shrinks proportionally to the distance from this line. *Bottom right* In the RS phase the $P(q)$ is a delta function centered in q_{EA}, that goes to zero as $h \to 0$

It follows from the Parisi ansatz that there is an underlying hierarchical structure in the organization of the states in the SG phase, that results in an ultrametric overlap space where $q^{ac} \geq \min(q^{ab}q^{bc})$[Mez84, Mez85, Ram86]. This can be seen by following the RSB process as a tree (Fig. 1.5). At the RS level all the states have the same overlap q_0, this represents the root of the tree. After one step of replica symmetry breaking the replicas part in two groups. Replicas within the same group share have overlap q_1, otherwise it is $q_0 < q_1$, and so on for further steps of RSB. The overlap between two replicas α and β can be identified by returning back towards the root until the two states belong to the same group. For example, the overlap between states α and β in Fig. 1.5 is $q^{\alpha\beta} = q_1$. The ultrametricity condition is easily verified by picking three generic states.

The full RSB $P(q)$ is sign of a SG phase with a complex energy landscape and an infinitely large number of metastable states that are not related through evident symmetries: "The space of configurations consists of many valleys separated by high mountains (free energy barriers) whose height goes to infinity in the infinite-volume limit" [Par83]. The number of valleys is exponential in the number of spins N [Bra80a, Dom80, You81], and so is the time spent in a single valley, meaning that the dynamics of a SG are extremely slow, and when the system size goes to

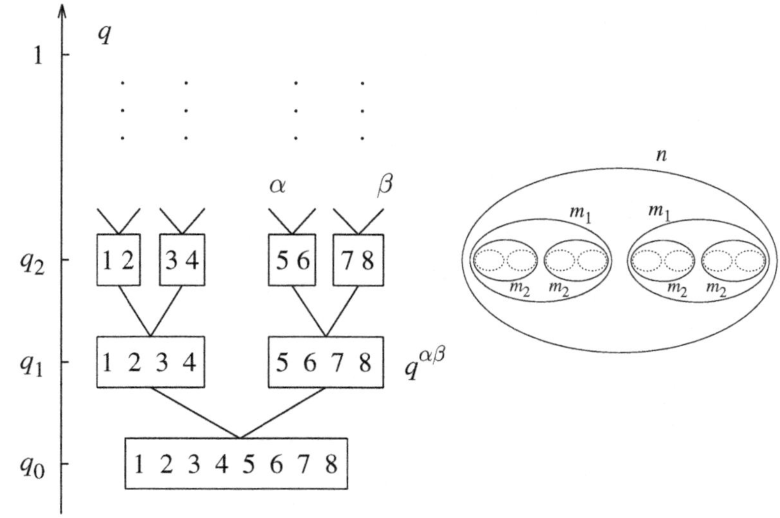

Fig. 1.5 RSB as branching process. The overlap between two states α and β can be seen as the first common level of RSB between α and β (*left*). Another way to visualize this is to represent the RSB process as an iterative subdivision in subsets (*right*), then the overlap between two states α and β is given by the smallest set containing both α and β. The n, m_1, m_2, . . . are the parameters that define the overlap matrix in the Parisi solution

infinity ergodicity is broken [Mac82] in the whole SG phase. This was made clear at first in the infinite-range model [Kir78], an extension of the SK model that mixes interactions between p spins (also called the p-spin model). The limit $p \to \infty$ of the p-spin model yields an exactly solvable model called the REM [Der81], where the probability of a state depends exclusively on its energy and not on the configuration itself.

Despite the Parisi solution of the SK model was physically consistent and confirmed by numerical simulations and other analytical methods (for example the cavity method [Mez86]), it contained some mathematical arbitrarities, some of which we already mentioned, that made it non-rigorous. It took over 20 years later before it was confirmed rigorously through a mathematical proof [Gue02, Gue03, Tal06]. Nonetheless, this mean field solution of the EA model posed a first hypothesis on the nature of the SG phase in finite dimensions. Just as the mean field solution of the Ising model, valid in infinite dimensions, is a good qualitative descriptor of the ferromagnetic transition, the SG phase in a lattice of size $L \times L \times L$ would be qualitatively similar to the one detected in the SK model. This means for instance that the SG phase would resist the application of a small magnetic field, the $P(q)$ would be non-trivial and the overlap space would be ultrametric. Also, in low dimensions the RSB the domains are expected to be space-filling, i.e. with a fractal dimension $d_s = d$, and it is possible to have excitations that involve a finite fraction, $O(L^d)$, of the total spins with a finite-energy cost.

This attractive[12] vision of how real SGs are is called *RSB scenario*.[13]

The droplet picture.

Stimulated by earlier numerical domain-wall renormalization group studies of low-dimensional SGs [Bray85, McM85], and inspired on a schematic scaling theory of SGs proposed by Mc Millan [McM84], Fisher and Huse proposed a new picture of the ordered phase in SGs [Fis86], called *droplet picture* [Fis87, Hus87, Fis88a, Fis88b]. The theory, that derives from a Migdal-Kadanoff approximation [Mig75, Kad76] on the EA model [And78], exact in one dimension, describes the SG phase of low-dimensional SGs as a "ferromagnet in disguise",[14] with only two pure states, with order parameter $q = \pm q_{\rm EA}$. Within a pure state, phase coexistence occurs in form of low lying excitations (droplets) of spins in the subdominant state. The boundaries of these domains are not fixed, but move around due to the disorder, exploiting unsatisfied links and avoiding the strongly satisfied ones. The effect is that the droplets are non-convex, and their boundary scales as L^{d_s}, with $d-1 \le d_s < d$, so not space-filling. The fundamental ansatz, inspired by an earlier argument from Anderson and Pond in the aforementioned Migdal-Kadanoff approach [And78], is that the free-energy cost of the lowest-energy excitations of linear size ℓ is

$$F_\ell \sim \gamma(T)\ell^\theta, \tag{1.19}$$

where θ is the stiffness coefficient, $0 < \theta < (d-1)/2$ and γ is the stiffness modulus. A direct implication is that an infinite energy would be necessary to excite a finite fraction ($\ell \sim L$) of the total number of spins, so only small excitations ($\ell \ll L$) are supported.

In the droplet picture the stiffness coefficient controls the decay of the correlations that go as

$$C(|\boldsymbol{x}-\boldsymbol{y}|) = \overline{\langle s_x s_y\rangle^2} - \overline{\langle s_x\rangle^2 \langle s_y\rangle^2} \sim \frac{1}{|\boldsymbol{x}-\boldsymbol{y}|^\theta}, \tag{1.20}$$

that entails $\overline{q^2} - \overline{q}^2 \to 0$, and therefore the overlap distribution is a delta function, $P(q) = \delta(q - q_{\rm EA})$.

One last remarkable feature of the droplet theory is that the energy barrier for flipping a droplet in a field h scales as $L^\theta - hL^{d/2}$. Because of the bound $\theta < (d-1)/2$, the SG phase is unstable to the presence of any magnetic field. This prediction in

---

[12] *"God used beautiful mathematics in creating the world"*, Paul Dirac, as quoted in *The Cosmic Code : Quantum Physics As The Language Of Nature* (1982) by Heinz R. Pagels, p. 295; also in *Paul Adrien Maurice Dirac : Reminiscences about a Great Physicist* (1990) edited by Behram N. Kursunoglu and Eugene Paul Wigner, p. xv.

[13] For a detailed review on the RSB scenario see [Mar00b]. See also [Par96].

[14] Ferromagnets in disguise can be obtained, for example, by performing a random gauge transformation on an ordered system [Nis01], as it is done in the Mattis model [Mat76].

particular is in contrast with the RSB theory, that predicts a de Almeida-Thouless (dAT) line for $h > 0$.[15]

It is still matter of debate whether which of the two dominant theories for the SG phase, the droplet and the RSB scenario, describes well the SG phase [Moo11, Par12b, Yeo12, Yuc12, Bil13, Yuc13]. It is predominantly accepted that the RSB scenario is valid for dimensions greater than the upper critical dimension $d_u = 6$, and that the droplet picture is exact in $d = 1$.

A different order parameter.

The reason why it is hard to understand the SG in real-life (three-dimensional) SGs could be that we are not looking at the most useful order parameter [Con03, Con05a, Con06].

From a purely mathematical perspective, in the SK model the square of overlap (1.8) represents the covariance of Hamiltonian (1.9). On the other side, in a finite-dimensional EA model, the covariance of Hamiltonian (1.5) is given by the square of the link overlap

$$q_{\text{link}}^2 = \frac{1}{Nz} \sum_{xy}^{N} \sum_{\mu=1}^{d} q_x q_y \tag{1.21}$$

where $q_x = s_x^{(a)} \cdot s_x^{(b)}$ and z is the connectivity.

Overlap and link overlap are the same in the SK model, but in finite-dimensional lattices the two behave differently, as, for instance, under an inversion of all the spins the change in q is $O(L^d)$, while in the case of the link overlap the only changes are in the links that cross the domain surfaces, so the variation is $O(L^{d_s})$.

Droplet and RSB theories have different predictions for the relation between q and q_{link}. In the droplet picture, where the surface-volume ratio vanishes for large systems, q_{link} should be constant, with no correlation with q. On the other side, in the RSB scenario the surfaces are space-filling, so there should be a correlation between q_{link} and q, implying that also $P(q_{\text{link}})$ is non-trivial.

1.2 Observables in Simulations

The reason why numerical simulations became so popular in the last decades is that they are able to give a perspective to physical phenomena orthogonal to the one coming from analytical work and experiments. It is often not possible to validate a model, nor to make predictions that experimentalists can use by using only analytical tools. A numerical simulation can take advantage of the knowledge of the Hamiltonian to test it straightforwardly. As an advantage with respect to experiments, computer simulations are able to measure a large set of observables, mostly microscopic, that

[15] In Chap. 2 we will try to see whether there is or not a phase transition in a field, that would discriminate the (in)correct theory.

are not accessible on real samples. The conjunction of these three aspects of research makes scientific advance much more effective. In this chapter we will discuss most of the observables that we kept track of in our simulations and analyses.

Some notation.

Most of the work presented in this thesis comes from numerical simulations on systems of N spins, both in regular d-dimensional cubic lattices of size $L^d = N$ (Chaps. 2, 3, 4, 6), and in fully connected networks (Chap. 5). Each spin $\vec{s}_{\boldsymbol{x}}$ occupies a position $\boldsymbol{x}$ and has m components, $\vec{s}_{\boldsymbol{x}} = (s_{\boldsymbol{x},1}, s_{\boldsymbol{x},2}, \ldots, s_{\boldsymbol{x},m})$. If $m = 1$ we call them *Ising spins* and often remove the vector symbol, $s_{\boldsymbol{x}}$. If $m = 2$ they are *XY spins*, while if $m = 3$ we call them *Heisenberg spins*. The set of all the spins $\vec{s}_{\boldsymbol{x}}$ of the system is denoted with a ket, $|\vec{s}\rangle$, and constitutes a *configuration*.

Through MC simulations we thermalize the system at a temperature T, taking them to follow the Boltzmann distribution

$$P(|s\rangle) \sim \mathrm{e}^{-\beta\mathcal{H}(|s\rangle)}, \tag{1.22}$$

where $\mathcal{H}$ is the model's Hamiltonian and $\beta = 1/k_{\mathrm{B}}T = 1/T$ is the inverse temperature, as we set to one the Boltzmann constant, $k_{\mathrm{B}} = 1$.

Once the system is thermalized, one can take thermal averages of any measurable observable $\mathcal{O}$, that we denote with $\langle\mathcal{O}\rangle$. The averages over the disorder, instead, are indicated with an over line $\overline{\mathcal{O}}$. To make the notation lighter, we use $E(\mathcal{O})$ when both averages are performed, $E(\mathcal{O}) \equiv \overline{\langle\mathcal{O}\rangle}$.

It can be useful to define a scalar product between two configurations $|s\rangle$ and $|s'\rangle$, for which we use again Dirac's notation

$$\langle s|\, s'\rangle = \sum_{\boldsymbol{x}}^{N} \vec{s}_{\boldsymbol{x}} \cdot \vec{s}\,'_{\boldsymbol{x}}. \tag{1.23}$$

It is straightforward to define the 1- and 2-norms in this space

$$\|s\|_1 = \sum_{\boldsymbol{x}}^{N} |\vec{s}_{\boldsymbol{x}}|, \tag{1.24}$$

$$\|s\|_2 = \sum_{\boldsymbol{x}}^{N} |\vec{s}_{\boldsymbol{x}}|^2 = \langle s|\, s\rangle. \tag{1.25}$$

Now that the notation is defined, we can proceed describing the set of observables $\mathcal{O}$ that we measured in our simulations, that can be used to validate theories and physical scenarios.

1.2.1 Overlaps

We will use two replicas in order to create gauge-invariant observables [Mez87]. To identify different replicas we use the superscripts $^{(a)}, ^{(b)}, ^{(c)}$ and $^{(d)}$. The definition of overlap we use depends on the model we consider and on its symmetries.

Ising overlap.

With Ising spins $s_x = \pm 1$ we can define the local overlap as

$$q_x = s_x^{(a)} s_x^{(b)}, \tag{1.26}$$

from which we can create the global overlap

$$q = \frac{1}{N} \sum_x^N q_x = \frac{1}{N} \langle s^{(a)} | s^{(b)} \rangle, \tag{1.27}$$

where we used notation (1.23).

Tensorial overlap When the spins are m-component vectors $\vec{s}_x = (s_{x,1}, s_{x,2}, \ldots, s_{x,m})$ and $\mathcal{H}$ displays an $O(m)$ symmetry it is convenient to define a rotationally invariant overlap.

We define the tensorial site overlap is defined as

$$\tau_{\alpha\beta}(\boldsymbol{x}) = s_{\boldsymbol{x},\alpha}^{(a)} s_{\boldsymbol{x},\beta}^{(b)}, \tag{1.28}$$

where $\alpha, \beta = 1, \ldots, m$ indicate the components of the vector. Notice that $\tau_{\alpha\beta}(\boldsymbol{x})$ is not Hermitian, since

$$\tau_{\alpha\beta}(\boldsymbol{x})^\dagger = \tau_{\beta\alpha}(\boldsymbol{x}) = s_{\boldsymbol{x},\beta}^{(a)} s_{\boldsymbol{x},\alpha}^{(b)}. \tag{1.29}$$

The order parameter is the overlap tensor [Fer09b]:

$$Q_{\alpha\beta} = \frac{1}{N} \sum_{\boldsymbol{x}} \tau_{\alpha\beta}(\boldsymbol{x}). \tag{1.30}$$

This quantity is not rotationally invariant, and since it is a tensor it is not easy to deal with, so we use the square overlap [Bin86, Col95]

$$\begin{aligned} Q^2 &= \mathrm{Tr}\left[Q Q^\dagger\right] \\ &= \frac{1}{N^2} \sum_{\boldsymbol{x},\boldsymbol{y}} \mathrm{Tr}\left[\tau(\boldsymbol{x}) \tau(\boldsymbol{y})^\dagger\right] \\ &= \frac{1}{N^2} \sum_{\boldsymbol{x},\boldsymbol{y}} (\vec{s}_{\boldsymbol{x}}^{(a)} \cdot \vec{s}_{\boldsymbol{y}}^{(a)})(\vec{s}_{\boldsymbol{x}}^{(b)} \cdot \vec{s}_{\boldsymbol{y}}^{(b)}), \end{aligned} \tag{1.31}$$

that is $O(m) \times O(m)$ invariant (rotational invariance for replica a and replica b). Even though the Q^2 defined in Eq. (1.31) is a square overlap, we will be calling it overlap when referring to it.

The self overlap Q^2_{self} is defined analogously, by taking (a) = (b) in the previous definitions. Notice that the self overlap is not identically equal to 1. It is easy to see, for example, that at infinite temperature, in the thermodynamic limit it is equal to $Q^2_{\text{self}}(T = \infty; L = \infty) = 1/m$ (see for example the Appendix of [BJ11]).

Scalar overlap With vector spins, if the Hamiltonian is not rotationally invariant the overlap can be expressed straightforwardly through the scalar product between spins of different replicas. The site overlap would be

$$q_{\text{SG},x} = \vec{s}_x^{\,\text{a}} \cdot \vec{s}_x^{\,\text{b}}, \tag{1.32}$$

and the global overlap

$$q_{\text{SG}} = \frac{1}{N} \sum_x^N q_{\text{SG},x}. \tag{1.33}$$

We will be calling q_{SG} the SG overlap, to differentiate it from the CG overlap Q_{CG}, defined in the next paragraph.

Chiral overlap With vector spins it is possible to define the chirality, an observable whose importance we will discuss in Chap. 3. It represents the amplitude and handedness of the alignment of the spins along the axis μ, and is expressed with the mixed product of three consecutive spins

$$\zeta_{x,\mu} = \vec{s}_{x+e_\mu} \cdot (\vec{s}_x \times \vec{s}_{x-e_\mu}), \quad \mu = 1, \ldots, d, \tag{1.34}$$

where e_μ is the unitary vector along the μ direction. We can see it as the oriented volume of the parallelepiped we can construct with the three spins. The CG overlap is defined similarly to the SG one,

$$\kappa_{x,\mu} = \zeta^{(\text{a})}_{x,\mu} \zeta^{(\text{b})}_{x,\mu}, \tag{1.35}$$

but in this case we also sum over the d equivalent directions μ

$$q_{\text{CG}} = \frac{1}{Nd} \sum_{x,\mu}^{d,N} \kappa_{x,\mu}. \tag{1.36}$$

Link overlap We also measured the link overlaps, that were shown to be equivalent to the overlaps in the description of the low temperature phase [Con05b, Con06]. In the case of Ising spins the link overlap is

$$q_{\text{link}}^2 = \frac{1}{Nd} \sum_{x}^{N} \sum_{\mu=1}^{d} q_x q_{x+e_\mu} \tag{1.37}$$

$$= \frac{1}{Nd} \sum_{x}^{N} \sum_{\mu=1}^{d} s_x^{(a)} s_{x+e_\mu}^{(a)} s_x^{(b)} s_{x+e_\mu}^{(b)},$$

while for vector spins

$$Q_{\text{link}}^2 = \frac{1}{Nd} \sum_{x}^{N} \sum_{\mu=1}^{d} q_{\text{link}}^{\mu\,2}(x), \tag{1.38}$$

$$q_{\text{link}}^{\mu\,2}(x) = \text{Tr}\left[\tau(x)\tau(x+\hat{e}_\mu)^\dagger\right] = \tag{1.39}$$

$$= (\vec{s}_x^{\,(a)} \cdot \vec{s}_{x+\hat{e}_\mu}^{\,(a)})(\vec{s}_x^{\,(b)} \cdot \vec{s}_{x+\hat{e}_\mu}^{\,(b)}),$$

which is a generalization of (1.37).

1.2.2 Scalar Correlators

For a given the wave vector $\boldsymbol{k}$ we can define the Fourier transforms of the overlap fields

$$\hat{q}_{\text{SG}}(\boldsymbol{k}) = \frac{1}{N} \sum_{x}^{N} q_x e^{i\boldsymbol{k}\cdot x} \tag{1.40}$$

$$\hat{q}_{\text{CG}}^{\mu}(\boldsymbol{k}) = \frac{1}{N} \sum_{x}^{N} \kappa_x e^{i\boldsymbol{k}\cdot x}, \tag{1.41}$$

that we use to build the wave-vector dependent susceptibilities as

$$\chi_{\text{SG}}(\boldsymbol{k}) = N\overline{\langle |q_{\text{SG}}(\boldsymbol{k})|^2\rangle}, \tag{1.42}$$

$$\chi_{\text{CG}}(\boldsymbol{k}) = N\overline{\langle |q_{\text{CG}}(\boldsymbol{k})|^2\rangle}. \tag{1.43}$$

Since the lattice is finite and has discrete spacings, in our simulations we measure $\hat{q}(\boldsymbol{k})$ for a specific set of wave vectors that we need to compute relevant observables. Calling $k_{\text{min}} = 2\pi/L$ the lowest wave number allowed by periodic boundary conditions, we seek

$$\boldsymbol{k}_n = (nk_{\text{min}}, 0, 0) \qquad n = 0, \ldots, L/2, \tag{1.44}$$

$$\boldsymbol{k}_{11} = (k_{\text{min}}, \pm k_{\text{min}}, 0), \tag{1.45}$$

and the permutations of their components.

We can then construct the susceptibilities $\chi_{SG} = \chi_{SG}(\mathbf{0})$ and $\chi_{CG} = \chi_{CG}(\mathbf{0})$ and the dimensionless cumulant R_{12} that will be useful to spot phase transitions with the finite-size scaling method (Sect. 1.3.4):

$$R_{12} = \frac{\chi(\boldsymbol{k}_1)}{\chi(\boldsymbol{k}_{11})}, \tag{1.46}$$

where we averaged over all the possible permutations of the components of $\boldsymbol{k}_1$ and $\boldsymbol{k}_{11}$.

We define the two-point correlation functions $C(\boldsymbol{x}, \boldsymbol{y}) = \langle q_x q_y \rangle$. When the system is translationally invariant, this correlation can be expressed as a function of the separation $\boldsymbol{r} = \boldsymbol{x} - \boldsymbol{y}$, being called $C(\boldsymbol{r})$. We compute $C(\boldsymbol{r})$ and its Fourier transform $\hat{C}(\boldsymbol{k})$ as

$$C(\boldsymbol{r}) = \frac{1}{N} \sum_{x}^{N} q_x q_{x+r}, \tag{1.47}$$

$$\hat{C}(\boldsymbol{k}) = \sum_{r} C(\boldsymbol{r}) e^{i\boldsymbol{k}\cdot(\boldsymbol{r})}, \tag{1.48}$$

and consequently $C(\boldsymbol{r})$ can be obtained back as the anti Fourier transform $C(\boldsymbol{r}) = \frac{1}{L} \sum_{k} \hat{C}(\boldsymbol{k}) e^{-i\boldsymbol{k}\cdot(\boldsymbol{r})}$. In appendix D.3 we discuss the numerical estimators of these quantities.

The wave-vector dependent susceptibilities are directly related to the correlation functions. Using Eqs. (1.40), (1.42) we have

$$\chi(\boldsymbol{k}) = N\left[\hat{q}_{SG}(\boldsymbol{k})\hat{q}_{SG}(\boldsymbol{k})^*\right] = \tag{1.49}$$

$$= \frac{1}{N} \sum_{x}^{N} q_x e^{i\boldsymbol{k}\cdot\boldsymbol{x}} \sum_{y}^{N} q_y e^{-i\boldsymbol{k}\cdot\boldsymbol{y}} = \tag{1.50}$$

$$= \frac{1}{N} \sum_{x,y}^{N} C(\boldsymbol{x}, \boldsymbol{y}) e^{i\boldsymbol{k}\cdot(\boldsymbol{x}-\boldsymbol{y})} = \tag{1.51}$$

that in the presence of translational invariance and recalling Eq. (1.48) becomes

$$= \frac{1}{N} \sum_{x}^{N} \sum_{r}^{N} C(\boldsymbol{r}) e^{i\boldsymbol{k}\cdot(\boldsymbol{r})} = \hat{C}(\boldsymbol{k}). \tag{1.52}$$

This means that we can measure correlation functions both in the real and in the Fourier space, depending on which of the procedures is more convenient numerically.

The point-to-plane correlation functions are computed from the Fourier transform of the fields,

$$C(r) = \frac{1}{L}\sum_{n=0}^{L-1} \mathrm{e}^{-\mathrm{i}\boldsymbol{r}\cdot\boldsymbol{k}_n}\chi(\boldsymbol{k}_n) \equiv \sum_{y,z} C(x=r,y,z), \tag{1.53}$$

where r is the modulus of the distance. Equation (1.53) is equivalent if we align the wave vector along any of the three coordinate axes, so we average over these choices.

In Chap. 2 we will use similar procedures to construct correlation functions with four replicas instead of two.

1.2.3 Tensorial Correlation Functions

We will be measuring both point and plane correlation functions. The point correlation function is

$$C^{(\text{point})}(r) = \frac{1}{Nd}\sum_{\mu=1}^{d}\sum_{\boldsymbol{x}}^{N} \mathrm{Tr}[\tau(\boldsymbol{x})\tau(\boldsymbol{x}+\hat{e}_\mu r)^\dagger], \tag{1.54}$$

where $\mu = 1$ (or x), 2 (or y), 3 (or z) is a coordinate axis, and e_μ is the unitary vector in that direction. We also use plane correlation functions because they decay slower and have a better signal-to-noise ratio. If we denominate the plane-overlap tensor as the mean overlap tensor over a plane

$$P^x_{\alpha\beta}(x) = \frac{1}{L^2}\sum_{y,z=0}^{L-1} \tau_{\alpha\beta}(x,y,z), \tag{1.55}$$

we can define the plane correlation function as

$$C^{(\text{plane})}(r) = \frac{1}{Ld}\sum_{\mu=1}^{d}\sum_{x=0}^{L-1} \mathrm{Tr}[P^\mu(x)P^\mu(x+r)^\dagger]. \tag{1.56}$$

These tensorial definitions of $C(r)$ are $O(m)\times O(m)$ invariant.

The link-overlap correlation functions are

$$C^{(\text{point})}_{\text{link}}(r) = \frac{1}{Nd^2}\sum_{\mu,\nu=1}^{d}\sum_{\boldsymbol{x}}^{N} q^{\nu\,2}_{\text{link}}(\boldsymbol{x})q^{\nu\,2}_{\text{link}}(\boldsymbol{x}+r\hat{e}_\mu), \tag{1.57}$$

$$C^{(\text{plane})}_{\text{link}}(r) = \frac{1}{Ld}\sum_{\mu=1}^{d}\sum_{x=0}^{L-1} P_{\text{link}}(x)P_{\text{link}}(x+r), \tag{1.58}$$

with

$$P_{\text{link}}^{x}(x) = \frac{1}{L^2 d} \sum_{\nu=1}^{d} \sum_{y,z=0}^{L-1} q_{\text{link}}^{\nu\,2}(x, y, z). \tag{1.59}$$

One could in principle choose to subtract from those correlators the equilibrium link overlap, to obtain connected correlators, since the link overlap is non-zero also in the paramagnetic phase.

1.2.4 Four-Replica Correlators

We will be working with Ising spins under an applied magnetic field $h > 0$. In this situation the order parameter q_{EA} is not zero even in the paramagnetic phase. This implies that we cannot construct connected correlation functions by means of only two replicas. Therefore, for each sample we simulated 4 different replicas, in order to be able to compute connected correlation functions that go to zero at infinite distance. In appendix B we give more details and show that the most informative connected correlator we can construct with 4 replicas is the replicon propagator [Alm78b, Dom06]

$$G_{\text{R}}(\boldsymbol{r}) = \frac{1}{N} \sum_{\boldsymbol{x}} \overline{(\langle s_{\boldsymbol{x}} s_{\boldsymbol{x}+\boldsymbol{r}} \rangle - \langle s_{\boldsymbol{x}} \rangle \langle s_{\boldsymbol{x}+\boldsymbol{r}} \rangle)^2}. \tag{1.60}$$

To compute G_{R} we calculate the 4-replica field

$$\Phi_{\boldsymbol{x}}^{(\text{ab;cd})} = \frac{1}{2}(s_{\boldsymbol{x}}^{(\text{a})} - s_{\boldsymbol{x}}^{(\text{b})})(s_{\boldsymbol{x}}^{(\text{c})} - s_{\boldsymbol{x}}^{(\text{d})}), \tag{1.61}$$

where the indexes a, b, c, d indicate strictly different replicas. Notice that

$$\left\langle \Phi_{\boldsymbol{x}}^{(\text{ab;cd})} \Phi_{\boldsymbol{y}}^{(\text{ab;cd})} \right\rangle = (\langle s_{\boldsymbol{x}} s_{\boldsymbol{x}+\boldsymbol{r}} \rangle - \langle s_{\boldsymbol{x}} \rangle \langle s_{\boldsymbol{x}+\boldsymbol{r}} \rangle)^2, \tag{1.62}$$

so we obtain G_{R} by taking also the average over the samples

$$E(\Phi_{\boldsymbol{x}}^{(\text{ab;cd})} \Phi_{\boldsymbol{y}}^{(\text{ab;cd})}) = G_{\text{R}}(\boldsymbol{x} - \boldsymbol{y}). \tag{1.63}$$

Here, and everywhere there is more than one possible permutation of the replica indices, we average over all of them to gain statistics.

From this point on everything is formally like the two-replica construction, using $\Phi_{\boldsymbol{x}}$ instead of $q_{\boldsymbol{x}}$ to construct the susceptibilities $\chi(\boldsymbol{k})$. For example correlations in the Fourier space are defined by Fourier-transforming $\Phi_{\boldsymbol{x}}^{(\text{ab;cd})}$, so the wave-vector dependent replicon susceptibility is expressed as

$$\chi_{\mathrm{R}}(\boldsymbol{k}) = \frac{1}{N} E(|\hat{\Phi}_{\boldsymbol{k}}^{(\mathrm{ab;cd})}|^2), \quad \hat{\Phi}_{\boldsymbol{k}}^{(\mathrm{ab;cd})} = \sum_{\boldsymbol{x}}^{N} \mathrm{e}^{\mathrm{i}\boldsymbol{k}\cdot\boldsymbol{x}} \Phi_{\boldsymbol{x}}^{(\mathrm{ab;cd})}. \tag{1.64}$$

Point-to-plane correlation functions are computed through Eq. (1.53).

1.2.5 Correlation Lengths

The correlation length is the average distance weighed with the $C(r)$. We will be constructing second-moment correlation lengths for point and plane correlations

$$\xi_2^{(\mathrm{point})} = \sqrt{\frac{\int_0^{L/2} C^{(\mathrm{point})}(r) r^4 dr}{\int_0^{L/2} C^{(\mathrm{point})}(r) r^2 dr}}, \tag{1.65}$$

$$\xi_2^{(\mathrm{plane})} = \sqrt{\frac{\int_0^{L/2} C^{(\mathrm{plane})}(r) r^2 dr}{\int_0^{L/2} C^{(\mathrm{plane})}(r) dr}}. \tag{1.66}$$

The difference in the definitions is due to the presence of a Jacobian term when we want to integrate the point correlation function over the space. These two lengths would be proportional by a factor $\sqrt{6}$ if they had the same purely exponential correlation function. Note that $\xi_2^{(\mathrm{point})}$ and $\xi_2^{(\mathrm{plane})}$ are proper estimators of a correlation length only when the correlation functions $C^{(\mathrm{point})}(r)$ and $C^{(\mathrm{plane})}(r)$ are connected (i.e. they go to zero for large r). Otherwise, in principle they could be used to individuate if a quench penetrated in the SG phase. In fact, depending on m a quench will drive us in a ferromagnetic or in a SG phase. Our correlation functions are connected in the SG phase, but they are not in a ferromagnetic state. Consequently, a cumulant such as ξ_L/L—being ξ_L the correlation length measured in a lattice of size L—will diverge as $L^{\theta/2}$ (see Ref. [AB10a] for a definition of θ and an explanation of this behavior) when m is too large for a SG phase, it will converge as $1/L$ if the quench penetrates in the SG phase, and it will be of order 1 right at the critical m, m_{SG}, that is probably not integer, so not exactly locatable.

When the correlation function decays very quickly and the noise becomes larger than the signal, one could measure negative values of $C(r)$, that would be amplified by the factors r^2 and r^4 in the integrals. This would imply very large errors in ξ, or even the square root of a negative number. To overcome this problem, we truncated the correlation functions when they became less than three times the error [Bel09a]. This procedure introduces a small bias, but reduces drastically the statistical error. Furthermore, the plane correlation function required the truncation much more rarely, therefore we compared the behaviors as a consistency check.

As shown in the appendices of [BJ11], in the thermodynamic limit ($k_{\min} = 0$) the second moment correlation length can be re-expressed as

$$\xi_L = \frac{1}{2\sin(k_{\min}/2)}\sqrt{\frac{\chi(\mathbf{0})}{\chi(\boldsymbol{k}_{\min})} - 1}. \tag{1.67}$$

being $\boldsymbol{k}_{\min} = (2\pi/L, 0, 0)$ or permutations. The sub-index L stresses the dependence the linear size of the lattice (recall that $k_{\min}$ depends on L). This same definition can be used with any of the observables defined in the previous section (SG susceptibility, SG susceptibility, replicon susceptibility, ...). This quantity will be used only using with plane correlations, since integrating over all the directions in the lattice to calculate $\chi(\boldsymbol{k}_{\min})$ is a cumbersome and imprecise task. When computing ξ_{CG}, one can choose μ parallel or orthogonal to the wave vector $\boldsymbol{k}_{\min}$. As it was already observed in [Fer09b], there is no apparent difference between the two options, so we averaged over all the values of μ to enhance our statistics.

The definitions of the link correlation lengths $\xi_{2,\mathrm{link}}^{(\mathrm{point})}$ and $\xi_{2,\mathrm{link}}^{(\mathrm{plane})}$ can be obtained from Eqs. (1.65) and (1.66), by substituting the spin with link correlation functions.

1.3 Phase Transitions with a Diverging Length Scale

The topics treated in this section have been very successful in describing phase transitions and are very well consolidated tool since the 1970s. Our scope here is not to give an extended treatment, that can be found elsewhere (see e.g. [Ma76, Bin86, Hua87, Car96, Ami05]), but to refresh the reader's memory on some concepts that we will be using throughout this dissertation.

1.3.1 Second-Order-Like Phase Transitions

The phenomenology of the spin-glass transitions we will treat is similar to that of a second-order phase transition. In this section we will assume Ising spins, but the description is the same with m-component spins. The coherence length ξ, that we can define through the long-distance decay of two-point correlation function,

$$\langle s_{x+r} s_x \rangle \overset{|r|\to\infty}{\sim} \mathrm{e}^{-|r|/\xi}, \tag{1.68}$$

diverges in power law as we approach the critical point

$$\xi \propto |t|^{-\nu}. \tag{1.69}$$

In Eq. (1.69) we defined the reduced temperature $t = \frac{T-T_c}{T_c}$, and T_c is the critical temperature. Mind that the symbol t will represent the reduced temperature only in this chapter, while throughout the rest of the text it will indicate the time. The exponent ν characterizes the phase transition and sets its Universality class. The correlation length ξ is not the only diverging observable. To fully identify the type of phase transition we can define six critical exponents $\alpha, \beta, \gamma, \delta, \eta, \nu$ that describe the power law behavior of the observables that are relevant in our case.

The specific heat diverges as

$$C_{\mathrm{h}}(t) \sim |t|^{\alpha}. \tag{1.70}$$

The case $\alpha = 0$ can indicate a discontinuity or a logarithmic divergence.

The order parameter, for example the magnetization m or the overlap q, vanishes as

$$\hat{m}(t) \sim (-t)^{\beta} \tag{1.71}$$

when we approach the critical temperature from below.

The response to a small external field h, that we call susceptibility, diverges like

$$\chi(t) \sim |t|^{\gamma}. \tag{1.72}$$

If we are exactly at the critical point $t = 0$, for small fields h the order parameter behaves as

$$m(t = 0, h) \sim |h|^{1/\delta}, \tag{1.73}$$

and the correlation length decays with a power law

$$\langle s_{x+r} s_x \rangle \overset{|r| \to \infty}{\sim} |\boldsymbol{r}|^{-(d-2+\eta)}, \tag{1.74}$$

and we call η the anomalous dimension.

These critical exponents are constrained by a set of four independent *scaling relations*,

$$\begin{aligned} 2\beta + \gamma &= 2 + \alpha, \\ 2\beta\delta - \gamma &= 2 + \alpha, \\ \gamma &= \nu(2 - \eta), \\ \nu d &= 2 - \alpha, \end{aligned} \tag{1.75}$$

that reduce to two the number of independent exponents. The fourth of Eq. (1.75) relates the exponents to the dimension of space. It is called *hyperscaling* relation and is valid only under the upper critical dimension d_u. From the hyperscaling relation one understands directly that the universality class must depend on dimensionality, since the critical exponents change with d.

1.3.2 Real-Space Coarse Graining

The coherence length ξ represents the size of patches of highly correlated spins. One can think that patches of size ξ interact one with the other. This concept works very well in ferromagnets [Ma76, Hua87, Ami05], but though plausible it is still not fully developed for disordered systems [Har76, Dot87, Dot01, Ang13]. Following this idea, since ξ is singular at the critical temperature, we can think to construct a block Hamiltonian that describes the interactions between patches of spins. Let us call b the linear size of these blocks. Then there will be $L^d b^{-d}$ blocks, each including b^d spins. The block variables σ_x can be defined as the mean spin in the block

$$\sigma_x = b^{-d} \sum_{y \in x}^{b^d} s_y, \tag{1.76}$$

where the sum runs over all the spins s_y that belong to the block σ_x. The probability distribution for the blocks of spins is

$$\begin{aligned} P'[\{\sigma\}] &= \left\langle \prod_x \delta\left(\sigma_x - b^{-d}\sum_{y \in x}^{b^d} s_y\right)\right\rangle_P \propto \\ &\propto \int e^{-\mathcal{H}[\{s\}]/T} \prod_x \delta\left(\sigma_x - b^{-d}\sum_{y \in x}^{b^d} s_y\right) ds_1 ds_2 \ldots ds_N \equiv \\ &\equiv e^{-\mathcal{H}_{\text{block}}[\{\sigma\}]/T}, \end{aligned} \tag{1.77}$$

where with $\langle \ldots \rangle_P$ we indicate the average using the equilibrium distribution P of the spins s_y, $P = \mathcal{Z}^{-1} e^{-\mathcal{H}[\{s\}]/T}$, being $\mathcal{Z}$ the partition function. $\mathcal{H}_{\text{block}}$ is the block Hamiltonian deriving from the coarsening we made, and is equivalent to the original Hamiltonian as long as we are interested in spatial resolutions larger than b. This is our case, since we want to use this procedure to describe diverging length scales. Once we constructed the blocks once, we can obviously iterate the process, renormalizing each time dynamics variables and Hamiltonian.

1.3.3 Scaling Hypothesis and Widom Scaling

The scaling hypothesis, first conjectured by Widom [Wid65], is the reasonable assumption that if we have a phase transition with a diverging length ξ, then ξ is the only relevant length. It is model-independent and has been very effective in describing observations. The main idea is that the singular behavior is completely due to the long-range correlation of spin fluctuations near T_c.

To formalize this setting, we assume that when we coarsen the lattice in block variables the free energy remains unchanged, $\mathcal{F} = \mathcal{F}_{\text{coarse}}$: even though our model is short ranged, we are only interested in the long-range correlations that arise from being at criticality. The renormalized temperature $\tilde{t}$ and field $\tilde{h}$ will have to be rescaled in a consonous way. This rescaling can be written as

$$\begin{cases} \tilde{t} = t\, b^{y_t} \\ \tilde{h} = h\, b^{y_h}, \end{cases} \tag{1.78}$$

where y_t and y_h are generic exponents that describe the rescaling. Using Eq. (1.78) and $F = F_{\text{coarse}}$ we have that the intensive free energy scales as

$$f(t,h) = b^{-d} f(\tilde{t},\tilde{h}) = b^{-d} f(t\, b^{y_t}, h\, b^{y_h}). \tag{1.79}$$

To obtain y_t and y_h as a function of the critical exponents we study the behavior of the magnetization m, that we can obtain by deriving f by the magnetic field:

$$m(t,h) = \frac{\partial f(t,h)}{\partial h} = b^{-d} \frac{\partial f(t\, b^{y_t}, h\, b^{y_h})}{\partial h} = b^{y_h - d} m(t\, b^{y_t}, h\, b^{y_h}). \tag{1.80}$$

Since b is an arbitrary scaling parameter, we can set it to grow as any diverging function of $\tilde{t}$ or $\tilde{h}$. If we place ourselves in the zero-field limit $h = 0$ it is convenient to choose $b = (-t)^{-1/y_t}$, so Eq. (1.80) becomes

$$m(t,0) = (-t)^{(d-y_h)/y_t} m(-1,0). \tag{1.81}$$

Remembering the definition of the critical exponent β, that defines that approaching the critical point from below the magnetization goes to zero as $m(t) \sim (-t)^{\beta}$, we can determine the constraint $\beta = (d - y_h)/y_t$.

We can also study the behavior of the system along the critical curve $t = 0$. A helpful choice of b is then $b = h^{-1/y_h}$, in such a way that

$$m(0,h) = (h)^{(d-y_h)/y_h} m(0,1). \tag{1.82}$$

This time we use the definition of δ, that for small h sets the behavior of m along the critical line as $m(0,h) \sim h^{1/\delta}$, and obtain the constraint $\delta = y_h/(d - y_h)$.

Using Eq. (1.75) it becomes possible to reconstruct all the other critical exponents.

1.3.4 Finite-Size Scaling

Simulations near T_c in a lattice of linear size L are usually far from the thermodynamic, due to the extreme growth of the correlation length. Finite-size scaling (FSS) techniques let us measure properties of the thermodynamic limit by using L as a scaling variable, just like we did with the parameter b in the previous paragraphs. It was proposed by Nightingale [Nig75] and developed by Binder [Bin82], and it is nowadays the method of choice to study this type of phase transitions (see e.g. [Bin86, Bal96, Bal98a, Bal00, Lee03, Cam06, Jor06, Leu08, Jor08b, Has08, Fer09c, Bn12b, BJ13, BJ14d, Lul15] for applications of FSS in the field of SGs).

1.3.4.1 Spotting the Transition

If an observable $\mathcal{O}$ diverges at the critical temperature as $\mathcal{O} \propto |t|^{x_{\mathcal{O}}}$, then its thermal average close to the critical point can be expressed like

$$\begin{aligned}\langle \mathcal{O}(L,T)\rangle = L^{x_{\mathcal{O}}/\nu}\Big[& f_{\mathcal{O}}(L^{1/\nu}(t)) \\ & + L^{-\omega} g_{\mathcal{O}}\big(L^{1/\nu}t\big) \\ & + L^{-2\omega} h_{\mathcal{O}}\big(L^{1/\nu}t\big) + \cdots \Big],\end{aligned} \tag{1.83}$$

where $f_{\mathcal{O}}, g_{\mathcal{O}}$ and $h_{\mathcal{O}}$ are analytic scaling functions for observable $\mathcal{O}$, while ν is defined in Eq. (1.69). The exponent $\omega > 0$ is the largest irrelevant exponent. It is universal, and it expresses the corrections to the dominant scaling. The lower dots, ..., stand for subleading corrections to scaling.

The case $\mathcal{O} = \xi_L(T)/L$ is of special interest, since ν is the critical exponent for the correlation length. Then, Eq. (1.83) becomes in this case, up to the leading-order,

$$\frac{\xi_L}{L} = f_\xi\big(L^{1/\nu}(t)\big) + \cdots . \tag{1.84}$$

Therefore, we can identify $T = T_c$ ($t = 0$) as the temperature where the curves $\xi_L(T)/L$ cross for all L for sufficiently large L. The same reasoning is valid also for R_{12}, defined in Eq. (1.46)

$$R_{12} = f_R\big(L^{1/\nu}(t)\big) + \cdots , \tag{1.85}$$

so R_{12} as well can be used to identify the phase transition, and has the feature of not depending on the susceptibility.

The cumulant R_{12} (recall Fig. 2.1) was introduced in [Bn12a] to estimate the critical temperature bypassing pathologies on $\chi(\mathbf{0})$ due to the fact that the overlap is non-zero in the paramagnetic phase [Leu09].

Note that the value of ξ_L/L and R_{12} at the crossing tends to a non-trivial universal quantity (see also footnote in Sect. 1.3.5):

$$\left.\frac{\xi_L}{L}\right|_{T^{L,2L}} = \left.\frac{\xi}{L}\right|_{L=\infty} + A_\xi L^{-\omega} + \cdots , \tag{1.86}$$

$$R_{12}|_{T^{L,2L}} = R_{12}|_{L=\infty} + A_R L^{-\omega} + \cdots . \tag{1.87}$$

If we let $T^{L,2L}$ be the temperature where $\xi_L(T)/L$ crosses $\xi_{2L}(T)/(2L)$, this regime is reached once the $T^{L,2L}$ has converged. Yet, if ω is small, our lattice sizes may not be large enough, so we will have to take in account the aforementioned corrections to scaling. Including the second-order corrections, the approach of the crossing temperature $T^{L,2L}$ to the asymptotic value T_c can be written as

$$T^{L,2L} - T_c = AL^{-(\omega+1/\nu)} + BL^{-(2\omega+1/\nu)} + \cdots , \tag{1.88}$$

where A and B are non-universal scaling amplitudes.

1.3.4.2 Critical Exponents

To compute the critical exponents ν and η we use the quotients' method, taking the quotient of the same observable between different lattice sizes L and $2L$. At the temperature $T^{L,2L}$ we get:

$$\frac{\overline{\langle \mathcal{O}_{2L}(T^{L,2L})\rangle_J}}{\overline{\langle \mathcal{O}_L(T^{L,2L})\rangle_J}} = 2^{x_\mathcal{O}/\nu} + A_{x_\mathcal{O}} L^{-\omega} + \cdots . \tag{1.89}$$

Again, $A_{x_\mathcal{O}}$ is a non-universal amplitude, while the dots stand for subleading corrections to scaling. Therefore, if $\mathcal{O}$ is the thermal derivative of ξ, we can compute the ν critical exponent through the relation

$$\frac{d\xi_{2L}(T^{L,2L})/dT}{d\xi_L(T^{L,2L})/dT} = 2^{1+1/\nu} + A_\nu L^{-\omega} + \cdots . \tag{1.90}$$

To calculate η we use the susceptibility, as $\chi \propto |T - T_c|^{-\gamma} \sim L^{\gamma/\nu}$. Since for the scaling relations (1.75) $2 - \eta = \gamma/\nu$, the susceptibility at the critical temperature scales as

$$\chi_L \sim L^{2-\eta}, \tag{1.91}$$

so the exponent η can be calculated by taking the quotient between sizes $2L$ and L

$$\frac{\chi_{2L}(T^{L,2L})}{\chi_L(T^{L,2L})} = 2^{2-\eta} + A_\eta L^{-\omega} + \cdots . \tag{1.92}$$

Due to the scaling relations (1.75) determining the two exponents η and ν is enough to be able to estimate them all.

1.3.5 Universality and Renormalization Group Flow

The RG assumption is that the coarse-graining transformation (1.76) will transform smoothly the free energy (Eq. (1.79)), that will converge to aFP in the space of the rescaled parameters ($\tilde{t}$ and $\tilde{h}$ in the case of Eq. (1.79)). That is, when the system is looked at large enough scales, the whole behavior of the system will be given

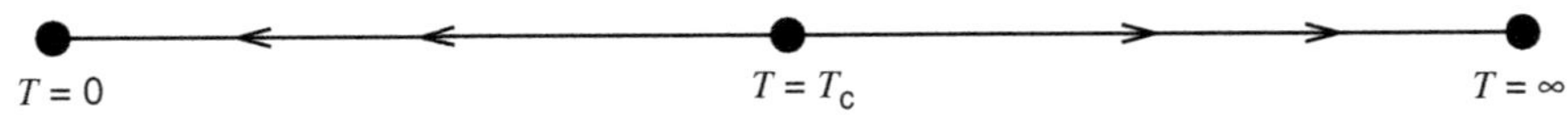

Fig. 1.6 RG flow in the Ising model. The only control parameter is the temperature. There are two stable FP at zero and infinite temperature, and one unstable FP at the critical temperature T_c. The arrows represent the direction of the flow

by the FP, that depends in a complicated way on physical parameters such as the temperature T, the magnetic field h, etc.

Now, the physical parameters can be adjusted in experiments, and can be imposed in calculations, in order to tune the regime in which the system finds itself. In the space of the (rescaled) parameters, a FP will attract the RG trajectories that start in a finite region around it. This region is often a hypersurface in the space of the scaling variables. Since all the trajectories of the hypersurface converge to the same FP, in the infinite-size limit all these starting points will share the same behavior. More precisely, the value to which the observables converge will be the same, such as $\xi|_{L=\infty}$ and $R_{12}|_{L=\infty}$,[16] and the way they converge to this quantity also will coincide, so the critical exponents will be the same. The set of all the quantities that are set by the FP is called *universality class*. In principle, different models can fall in the same universality class as long as they are dominated by the same FP.

Let us take as an example the Ising model [Hua87] with $d > 1$ spatial dimensions, that has the temperature as only control parameter, and displays a second-order phase transition at a temperature T_c. In this case the parameter space is 1 dimensional, so the critical hypersurface is a point. There is a zero-temperature and an infinite-temperature stable FP, respectively governing the behavior of the ferromagnetic and of the paramagnetic phases. By stable we mean that the FP is attractive, and RG trajectories starting from a neighborhood finish in those FPs. The two are separated by a FP at T_c that represents the critical point (Fig. 1.6). Any RG trajectory starting at $T > T_c$ will converge to the $T = \infty$ FP after a large enough number of coarse-graining steps. Equivalently, the behavior at $T < T_c$ is described by the zero-

[16] For systems belonging to the same universality class the correlation function scales as

$$C(\boldsymbol{r}, L) = \frac{1}{L^{2-\eta-d}} f_C\left(\frac{\boldsymbol{r}}{L}\right). \tag{1.93}$$

The scaling function f_C is depends on the geometry of the system (ratio between the sides, type of boundary conditions, etc.), but not on the Hamiltonian (as long as it is dominated by the same FP). If we take the ratio between the Fourier transforms of the correlation function $R_{12} = \frac{\chi(\boldsymbol{k}_1)}{\chi(\boldsymbol{k}_{11})}$, the divergences even out an it tends to a constant value. Similarly, for large L, $\frac{\xi_L}{L} = \frac{1}{2L\sin(\pi/L)}\sqrt{\frac{\chi(\boldsymbol{0})}{\chi(\boldsymbol{k}_1)} - 1}$ tends to a constant value, since $L\sin(\pi/L) \to \pi$.

Even though $\xi|_{L=\infty}$ is universal, its value is not very interesting, since it diverges. To obtain some non-trivial limit, we can divide it by some power of L. The ratio $\left.\frac{\xi_L}{L^A}\right|_{L=\infty}$ has three limits, two of which are trivial. If $A > 1$ we get $\left.\frac{\xi_L}{L^A}\right|_{L=\infty} = 0$, while if $A < 1$ then $\left.\frac{\xi_L}{L^A}\right|_{L=\infty} = \infty$, no matter the universality class of the phase transition. Only $A = 1$ gives therefore a useful indicator of the universality class, since $\frac{\xi_L}{L}$ tends to a finite value.

temperature FP after the system is coarse-grained enough. Moreover, the fact that the ferromagnetic phase is described by a FP at zero temperature means that neglecting thermal fluctuations is a fair way to treat this phase. Being the critical FP unstable, the only way for a trajectory to converge to it is if it starts at $T = T_c$.

The "speed" of the rescaling is proportional to the distance from the critical temperature t [recall the first of (1.78)], so the closer we are to T_c the longer it will take to reach the FP. Suppose that starting from $T > T_c$ we want to reach a correlation length $\xi_0 = O(1)$, this will take an amount $n(T)$ of coarse-graining steps, so $\xi(T)b^{-n(T)} = \xi_0$. The smaller $|t|$, the higher $n(T)$. So, as $T \to T_c$, $n(T) \to \infty$, meaning that $\xi(T_c) \to \infty$, representing a critical point.

We stress that as long as $t \simeq 0$, it will take a very large number of coarsening steps before the behavior of the system (for example the size of the correlation length) start to appear more similar to that of the stable infinite-temperature FP (to which it will eventually converge) than to that of the unstable critical FP.

1.3.5.1 Crossover Behaviors

As pointed out in the previous section, when we find ourselves very close to a critical (unstable) FP, the system will show for a long time (in terms of coarse-graining steps) echoes of that FP's behavior.

To tackle the role of crossover behaviors we make an explicit example. Let us take in account an $m = 3$ Heisenberg magnet with single-ion uniaxial anisotropy and nearest-neighbor interactions. The Hamiltonian is

$$\mathcal{H}_{si} = -\frac{1}{2} \sum_{|x-y|=1} \vec{s}_x \cdot \vec{s}_y - D \sum_x s_{x,z}^2, \tag{1.94}$$

where $s_{x,z}$ is the z component of spin $\vec{s}_x$. The anisotropy term D splits the Heisenberg $O(3)$ symmetry into a direct product of an XY $O(2)$ and an Ising symmetry. When $D = 0$ the symmetry of the model is $O(3)$, and the critical behavior is governed by a Heisenberg FP. When $D \to +\infty$ the z component is infinitely favored, only configurations with $s_{x,z} = \pm 1$ ($\forall x$) are eligible, and the system falls in the Ising universality class. When $D \to -\infty$ the z component is infinitely suppressed, only configurations with $s_{x,z} = 0$ are allowed, and the critical behavior is XY. Thus, in the RG flow diagram that we can draw in the (T, D) plane, there will be three fixed points, Ising, XY and Heisenberg. Figure 1.7 gives a qualitative picture of what the phase diagram could look like. Two critical lines will part from the $D = 0$ fixed point. It is reasonable that the XY and Ising universality classes for $D \neq 0$ are maintained along the whole area of the phase diagram where the symmetries are broken, so the Ising and XY fixed points will be attractive along the critical lines.

Now, assume we are in a situation of small positive anisotropy D. Depending on the temperature we will find ourselves in some part of the dashed line drawn in Fig. 1.7. We can start our RG flow, for example, from point A, deep in the ferro-

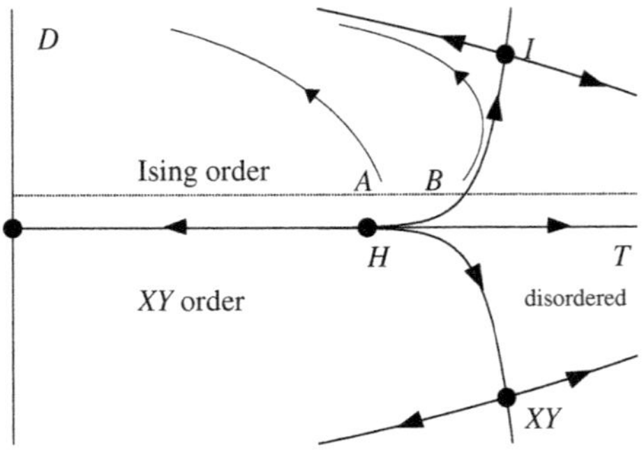

Fig. 1.7 Phase diagram of the Heisenberg model with uniaxial anisotropy. When the anisotropy parameter D is positive the low-T phase has Ising order, and the critical line is dominated by the Ising FP (I). Equivalently, for $D < 0$ the order is XY-like, and the critical behavior is XY-like. Only when $D = 0$ the critical behavior is Heisenberg-like. The arrows show the direction of the RG flow. See main text for more discussions of the figure

magnetic phase, or from point B, still in the ordered phase, but very close to the critical line. Both trajectories will eventually finish in the fixed point that describes the ordered phase. Yet, a trajectory starting from A will head directly towards the low-temperature fixed point. The one that begins on B instead, will pass very close to the Ising FP, and since it is a FP it will spend a lot of time near it. As we argued in the previous section, this amount of time diverges as point B approaches the critical line.

This implies that despite the Ising FP attracts trajectories that come away from the Heisenberg fixed point, when one explores the phase diagram with numerical RG methods his measurements might be biased by pure echoes of the more unstable Heisenberg FP. On the present example there are several ways to try to avoid this, such as (1) using a large D, (2) working very close to the critical temperature, (3) or tuning the starting point after having performed some RG steps (i.e. working on very large lattices). Unfortunately these measures are seldom adoptable. In fact (1) the critical line could exist only for small D, so too large anisotropies would hurl us in the paramagnetic phase, far from the critical line. This is not the case in this example (in Chap. 3 we will succesfully use strong anisotropies to study the critical behavior of the model), but is it, by instance, the case when we deal with SGs in a field. In that case the control parameter is the field h instead of D, and the critical line is dominated by a fixed point at $T = 0$ and finite field [You97] (recall Fig. 1.2). Working at large h would yes take us far from the echoes of the $h = 0$ fixed point, but there is a large risk of overshootings that would make the critical line invisible, which is a big problem especially if we are not sure whether it exists or not (see Chap. 2). One could then rely on working very close to the critical line (2), but this is very hard task when the position (or even the existence) of the critical line is unknown, or try to use extremely large lattices, that in SGs is rarely feasible because of their very sluggish dynamics [Bel08b]. The solution is to try to tune these three factors in the best possible way and to pay special attention, during the analysis, to the crossover echo effects.

For more quantitative explanations on this and other crossover behaviors see e.g. [Fis74] and [Car96] (where the previous example is taken from).

1.3.5.2 A Note on the Distribution of the Couplings

The first SG Hamiltonian, proposed in [And70, Edw75] (see Sect. 1.1.2), depended on a set of coupling J_{xy} that followed a distribution of quenched couplings $P(J_{xy})$. In [And70] $P(J_{xy})$ was meant to follow roughly the RKKY distribution, but already in [Edw75] this idea was abandoned in favor of a Gaussian distribution, for sake of simplicity. The EA model described successfully the phenomenology of the spin glass, therefore it was kept. There is no solid argument stating that the $P(J_{xy})$ should be Gaussian rather than, for instance, bimodal. The reason of this approach is often seen in theoretical physics: one simplifies the model as much as it is possible, trying to keep track of only the most fundamental traits, so Edwards and Anderson hypothesized that it was important that $P(J_{xy})$ imply frustration, but it did not have to be necessarily the real one (the one that would descend from a renormalization of the RKKY couplings), provided that the $P(J_{xy})$ is "decent" enough, and for example it has a finite variance.[17]

This said, a very large amount of Hamiltonians were proposed after the EA model, and all of them tried to pick the fundamental aspects, such as disorder, symmetries and range of the interactions, and to neglect what seemed to be unimportant, such as the exact distribution of the couplings. The Gaussian pdf has often been chosen, but depending on the context other distributions were used as well.

That these models belong to the same universality class no matter the $P(J_{xy})$ is a natural hypothesis in SG theory. If it were contradicted there would be no reason to choose one distribution over another, and all the results obtained by SG theory would have a very limited impact.

The general feeling in the SG community has always been, indeed, that the precise distribution of the couplings is an unimportant feature in their description, despite no proof has been given yet. Some doubts arose from numerical works in which different critical exponents were measured (e.g. [Ber95]), but recent careful literature suggests that it is a matter of finite-size effects, and when scaling corrections are taken in account the universality is confirmed.

This independence from microscopic details like the disorder distribution has been found for spin glasses [Hem84, Jor06, Kat06, Has08, Jor08a], but also for other disordered systems such as the Random Field Ising model [Fyt13], or disordered ferromagnets (either site [Bal98b] or bond [Ber04a, Mal12] diluted).

[17] Even though in this text we did not treat ppdf with non-zero mean, the mean of $P(J_{xy})$is generally taken as a parameter [Mez87].

References

[Adk74] K. Adkins, N. Rivier, J. Phys. Colloques **35**, 237–240 (1974). doi:10.1051/jphyscol:1974443

[AB10a] R. Alvarez Baños, A. Cruz, L.A. Fernandez, J.M. Gil-Narvion, A. Gordillo-Guerrero, M. Guidetti, A. Maiorano, F. Mantovani, E. Marinari, V. Martin-Mayor, J. Monforte-Garcia, A. Muñoz Sudupe, D. Navarro, G. Parisi, S. Perez-Gaviro, J.J. Ruiz-Lorenzo, S.F. Schifano, B. Seoane, A. Tarancon, R. Tripiccione, D. Yllanes (Janus Collaboration), J. Stat. Mech. **2010**, P06026 (2010). doi:10.1088/1742-5468/2010/06/P06026. arXiv:1003.2569

[Ami05] D.J. Amit, V. Martin-Mayor, in *Field Theory, the Renormalization Group and Critical Phenomena*, 3rd edn. (World Scientific, Singapore, 2005). doi:10.1142/9789812775313_bmatter. http://www.worldscientific.com/worldscibooks/10.1142/5715

[And70] P. Anderson, Mat. Res. Bull. **5**, 549 (1970). ISSN 0025-5408. doi:10.1016/0025-5408(70)90096-6

[And78] P.W. Anderson, C.M. Pond, Phys. Rev. Lett. **40**, 903–906 (1978). doi:10.1103/PhysRevLett.40.903. http://link.aps.org/doi/10.1103/PhysRevLett.40.903

[Ang13] M. Angelini, F. Parisi, G. Ricci-Tersenghi, Phy. Rev. B **87**, 134201 (2013). doi:10.1103/PhysRevB.87.134201. arXiv:1111.6869

[Ang95] C.A. Angell, Science **267**, 1924–1935 (1995). doi:10.1126/science.267.5206.1924; http://www.sciencemag.org/content/267/5206/1924.full.pdf; http://www.sciencemag.org/content/267/5206/1924.abstract

[BJ11] M. Baity-Jesi, in *Energy landscape in three-dimensional Heisenberg spin glasses*. Master's thesis, Sapienza, Universitá di Roma, Rome, Italy (2011). arXiv:1503.08409

[BJ12] M. Baity-Jesi, R. A. Baños, A. Cruz, L.A. Fernandez, J.M. Gil-Narvion, A. Gordillo-Guerrero, M. Guidetti, D. Iniguez, A. Maiorano, F. Mantovani, E. Marinari, V. Martin-Mayor, J. Monforte-Garcia, A. Munoz Sudupe, D. Navarro, G. Parisi, M. Pivanti, S. Perez-Gaviro, F. Ricci-Tersenghi, J.J. Ruiz-Lorenzo, S.F. Schifano, B. Seoane, A. Tarancon, P. Tellez, R. Tripiccione, D. Yllanes, Eur. Phys. J. Spec. Topics **210**, 33 (2012). doi:10.1140/epjst/e2012-01636-9. arXiv:1204.4134

[BJ13] M. Baity-Jesi, R.A. Baños, A. Cruz, L.A. Fernandez, J.M. Gil-Narvion, A. Gordillo-Guerrero, D. Iniguez, A. Maiorano, F. Mantovani, E. Marinari, V. Martin-Mayor, J. Monforte-Garcia, A. Muñoz Sudupe, D. Navarro, G. Parisi, S. Perez-Gaviro, M. Pivanti, F. Ricci-Tersenghi, J. J. Ruiz-Lorenzo, S. F. Schifano, B. Seoane, A. Tarancon, R. Tripiccione, D. Yllanes (Janus Collaboration). Phys. Rev. B **88**, 224416 (2013). doi:10.1103/PhysRevB.88.224416. arXiv:1310.2910

[BJ14c] M. Baity-Jesi, R.A. Baños, A. Cruz, L.A. Fernandez, J.M. Gil-Narvion, A. Gordillo-Guerrero, D. Iniguez, A. Maiorano, F. Mantovani, E. Marinari, V. Martin-Mayor, J. Monforte-Garcia, A. Muñoz Sudupe, D. Navarro, G. Parisi, S. Perez-Gaviro, M. Pivanti, F. Ricci-Tersenghi, J.J. Ruiz-Lorenzo, S.F. Schifano, B. Seoane, A. Tarancon, R. Tripiccione, D. Yllanes (Janus Collaboration): Comp. Phys. Comm. **185**, 550–559 (2014). doi:10.1016/j.cpc.2013.10.019. arXiv:1310.1032

[BJ14d] M. Baity-Jesi, L.A. Fernandez, V. Martin-Mayor, J.M. Sanz, Phys. Rev. **89**, 014202 (2014). doi:10.1103/PhysRevB.89.014202. arXiv:1309.1599

[Bal00] H.G. Ballesteros, A. Cruz, L.A. Fernandez, V. Martin-Mayor, J. Pech, J.J. Ruiz-Lorenzo, A. Tarancon, P. Tellez, C.L. Ullod, C. Ungil, Phys. Rev. B **62**, 14237–14245 (2000). doi:10.1103/PhysRevB.62.14237. arXiv:cond-mat/0006211

[Bal98a] H.G. Ballesteros, L.A. Fernandez, V. Martin-Mayor, A. Muñoz Sudupe, G. Parisi, J.J. Ruiz-Lorenzo. Nucl. Phys. B **512**, 681 (1998)

[Bal98b] H.G. Ballesteros, L.A. Fernandez, V. Martin-Mayor, A. Muñoz Sudupe, G. Parisi, J.J. Ruiz-Lorenzo. Phys. Rev. B **58**, 2740 (1998). doi:10.1103/PhysRevB.58.2740

[Bal96] H.G. Ballesteros, L.A. Fernandez, V. Martin-Mayor, A. Muñoz, Sudupe. Phys. Lett. B **378**, 207 (1996). doi:10.1016/0370-2693(96)00358-9. arXiv:hep-lat/9511003

[Bn12a] R.A. Baños, A. Cruz, L.A. Fernandez, J.M. Gil-Narvion, A. Gordillo-Guerrero, M. Guidetti, D. Iniguez, A. Maiorano, E. Marinari, V. Martin-Mayor, J. Monforte-Garcia, A. Muñoz Sudupe, D. Navarro, G. Parisi, S. Perez-Gaviro, J.J. Ruiz-Lorenzo, S.F. Schifano, B. Seoane, A. Tarancon, P. Tellez, R. Tripiccione, D. Yllanes, Proc. Natl. Acad. Sci. USA **109**, 6452 (2012). doi:10.1073/pnas.1203295109

[Bn12b] R.A. Baños, L.A. Fernandez, V. Martin-Mayor, A.P. Young, Phys. Rev. B **86**, 134416 (2012). doi:10.1103/PhysRevB.86.134416. arXiv:1207.7014

[Bar73] J. Barker, R. Watts, Molecular Physics **26**, 789–792 (1973). doi:10.1080/00268977300102101

[Bec71] P. Beck, Metall. Trans. **2**, 2015 (1971). ISSN 0026-086X. doi:10.1007/BF02917527. http://link.springer.com/article/10.1007/BF02917527

[Bel08a] F. Belletti, M. Cotallo, A. Cruz, L.A. Fernandez, A. Gordillo, A. Maiorano, F. Mantovani, E. Marinari, V. Martin-Mayor, A. Muñoz Sudupe, D. Navarro, S. Perez-Gaviro, J.J. Ruiz-Lorenzo, S.F. Schifano, D. Sciretti, A. Tarancon, R. Tripiccione, J.L. Velasco (Janus Collaboration), Comp. Phys. Comm. **178**, 208–216 (2008). doi:10.1016/j.cpc.2007.09.006. arXiv:0704.3573

[Bel09a] F. Belletti, A. Cruz, L.A. Fernandez, A. Gordillo-Guerrero, M. Guidetti, A. Maiorano, F. Mantovani, E. Marinari, V. Martin-Mayor, J. Monforte, A. Muñoz Sudupe, D. Navarro, G. Parisi, S. Perez-Gaviro, J.J. Ruiz-Lorenzo, S.F. Schifano, D. Sciretti, A. Tarancon, R. Tripiccione, D. Yllanes (Janus Collaboration). J. Stat. Phys. **135**, 1121 (2009). doi:10.1007/s10955-009-9727-z. arXiv:0811.2864

[Bel06] F. Belletti, F. Mantovani, G. Poli, S.F. Schifano, R. Tripiccione, I. Campos, A. Cruz, D. Navarro, S. Perez-Gaviro, D. Sciretti, A. Tarancon, J.L. Velasco, P. Tellez, L.A. Fernandez, V. Martin-Mayor, A. Muñoz Sudupe, S. Jimenez, A. Maiorano, E. Marinari, J.J. Ruiz-Lorenzo (Janus Collaboration). Comput. Sci. Eng. **8**, 41 (2006)

[Bel08b] F. Belletti, M. Cotallo, A. Cruz, L.A. Fernandez, A. Gordillo-Guerrero, M. Guidetti, A. Maiorano, F. Mantovani, E. Marinari, V. Martin-Mayor, A.M. Sudupe, D. Navarro, G. Parisi, S. Perez-Gaviro, J.J. Ruiz-Lorenzo, S.F. Schifano, D. Sciretti, A. Tarancon, R. Tripiccione, J.L. Velasco, D. Yllanes, Janus Collaboration. Phys. Rev. Lett. **101**, 157201 (2008). doi:10.1103/PhysRevLett.101.157201. arXiv:0804.1471

[Ber04a] P.-E. Berche, C. Chatelain, B. Berche, W. Janke, Euro. Phys. J. B **38**, 463 (2004)

[Ber95] L. Bernardi, I.A. Campbell, Phys. Rev. B **52**, 12501–12504 (1995). doi:10.1103/PhysRevB.52.12501. http://link.aps.org/doi/10.1103/PhysRevB.52.12501

[Bia12] I. Biazzo, A. Braunstein, R. Zecchina, Phys. Rev. E **86**, 026706 (2012). doi:10.1103/PhysRevE.86.026706; http://link.aps.org/doi/10.1103/PhysRevE.86.026706

[Bil13] A. Billoire, L.A. Fernandez, A. Maiorano, E. Marinari, V. Martin-Mayor, G. Parisi, F. Ricci-Tersenghi, J.J. Ruiz-Lorenzo, D. Yllanes, Phys. Rev. Lett. **110**, 219701 (2013). doi:10.1103/PhysRevLett.110.219701. arXiv:1211.0843

[Bin82] K. Binder, Phys. Rev. A **25**, 1699 (1982)

[Bin86] K. Binder, A.P. Young, Rev. Mod. Phys. **58**, 801 (1986)

[Bla78] A. Blandin, J. Phys. Colloques **39**, 1499–1516 (1978). doi:10.1051/jphyscol:19786593. http://jphyscol.journaldephysique.org/articles/jphyscol/abs/1978/06/jphyscol197839C6593/jphyscol197839C6593.html

[Bla93] J. Blanshard, P. Lillford, *The Glassy State in Foods*, 1st edn. (Nottingham University Press, Nottingham, 1993)

[Blo55] N. Bloembergen, T.J. Rowland, Phys. Rev. **97**, 1679–1698 (1955). doi:10.1103/PhysRev.97.1679. http://link.aps.org/doi/10.1103/PhysRev.97.1679

[Bray85] A.J. Bray, M.A. Moore, Phys. Rev. B **31**, 631–633 (1985). doi:10.1103/PhysRevB.31.631. http://link.aps.org/doi/10.1103/PhysRevB.31.631

[Bra78] A.J. Bray, M.A. Moore, Phys. Rev. Lett. **41**, 1068–1072 (1978). doi:10.1103/PhysRevLett.41.1068. http://link.aps.org/doi/10.1103/PhysRevLett.41.1068

[Bra79] A. Bray, M. Moore, J. Phys. C: Solid State Phys. **12**, 79 (1979). doi:10.1088/0022-3719/12/1/020. http://stacks.iop.org/0022-3719/12/i=1/a=020

[Bra80a] A. Bray, M. Moore, J. Phys. C: Solid State Phys. **13**, L469 (1980). doi:10.1088/0022-3719/13/19/002. http://stacks.iop.org/0022-3719/13/i=19/a=002

[Bro59] R. Brout, Phys. Rev. **115**, 824–835 (1959). doi:10.1103/PhysRev.115.824. http://link.aps.org/doi/10.1103/PhysRev.115.824

[Cam06] I. Campos, M. Cotallo-Aban, V. Martin-Mayor, S. Perez-Gaviro, A. Tarancon, Phys. Rev. Lett. **97**, 217204 (2006). doi:10.1103/PhysRevLett.97.217204

[Can72] V. Cannella, J.A. Mydosh, Phys. Rev. B **6**, 4220–4237 (1972). doi:10.1103/PhysRevB.6.4220. http://link.aps.org/doi/10.1103/PhysRevB.6.4220

[Car96] J. Cardy, in *Scaling and Renormalization in Statistical Field Theory*, vol. 5, *Lecture Notes in Physics* (P. Goddard and J. Yeomans, Cambridge University Press, Cambridge, 1996). ISBN 0521499593

[Cav09] A. Cavagna, Phys. Rep. **476**, 51–124 (2009). arXiv:0903.4264

[Cha14] P. Charbonneau, J. Kurchan, G. Parisi, P. Urbani, F. Zamponi, Nat. Comm. **5**, 3725 (2014). doi:10.1038/ncomms4725. arXiv:1404.6809

[Che77] J.-H. Chen, T.C. Lubensky, Phys. Rev. B **16**, 2106–2114 (1977). doi:10.1103/PhysRevB.16.2106. http://link.aps.org/doi/10.1103/PhysRevB.16.2106

[Col95] B. Coluzzi, J. Phys. A: Math. Gen. **28**, 747 (1995)

[Con06] P. Contucci, C. Giardinà, C. Giberti, C. Vernia, Phys. Rev. Lett. **96**, 217204 (2006).doi:10.1103/PhysRevLett.96.217204

[Con03] P. Contucci, J. Phys. A: Math. Gen. **36**, 10961 (2003). doi:10.1088/0305-4470/36/43/020. arXiv:cond-mat/0302500, http://stacks.iop.org/0305-4470/36/i=43/a=020

[Con05a] P. Contucci, C. Giardinà, Ann. Henri Poincare **6**, 915 (2005)

[Con05b] P. Contucci, C. Giardinà, Phys. Rev. B **72**, 014456 (2005). doi:10.1103/PhysRevB.72.014456

[Cro98] J. Crowe, J. Carpenter, L. Crowe, Annu. Rev. Physiol. **60**, 73–103 (1998). doi:10.1146/annurev.physiol.60.1.73. PMID: 9558455; doi:10.1146/annurev.physiol.60.1.73

[Alm78a] J. de Almeida, R. Jones, J. Kosterlitz, D. Thouless, J. Phys. C: Solid State Phys. **11**, L871 (1978). doi:10.1088/0022-3719/11/21/005. http://stacks.iop.org/0022-3719/11/i=21/a=005

[Alm78b] J.R.L. de Almeida, D.J. Thouless, J. Phys. A: Math. Gen. **11**, 983 (1978). doi:10.1088/0305-4470/11/5/028. http://stacks.iop.org/0305-4470/11/i=5/a=028

[Dom80] C. de Dominicis, M. Gabay, T. Garel, H. Orland, J. Phys. France **41**, 923–930 (1980). doi:10.1051/jphys:01980004109092300. http://dx.doi.org/10.1051/jphys:01980004109092300

[Dom06] C. de Dominicis, I. Giardina, *Random Fields and Spin Glasses: a field theory approach* (Cambridge University Press, Cambridge, 2006)

[Nob59] J. de Nobel, F. Chatenier, Physica **25**, 969–979 (1959). ISSN 0031-8914. doi:10.1016/0031-8914(59)90018-7. http://www.sciencedirect.com/science/article/pii/0031891459900187

[Der81] B. Derrida, Phys. Rev. B **24**, 2613–2626 (1981). doi:10.1103/PhysRevB.24.2613. http://link.aps.org/doi/10.1103/PhysRevB.24.2613

[Dot87] V. Dotsenko, J. Phys. C: Solid State Phys. **20**, 5473 (1987). doi:10.1088/0022-3719/20/33/005. http://stacks.iop.org/0022-3719/20/i=33/a=005

[Dot01] V. Dotsenko, *Introduction to the Replica Theory of Disordered Statistical Systems* (Cambridge University Press, Cambridge, 2001)

[Dru07] D. Drung, C. Assmann, J. Beyer, A. Kirste, M. Peters, F. Ruede, T. Schurig, IEEE Trans. Appl. Supercond. **17**, 699–704 (2007). ISSN 1051-8223. doi:10.1109/TASC.2007.897403

[Edw76] S.F. Edwards, P.W. Anderson, J. Phys. F **6**, 1927 (1976). doi:10.1088/0305-4608/6/10/022. http://stacks.iop.org/0305-4608/6/i=10/a=022

[Edw75] S.F. Edwards, P.W. Anderson, J. Phys. F: Metal Phys. **5**, 965 (1975). doi:10.1088/0305-4608/5/5/017. http://stacks.iop.org/0305-4608/5/i=5/a=017

[Fer15] L.A. Fernández, V. Martín-Mayor, Phys. Rev. B **91**, 174202 (2015). doi:10.1103/PhysRevB.91.174202. http://link.aps.org/doi/10.1103/PhysRevB.91.174202

[Fer09b] L.A. Fernandez, V. Martin-Mayor, S. Perez-Gaviro, A. Tarancon, A.P. Young, Phys. Rev. B **80**, 024422 (2009). doi:10.1103/PhysRevB.80.024422

[Fer09c] L.A. Fernandez, V. Martin-Mayor, D. Yllanes, Nucl. Phys. B **807**, 424–454 (2009)

[Fis75] K.H. Fischer, Phys. Rev. Lett. **34**, 1438–1441 (1975). doi:10.1103/PhysRevLett.34.1438. http://link.aps.org/doi/10.1103/PhysRevLett.34.1438

[Fis74] D. Fisher, Rev. Mod. Phys. **46**, 597 (1974)

[Fis86] D.S. Fisher, D.A. Huse, Phys. Rev. Lett. **56**, 1601 (1986). doi:10.1103/PhysRevLett.56.1601. http://link.aps.org/doi/10.1103/PhysRevLett.56.1601

[Fis87] D.S. Fisher, D.A. Huse, J. Phys. A: Math. Gen. **20**, L1005 (1987). doi:10.1088/0305-4470/20/15/013

[Fis88a] D.S. Fisher, D.A. Huse, Phys. Rev. B **38**, 373 (1988). doi:10.1103/PhysRevB.38.373

[Fis88b] D.S. Fisher, D.A. Huse, Phys. Rev. B **38**, 386 (1988). doi:10.1103/PhysRevB.38.386

[Fro40] H. Fröhlich, F.R.N. Nabarro, Proc. R. Soc. Lond. A: Math. Phys. Eng. Sci. **175**, 382–391 (1940). ISSN 0080-4630. doi:10.1098/rspa.1940.0064

[Fyt13] N.G. Fytas, V. Martin-Mayor, Phys. Rev. Lett. **110**, 227201 (2013). doi:10.1103/PhysRevLett.110.227201. arXiv:1304.0318

[Gol92] R. Goldstein, Z. Luthey-Schulten, P. Wolynes, Proc. Natl. Acad. Sci. **89**, 4918–4922 (1992)

[Gre95] A.L. Greer, Science **267**, 1947–1953 (1995). doi:10.1126/science.267.5206.1947; http://www.sciencemag.org/content/267/5206/1947

[Gue03] F. Guerra, Comm. Math. Phys.**233**, 1–12 (2003). doi:10.1007/s00220-002-0773-5. arXiv:cond-mat/0205123

[Gue02] F. Guerra, F. L. Toninelli, Commun. Math. Phys. **230**, 71–79 (2002). ISSN 0010-3616. doi:10.1007/s00220-002-0699-y. http://dx.doi.org/10.1007/s00220-002-0699-y

[Har76] A.B. Harris, T.C. Lubensky, J.-H. Chen, Phys. Rev. Lett. **36**, 415–418 (1976). doi:10.1103/PhysRevLett.36.415. http://link.aps.org/doi/10.1103/PhysRevLett.36.415

[Has08] M. Hasenbusch, A. Pelissetto, E. Vicari, J. Stat. Mech. **2008**, L02001 (2008). doi:10.1088/1742-5468/2008/02/L02001

[Her02] D. Hérisson, M. Ocio, Phys. Rev. Lett. **88**, 257202 (2002). doi:10.1103/PhysRevLett.88.257202; arXiv:cond-mat/0112378

[Hua87] K. Huang, *Statistical Mechanics*, 2nd edn. (Wiley, Hoboken, 1987)

[Hus87] D.A. Huse, D.S. Fisher, Phys. Rev. B **35**, 6841–6846 (1987). doi:10.1103/PhysRevB.35.6841. http://link.aps.org/doi/10.1103/PhysRevB.35.6841

[Imr75] Y. Imry, S.-K. Ma, Phys. Rev. Lett. **35**, 1399–1401 (1975). doi:10.1103/PhysRevLett.35.1399. http://link.aps.org/doi/10.1103/PhysRevLett.35.1399

[Jon98] K. Jonason, E. Vincent, J. Hammann, J.-P. Bouchaud, P. Nordblad, Phys. Rev. Lett. **81**, 3243 (1998). doi:10.1103/PhysRevLett.81.3243

[Jor08a] T. Jörg, H.G. Katzgraber, Phys. Rev. Lett. **101**, 197205 (2008). doi:10.1103/PhysRevLett.101.197205

[Jor08b] T. Jörg, H.G. Katzgraber, F. Krzakala, Phys. Rev. Lett. **100**, 197202 (2008). doi:10.1103/PhysRevLett.100.197202. arXiv:0712.2009

[Jor06] T. Jörg, Phys. Rev. B **73**, 224431 (2006). doi:10.1103/PhysRevB.73.224431

[Kad76] L. Kadanoff, Ann. Phys. **100**, 359–394 (1976). ISSN 0003-4916. doi:10.1016/0003-4916(76)90066-X

[Kas56] T. Kasuya, Prog. Theor. Phys. **16**, 45 (1956)

[Kat06] H.G. Katzgraber, M. Körner, A.P. Young, Phys. Rev. B **73**, 224432 (2006). doi:10.1103/PhysRevB.73.224432

[Kaw10] H. Kawamura, J. Phys.: Conf. Ser. **233**, 012012 (2010). doi:10.1088/1742-6596/233/1/012012. http://stacks.iop.org/1742-6596/233/i=1/a=012012

[Kir78] S. Kirkpatrick, D. Sherrington, Phys. Rev. B **17**, 4384–4403 (1978). doi:10.1103/PhysRevB.17.4384. http://link.aps.org/doi/10.1103/PhysRevB.17.4384

[Kle63] M.W. Klein, R. Brout, Phys. Rev. **132**, 2412–2426 (1963). doi:10.1103/PhysRev.132.2412. http://link.aps.org/doi/10.1103/PhysRev.132.2412

[Kum14] P.A. Kumar, R. Mathieu, P. Nordblad, S. Ray, O. Karis, G. Andersson, D.D. Sarma, Phys. Rev. X **4**, 011037 (2014). doi:10.1103/PhysRevX.4.011037; http://link.aps.org/doi/10.1103/PhysRevX.4.011037
[Lee03] L.W. Lee, A.P. Young, Phys. Rev. Lett. **90**, 227203 (2003). doi:10.1103/PhysRevLett.90.227203
[Leu08] L. Leuzzi, G. Parisi, F. Ricci-Tersenghi, J.J. Ruiz-Lorenzo, Phys. Rev. Lett. **101**, 107203 (2008). doi:10.1103/PhysRevLett.101.107203
[Leu09] L. Leuzzi, G. Parisi, F. Ricci-Tersenghi, J.J. Ruiz-Lorenzo, Phys. Rev. Lett. **103**, 267201 (2009). doi:10.1103/PhysRevLett.103.267201. arXiv:0811.3435
[Lul15] M. Lulli, G. Parisi, A. Pelissetto, Out-of-equilibrium measure of critical parameters for second-order phase transitions (2015). In preparation
[Ma76] S.-K. Ma, *Modern Theory of Critical Phenomena* (Westview Press, New York, 1976)
[Mac82] N.D. Mackenzie, A.P. Young, Phys. Rev. Lett. **49**, 301–304 (1982). doi:10.1103/PhysRevLett.49.301. http://link.aps.org/doi/10.1103/PhysRevLett.49.301
[Mal12] A. Malakis, A. Nihat Berker, N.G. Fytas, T. Papakonstantinou. Phys. Rev. E **85**, 061106 (2012)
[Mar00b] E. Marinari, G. Parisi, F. Ricci-Tersenghi, J.J. Ruiz-Lorenzo, F. Zuliani, J. Stat. Phys. **98**, 973 (2000). doi:10.1023/A:1018607809852. arXiv:cond-mat/9906076
[Mar06] G. Marsh, 50 years of reinforced plastic boats (2006). http://www.materialstoday.com/composite-applications/features/50-years-of-reinforced-plastic-boats/
[Mar60] W. Marshall, Phys. Rev. **118**, 1519–1523 (1960). doi:10.1103/PhysRev.118.1519. http://link.aps.org/doi/10.1103/PhysRev.118.1519
[Mar01] L.-M. Martinez, C.A. Angell, Nature **410**, 663–667 (2001). doi:10.1038/35070517
[Mat81] D. Mattis, in *Theory of Magnetism I: Statics and Dynamics*, 1st edn. (Springer, Berlin, 1981). doi:10.1007/978-3-642-83238-3
[Mat76] D. Mattis, Phys. Lett. A **56**, 421 (1976)
[May93] R. Mayer, in *Design with Reinforced Plastics: A Guide for Engineers and Designers*, 1st edn. (Springer Science & Business Media, Netherlands, 1993). ISBN 978-0-85072-294-9, 978-94-011-2210-8
[McM85] W.L. McMillan, Phys. Rev. B **31**, 340–341 (1985). doi:10.1103/PhysRevB.31.340. http://link.aps.org/doi/10.1103/PhysRevB.31.340
[McM84] W.L. McMillan, J. Phys. C: Solid State Phys. **17**, 3179 (1984). doi:10.1088/0022-3719/17/18/010
[Mez84] M. Mézard, G. Parisi, N. Sourlas, G. Toulouse, M. Virasoro, Phys. Rev. Lett. **52**, 1156 (1984). doi:10.1103/PhysRevLett.52.1156
[Mez86] M. Mézard, G. Parisi, M. Virasoro, Europhys. Lett. **1**, 77 (1986). doi:10.1209/0295-5075/1/2/006
[Mez87] M. Mézard, G. Parisi, M. Virasoro, *Spin-Glass Theory and Beyond* (World Scientific, Singapore, 1987)
[Mez85] M. Mézard, M. Virasoro, J. Physique **46**, 1293–1307 (1985). doi:10.1051/jphys:019850046080129300
[Mig75] A. Migdal, Zhurnal Eksperimentalnoi i teoreticheskoi fiziki (1975). ISSN 0044-4510
[Mit57] A.H. Mitchell, Phys. Rev. **105**, 1439–1444 (1957). doi:10.1103/PhysRev.105.1439. http://link.aps.org/doi/10.1103/PhysRev.105.1439
[Mon70] C.G. Montgomery, J.I. Krugler, R.M. Stubbs, Phys. Rev. Lett. **25**, 669–672 (1970). doi:10.1103/PhysRevLett.25.669. http://link.aps.org/doi/10.1103/PhysRevLett.25.669
[Moo11] M.A. Moore, A.J. Bray, Phys. Rev. B **83**, 224408 (2011). doi:10.1103/PhysRevB.83.224408. arXiv:1102.1675
[Myd93] J.A. Mydosh, *Spin Glasses: An Experimental Introduction* (Taylor and Francis, London, 1993)
[Nag79] S. Nagata, P.H. Keesom, H.R. Harrison, Phys. Rev. B **19**, 1633–1638 (1979). doi:10.1103/PhysRevB.19.1633; http://link.aps.org/doi/10.1103/PhysRevB.19.1633
[Nig75] M.P. Nightingale, Phys. A **83**, 561 (1975). doi:10.1016/0378-4371(75)90021-7

[Nis01] H. Nishimori, in *Statistical Physics of Spin Glasses and Information Processing* (Oxford University Press, Oxford, 2001). http://oxfordindex.oup.com/view/10.1093/acprof:oso/9780198509417.003.0004
[Ons36] L. Onsager, J. Am. Chem. Soc. **58**, 1486–1493 (1936). doi:10.1021/ja01299a050
[Owe56] J. Owen, M. Browne, W.D. Knight, C. Kittel, Phys. Rev. **102**, 1501–1507 (1956). doi:10.1103/PhysRev.102.1501
[Par95] G. Parisi, in *The Oscar Klein Centenary*, ed. by U. Lindström (World Scientific, Singapore, 1995). http://arxiv.org/abs/cond-mat/9411115
[Par79a] G. Parisi, Phys. Rev. Lett. **43**, 1754 (1979). doi:10.1103/PhysRevLett.43.1754
[Par83] G. Parisi, Phys. Rev. Lett. **50**, 1946 (1983). doi:10.1103/PhysRevLett.50.1946. http://journals.aps.org/prl/abstract/10.1103/PhysRevLett.50.1946
[Par96] G. Parisi, Recent rigorous results support the predictions of spontaneously broken replica symmetry for realistic spin glasses (1996). Reply to [New96]. arXiv:cond-mat/9603101
[Par80a] G. Parisi, J. Phys. A: Math. Gen. **13**, 1887 (1980). doi:10.1088/0305-4470/13/5/047. http://stacks.iop.org/0305-4470/13/i=5/a=047
[Par80c] G. Parisi, J. Phys. A: Math. Gen.**13**, L115–L121 (1980). ISSN 0305-4470. doi:10.1088/0305-4470/13/4/009
[Par79b] G. Parisi, Phys. Lett. **73A**, 203 (1979). doi:10.1016/0375-9601(79)90708-4
[Par80b] G. Parisi, J. Phys. A: Math. Gen. **13**, 1101 (1980). doi:10.1088/0305-4470/13/3/042
[Par12b] G. Parisi, T. Temesvári, Nucl. Phys. B **858**, 293 (2012). arXiv:1111.3313
[Ram86] R. Rammal, G. Toulouse, M.A. Virasoro, Rev. Mod. Phys. **58**, 765–788 (1986). doi:10.1103/RevModPhys.58.765. http://link.aps.org/doi/10.1103/RevModPhys.58.765
[Rud54] M. Ruderman, C. Kittel, Phys. Rev. **96**, 99 (1954)
[She07] D. Sherrington, in *Spin Glasses*, ed. by E. Bolthausen, A. Bovier, volume 1900 of *Lecture Notes in Mathematics* (Springer, Berlin, 2007), pp. 45–62. ISBN 978-3-540-40902-1. doi:10.1007/978-3-540-40908-3_2; arXiv:cond-mat/0512425
[She75a] D. Sherrington, S. Kirkpatrick, Phys. Rev. Lett. **35**, 1792–1796 (1975). doi:10.1103/PhysRevLett.35.1792. http://link.aps.org/doi/10.1103/PhysRevLett.35.1792
[She75b] D. Sherrington, W. Southern, J. Phys. F: Metal Phys. **5**, L49 (1975). doi:10.1088/0305-4608/5/5/003. http://stacks.iop.org/0305-4608/5/i=5/a=003
[Smi74] D.A. Smith, J. Phys. F: Metal Phys. **4**, L266 (1974). doi:10.1088/0305-4608/4/12/003. http://stacks.iop.org/0305-4608/4/i=12/a=003
[Sou77] B.W. Southern, A.P. Young, J. Phys. C: Solid State Phys. **10**, 2179 (1977). doi:10.1088/0022-3719/10/12/023. http://stacks.iop.org/0022-3719/10/i=12/a=023
[Tal06] M. Talagrand, Ann. Math. **163**, 221 (2006)
[Tho77] D.J. Thouless, P.W. Anderson, R.G. Palmer, Phil. Mag. **35**, 593–601 (1977). doi:10.1080/14786437708235992
[Tou77] G. Toulouse, Commun. Phys. **2**, 115 (1977)
[Hem79] J. van Hemmen, R. Palmer, J. Phys. A: Math. Gen. **12**, 563 (1979). http://stacks.iop.org/0305-4470/12/i=4/a=016
[Hem84] J. van Hemmen, A. Suto, Journal de physique Paris **45**, 1277–1281 (1984). ISSN 0302-0738. doi:10.1051/jphys:019840045080127700
[VV62] J.H. Van Vleck, Rev. Mod. Phys. **34**, 681–686 (1962). doi:10.1103/RevModPhys.34.681. http://link.aps.org/doi/10.1103/RevModPhys.34.681
[Vin97] E. Vincent, J. Hammann, M. Ocio, J.-P. Bouchaud, L.F. Cugliandolo, in *Complex Behavior of Glassy Systems*, ed. by M. Rubí, C. Pérez-Vicente, number 492 in Lecture Notes in Physics (Springer, 1997)
[Web13] J. Weber, R. Jack, V. Pande, J. Am. Chem. Soc. **135**, 5501–5504 (2013). doi:10.1021/ja4002663
[Wid65] B. Widom, J. Chem. Phys. **43**, 3898 (1965)
[Wol92] P.G. Wolynes, Spin Glass Ideas and the Protein Folding Problems. Chap. **8**, pp. 225–259 (1992). doi:10.1142/9789814415743_0008. http://www.worldscientific.com/doi/pdf/10.1142/9789814415743_0008, http://www.worldscientific.com/doi/abs/10.1142/9789814415743_0008

[Yeo12] J. Yeo, M.A. Moore, Phys. Rev. E **86**, 052501 (2012). arXiv:1208.3044

[Yos57] K. Yosida, Phys. Rev. **106**, 893–898 (1957). doi:10.1103/PhysRev.106.893. http://link.aps.org/doi/10.1103/PhysRev.106.893

[You97] A.P. Young, in *Spin Glasses and Random Fields*, ed. by A.P. Young (World Scientific, Singapore, 1997)

[You81] A. Young, J. Phys. C: Solid State Phys. **14**, L1085 (1981). doi:10.1088/0022-3719/14/34/004. http://stacks.iop.org/0022-3719/14/i=34/a=004

[You76] A.P. Young, R.B. Stinchcombe, J. Phys. C: Solid State Phys. **9**, 4419 (1976). doi:10.1088/0022-3719/9/24/012. http://stacks.iop.org/0022-3719/9/i=24/a=012

[You05] A. Young, Pramana. J. Phys. **64**, 1087–1096 (2005). doi:10.1007/BF02704170

[Yuc12] B. Yucesoy, H.G. Katzgraber, J. Machta, Phys. Rev. Lett. **109**, 177204 (2012). doi:10.1103/PhysRevLett.109.177204. arXiv:1206.0783, http://link.aps.org/doi/10.1103/PhysRevLett.109.177204

[Yuc13] B. Yucesoy, H.G. Katzgraber, J. Machta, Phys. Rev. Lett. **110**, 219702 (2013). doi:10.1103/PhysRevLett.111.219702. arXiv:1304.5210

[Zen51a] C. Zener, Phys. Rev. **81**, 440–444 (1951). doi:10.1103/PhysRev.81.440. http://link.aps.org/doi/10.1103/PhysRev.81.440

[Zen51b] C. Zener, Phys. Rev. **82**, 403–405 (1951). doi:10.1103/PhysRev.82.403. http://link.aps.org/doi/10.1103/PhysRev.82.403

[Zen51c] C. Zener, Phys. Rev. **83**, 299–301 (1951). doi:10.1103/PhysRev.83.299. http://link.aps.org/doi/10.1103/PhysRev.83.299

[Zim60] J. Zimmerman, J. Phys. Chem. Solids **17**, 52–56 (1960). An additional author, F.E. Hoare, appears in Edwards' citation of this article. http://www.sciencedirect.com/science/article/pii/0022369760901748

Part II
Criticality

Chapter 2
The Ising Spin Glass in a Field

This chapter is dedicated to the search of a would-be phase transition in a three-dimensional spin glass. The discussion will focus entirely on the analysis of the data and on the results. We want to stress, in this context, that the equilibrium MC simulations performed in [BJ14a] required huge numerical efforts. On the one hand, lower temperatures drastically increased thermalization times; on the other hand, the significance of the results depends upon the size of the systems we are able to simulate.

The problem of enhancing the reach of our simulations is faced by resorting to advanced algorithms and techniques, such as PT[1] and MSC,[2] but that is still not enough. It would not have been possible to attain the results published in [BJ14a] with the mere use of ordinary computational resources. We drew upon (HPC) on one side by making use of the JANUS dedicated computer to simulate the largest lattices, and on the other by simulating the smaller systems on a large CPU cluster, *Memento*.

2.1 The de Almeida-Thouless Line in Three Dimensions

In Sect. 1.1.2 we explained that the nature of the SG phase in three dimensions it is still matter of debate. The two dominant theories are the droplet picture and the RSB scenario, and they have different predictions on the presence of a SG phase in a field. In the droplet picture even the smallest applied magnetic field destroys the SG phase, while in the RSB scenario there is a DAT line $h_c(T)$ that separates the SG from the paramagnetic phase.

A rather obvious way out would be the experimental study of spin glasses in a field. Unfortunately, opposing indications have been gleaned over the existence of a phase transition [Jön05, Pet99, Pet02, Tab10].

[1] A short discussion on PT is given in Appendix A.1.

[2] In Appendix B.3 we describe how MSC was implemented in the analysis stage. Multi-spin coding MC in the simulations [Seo13] follows roughly the same principles than in the analyses.

M. Baity Jesi, *Spin Glasses*, Springer Theses, DOI 10.1007/978-3-319-41231-3_2

The RG approach to this problem also provides conflicting results. No FPs were found by enforcing that the number of replicas of the replicated field theory be zero [Bra80b]. However, FPs were found relaxing this condition and using the most general Hamiltonian [Tem02]. Reasoning along this line, in [Tem08] (see also [Par12b]) the DAT line was computed for d slightly below the upper critical dimension $d_u = 6$ (the upper critical dimension remains 6 when an external magnetic field is applied).

Equilibrium numerical simulations offer an alternative approach, which has already been effective in establishing that a phase transition does occur at zero field in the $d = 3$ Edwards–Anderson model [Pal99a, Bal00] (in agreement with experiments [Gun91]). The same strategy has been followed for $h > 0$, with negative results [You04, Jör08b]. Yet, this cannot be the whole story: Recent work in $d = 4$, hence below d_u, using a non-standard finite-size scaling method has found clear evidence for a DAT line [Bn12a]. Furthermore, one may try to interpolate between $d = 3$ and $d = 4$ by tuning long-range interactions in $d = 1$ chains [Kot83, Leu08]. This approach suggests that a DAT might be present in $d = 4$, but not in $d = 3$ [Lar13] (yet, see the criticism in [Leu13]).

The problem being still open, in [BJ14b] we undertook a dynamical study of the 3-dimensional EA spin glass with the JANUS dedicated computer [Bel06, Bel08a, Bel09b, AB10a, Bn11, BJ12]. We studied very large lattices ($L^3 = 80^3$), in wide time scales (from an equivalent of $\sim$1 ps to $\sim$0.01 s), and gathered both equilibrium and non-equilibrium data. We focused on the increase of relaxation times and found a would-be dynamical transition, but at a suspiciously high temperature. A subsequent examination of the correlation length found a growth faster than predicted by the droplet theory, and slower than what RSB would expect. We also examined the problem from a supercooled liquid point of view [Deb97, Deb01, Cav09, Cas05, Kir87, Kir89], motivated by the equivalence of universality classes between spin and structural glasses [Moo02, Ful13]. At any rate, the study of the possible critical divergence of the correlation length allowed us to give upper bounds $T^{\mathrm{up}}(h)$ to the possible transition line for the studied fields.

The impossibility to get concluding evidence in [BJ14b], may be due to the fact that we did not reach low enough temperatures (our simulations fell out of equilibrium at temperatures T significantly higher than $T^{\mathrm{up}}(h)$). In any case, a study of the equilibrium properties of the model is mandatory if one wants to understand the nature of the thermodynamic phases of the three-dimensional EA spin glass in a field.

In this dissertation we will not talk about the aforementioned out-of-equilibrium results [BJ14b]. We will instead focus on the result of equilibrium simulations performed on JANUS, using lattices up to $L = 32$ [BJ14a].[3] For further reference we recall that $T^{\mathrm{up}}(h = 0.1) = 0.8$ and $T^{\mathrm{up}}(h = 0.2) = T^{\mathrm{up}}(h = 0.3) = 0.5$. Analogously to what has been already found in mean-field spin glasses on the DAT line, we find extreme fluctuations in the model's behavior [Par12a]. We will propose a

[3]In [BJ14b] we studied a bimodal field, while in the work we present here h is constant. Notwithstanding, we will make comparisons with the bounds $T^{\mathrm{up}}(h)$ by matching $\overline{h^2}$ in both models.

method to tame these fluctuations, and we will find out that, although the average behavior does not show any sign of a phase transition, this is not true for the medians of our observables, where we have indications of a possible phase transition at a temperature $T_c \lesssim T^{up}(h)$.

2.2 Model and Simulations

2.2.1 The 3d Edwards–Anderson Model in a Field

We consider a $3d$ cubic lattice of linear size L with periodic boundary conditions. In each of the $N = L^3$ vertices of the lattice there is an Ising spin $s_x = \pm 1$. The spins interact uniquely with their nearest neighbors and with an external magnetic field h. The Hamiltonian is

$$\mathcal{H}_h = -\frac{1}{2} \sum_{|x-y|=1} J_{xy} s_x s_y - h \sum_x s_x, \tag{2.1}$$

where the couplings J_{xy}, which are constant during each simulation, take the values ± 1 with equal probability (quenched disorder). As already stated in Sect. 1.2.4, a given instance of the bonds J_{xy} and of the intensity of the magnetic field h define a *sample*. We will consider real *replicas* of each sample, i.e., systems with identical couplings J_{xy} and field h, but independent evolutions (for a recent discussion see [Bel09a, AB10a]). In this work we will use 4 replicas per sample.

2.2.2 The Simulations

For all our simulations we made use of PT.[4] The whole procedure was very similar to the one in [Bn12a].

The smaller lattices ($L = 6, 8, 12$) were simulated with MSC (C code with words of 128 bits, by means of streaming extensions) [New99, Bn12a, Seo13] on the *Memento* CPU cluster at BIFI. See details on MSC in Appendix B.3. The larger samples ($L = 16, 24, 32$) were simulated on the JANUS computer [Bel06, BJ12].

An EMCS consisted in 1 PT exchange every 10 Metropolis steps for the MSC samples, and 1 PT every 10 (HB) for the samples simulated on JANUS. Table 2.1 shows the relevant parameters of the simulations. The temperatures were equally spaced between $T_{\min}$ and $T_{\max}$. The intensities of the external magnetic field we chose are $h = 0.05, 0.1, 0.2$ and 0.4.

---

[4] See the short note in Appendix A.1.

Table 2.1 Parameters of the simulations

h	L	N_{samples}	$N_{\text{EMCS}}^{\text{min}}$	f_{max}	N_{τ}^{min}	N_T	T_{min}	T_{max}
0.05	6	25600	1.6×10^6	1	40.0	14	0.5	1.8
0.05	8	25600	3.2×10^6	16	40.0	14	0.5	1.8
0.05	12	25600	3.2×10^6	16	15.6	12	0.7	1.8
0.05	16	12800	1.28×10^7	128	20.1	24	0.6	1.75
0.05	24	6400	1.28×10^7	110	16.0	20	0.78	1.54
0.05	32	2400	6.4×10^7	256	14.3	30	0.805128	1.54872
0.1	6	25600	1.6×10^6	4	40.0	14	0.5	1.8
0.1	8	25600	3.2×10^6	16	40.0	14	0.5	1.8
0.1	12	25600	3.2×10^6	16	14.4	12	0.7	1.8
0.1	16	12800	1.28×10^7	256	27.9	24	0.6	1.75
0.1	24	3200	1.28×10^7	4097	14.3	24	0.66	1.58
0.1	32	1600	6.4×10^7	533	14.4	30	0.805128	1.54872
0.2	6	25600	1.6×10^6	1	40.0	14	0.5	1.8
0.2	8	25600	3.2×10^6	16	40.0	14	0.5	1.8
0.2	12	25600	3.2×10^6	64	25.4	12	0.7	1.8
0.2	16	12800	1.28×10^7	256	18.4	24	0.6	1.75
0.2	24	3200	1.28×10^7	512	16.1	24	0.66	1.58
0.2	32	1600	1.6×10^7	513	16.0	30	0.805128	1.54872
0.4	6	25600	1.6×10^6	1	40.0	14	0.5	1.8
0.4	8	25600	3.2×10^6	4	30.7	14	0.5	1.8
0.4	12	25600	3.2×10^6	16	14.1	12	0.7	1.8
0.4	16	3200	1.28×10^7	32	20.1	24	0.6	1.75
0.4	24	800	1.28×10^7	29	16.1	24	0.66	1.58
0.4	32	800	3.2×10^6	16	16.4	30	0.805128	1.54872

We report the magnetic field h, the lattice linear size L, the number of simulated samples N_{samples}, and the basic length of a simulation in EMCS $N_{\text{EMCS}}^{\text{min}}$. In each simulation we measured the exponential correlation time τ of the PT random walk in temperatures. When τ was too large to meet our thermalization requirements, we extended the length of each simulation by an extension factor f. We denote with f_{max} the greatest extension factor. We also give the minimum length of a simulation N_{τ}^{min} in units of τ. In all cases we imposed $N_{\tau}^{\text{min}} > 14$. Finally, we give the number of temperatures N_T we used for the PT, and the minimum and maximum temperatures T_{min} and T_{max}

To check whether the samples were thermalized we measured the exponential autocorrelation time of the PT random walk in temperatures τ [Fer09b, AB10a, Yl11, Bn12a]. We required the simulations to last at least 14τ. To do so without consuming computing time on already thermalized lattices, we assigned a minimum number of EMCS, $N_{\text{EMCS}}^{\text{min}}$, for all the samples, and extended by a factor $f > 1$ only the ones that did not meet the imposed thermalization criterion. In Table 2.1 we report $N_{\text{EMCS}}^{\text{min}}$, the maximum extension factor f_{max} of the simulations, and minimum number N_{τ}^{min} of EMCS in units of τ.

Equilibrium measurements were taken offline over the second half of each simulation. Independently of how much the simulations were extended, we saved $N_{\mathrm{m}} = 16$ equally time-spaced configurations and performed measurements on them. We measured four-replica observables. Therefore, for each sample it was possible to choose quadruplets of configurations, each from a different replica, in N_{m}^4 ways. Out of the N_{m}^4 possibilities, we chose randomly $N_{\mathrm{t}} = 1000$ combinations. In other words, each sample participated in the statistics with $N_{\mathrm{t}} = 1000$ measurements.

The errors were estimated with the jackknife method (Appendix E).

2.3 Giant Fluctuations and the Silent Majority

2.3.1 No Signs of a Phase Transition with Common Tools

A common way to locate a phase transition is to proceed as described in Sect. 1.3.4, by locating the temperature where the curves $\frac{\xi_L}{L}(T)$ and $R_{12}(T)$ of different lattice sizes cross. For sufficiently large systems, if the curves do not cross, there is no phase transition in the simulated temperature range.

In the present case, this type of analysis yields a clear result: there is no evidence of a crossing at the simulated temperatures, magnetic fields and sizes. This is clearly visible from Fig. 2.1, where the curves $\frac{\xi_L}{L}(T)$ and $R_{12}(T)$ should have some crossing point if we were in the presence of a phase transition. This is in complete qualitative agreement with earlier works on this model [You04, Jör08b].

2.3.2 A Hidden Behavior

Although $\frac{\xi_L}{L}(T)$ is smaller the larger the lattice size, the coherence length ξ_L grows significantly even for our largest lattice sizes. For example at $h = 0.2$, $T = 0.81$ we have $\xi_{16} = 6.09(4)$, $\xi_{24} = 7.63(9)$ and $\xi_{32} = 9.0(2)$. The noticeable size evolution implies that the asymptotic correlation length ξ_∞ is large compared with $L = 32$.

Also, we can examine the behavior of the spin-glass order parameter, the overlap q, by studying its distribution function $P(q)$. In the absence of a phase transition we would be in the paramagnetic phase, and $P(q)$ should be a delta function of a positive overlap q_{EA} (so in finite systems it should be Gaussian).

Instead, we can see from Fig. 2.2 that its distribution $P(q)$ has a very wide support, with tails that, for small enough magnetic fields, reach even negative values of q. This is precisely what was observed in the mean-field version of the model on the de Almeida-Thouless line, and it was attributed to the contribution of few samples [Par12a].

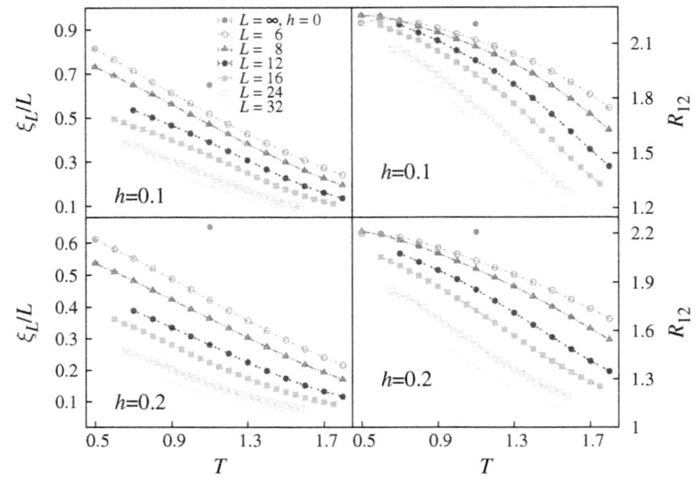

Fig. 2.1 The figures on the *left* show the standard correlation length ξ_L in units of the lattice size L as a function of the temperature T, for all our lattice sizes. The magnetic fields are $h = 0.1$ (*top*), and $h = 0.2$ (*bottom*). If the lattices are large enough, in the presence of a second-order phase transition, the curves are expected to cross at a finite temperature $T_c(h)$. The figures on the *right* show the cumulant R_{12}, which in the presence of a magnetic field is a better indicator of a phase transition [Bn12a], for the same magnetic fields. At zero field the heights of the crossings (which are universal quantities) are indicated with a point at $T_c = 1.1019(29)$. They are $\xi_L/L(h = 0; T_c) = 0.6516(32)$ and $R_{12}(h = 0; T_c) = 2.211(6)$ [BJ13]. In neither case we observe signs of a crossing at the simulated temperatures, nor can we state that the curves will cross at lower temperature. The reader might remark that the curve for $L = 32, h = 0.1$ is not as smooth as one would expect from parallel tempering simulations. The reason is twofold. On one side the number of simulated samples is much smaller than for $L < 32$, and on the other side temperature chaos, which is stronger the larger the lattice, is probably present [Fer13]

From these arguments it becomes reasonable to think that we may not be simulating large enough lattices to observe the asymptotic nature of the system and that there may be some hidden behavior that we are not appreciating.

2.3.3 *Giant Fluctuations*

In fact, we find out that the average values we measure are representative of only a small part of the data set. That is, the average of relevant observables (e.g., the spatial correlation function) only represents the small number of measurements that are dominating it. The rest of the measurements is not appreciated by using the average.

Clearly, standard finite-size scaling methods are not adequate to these systems, and we need to find a way to take into account *all* the measurements. Recalling the wide distributions of Fig. 2.2, it seems reasonable to sort our measurements according

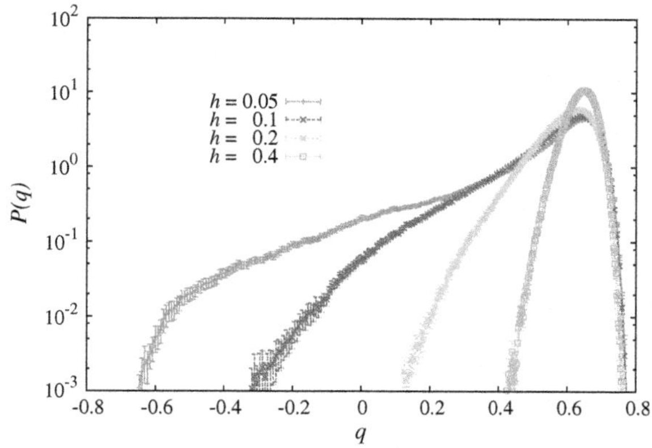

Fig. 2.2 The pdf $P(q)$ of the overlap q, for our largest lattices ($L = 32$) at the lowest simulated temperature ($T = 0.805128$), for all our magnetic fields ($h = 0.05, 0.1, 0.2, 0.4$), see Table 2.1. The order parameter in the EA model is the overlap q, and it is defined in the $[-1, 1]$ interval (see Sect. 1.2.1). The supports are wide, with exponential tails similar to those in the mean-field model at the DAT transition line [Par12a]

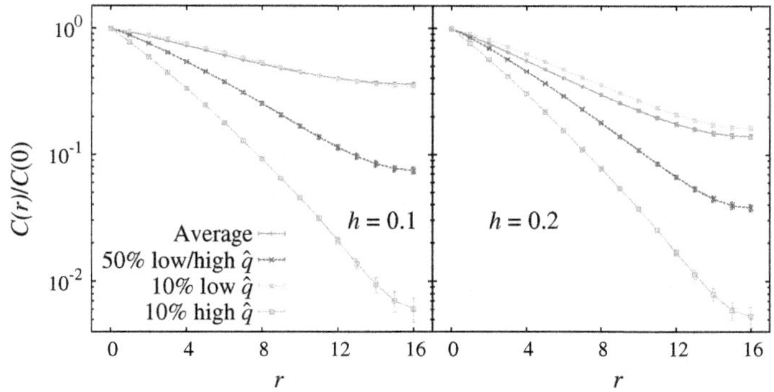

Fig. 2.3 Different instances of the normalised plane correlation function $C(r)$ (Eq. 1.5.3) for $L = 32$, $T = 0.805128$. The field is $h = 0.1$ on the *left*, and $h = 0.2$ in the *right* plot. We sort the measurements with the help of a conditioning variate $\hat{q}$ as described in Sect. 2.4. In this case $\hat{q}$ is the median overlap q_{med}. We show small sets of measurements. Namely, the ones with the 10 % lowest (*top curve*) and highest (*bottom curve*) $\hat{q}$ and those whose $\hat{q}$ corresponds to the median of the distribution of $\hat{q}$ (50 % lowest/highest $\hat{q}$). This sorting reveals extreme differences in the *fauna* of measurements. The average and median of the correlation functions are very different. The average is very similar to the 10 % lowest ranked measures, i.e., it is only representative of a very small part of the data. We normalise $C(r)$ by dividing by $C(0)$ because we measure point-to-plane correlation functions (Eq. 1.5.3). The correlation functions have zero slope at $r = L/2$ due to the periodic boundary conditions

to some conditioning variable $\hat{q}$ related to the overlaps between our replicas (see Sect. 2.4). This way, we find out that the average values we measure are given by only a small part of the measurements. For example in Fig. 2.3 we show the correlation function $C(r)$. We plot 4 estimators of $C(r)$: the average (which is the standard quantity studied in almost all, if not all, previous work), the $C(r)$ that corresponds to the median of the $\hat{q}$ distribution, and the measurements with the 10 % highest (lowest) value of $\hat{q}$. We see that the average is very close to the 10 % lowest $\hat{q}$, and very far from the two other curves. So, when we plot the average curve, we are only representing the behavior of that small set of data.

Therefore, if we want to understand the behavior of the *whole* collection of measurements, we have to be able to find some criterion to sort them and analyse them separately.

2.4 Conditional Expectation Values and Variances

2.4.1 The Conditioning Variate

As we pointed out in Sect. 2.3, the behavior of the system is dominated by a very small number of measurements.

This means that the average over all the measurements of an observable does not describe the typical behavior of the system. Furthermore, the behavior of the measurements that contribute less to the full averages is qualitatively different from the one of those who give the main contribution (see Fig. 2.3 and later on Sect. 2.9).

We want to classify our measurements in a convenient way, in order to be able to separate different behaviors, and analyse them separately. To this goal, we replace normal expectation values $E(\mathcal{O})$ of a generic observable $\mathcal{O}$, with the expectation value $E(\mathcal{O}|\hat{q})$ conditioned to another random variable $\hat{q}$. Perhaps for lack of imagination $\hat{q}$ will be named CV. For each instance of $\mathcal{O}$ we monitor also the value of $\hat{q}$, and we use it to label $\mathcal{O}$. Hopefully, there will be some correlation.

The conditional expectation value is defined as the average of $\mathcal{O}$, restricted to the measurements i (out of the $\mathcal{N}_\mathrm{m} = N_\mathrm{t} N_\mathrm{samples}$ total measurements) that simultaneously yield $\mathcal{O}_i$ and $\hat{q}_i$ [so we are actually talking about couples of simultaneous measurements $(\mathcal{O}_i, \hat{q}_i)$] in a small interval around $\hat{q} = c$,

$$E(\mathcal{O}|\hat{q} = c) = \frac{E\left[\mathcal{O}_i \mathcal{X}_{\hat{q}=c}(\hat{q}_i)\right]}{E\left[\mathcal{X}_{\hat{q}=c}(\hat{q}_i)\right]}. \tag{2.2}$$

where we have used the characteristic function

$$\mathcal{X}_c(\hat{q}_i) = \begin{cases} 1, & \text{if } |c - \hat{q}_i| < \epsilon \sim \frac{1}{\sqrt{V}} \\ 0, & \text{otherwise.} \end{cases} \tag{2.3}$$

In Appendix C we give technical details on the choice of ϵ. To make notation lighter, in the rest of the paper we will replace $E(\mathcal{O}|\hat{q} = c)$ with $E(\mathcal{O}|\hat{q})$.

The traditional expectation value $E(\mathcal{O})$ can be recovered by integrating over all the possible values of the CV $\hat{q}$:

$$E(\mathcal{O}) = \int d\hat{q}\ E(\mathcal{O}|\hat{q})P(\hat{q}), \quad P(\hat{q}) = E[\mathcal{X}_{\hat{q}}], \tag{2.4}$$

where $P(\hat{q})$ is the probability distribution function of the CV.

We remark that the concept of CV is fairly similar to the one of control-variate. Yet, the latter was formalised slightly differently, and with the objective of enhancing the precision of the measures [Fer09a]. In [AB10a, AB10b] a procedure very similar to the present one was followed, but the aim was constructing clustering correlation functions, while in our case the CV is used to analyse *separately* different behaviors outcoming from the same global data set, so that a sensible finite-size scaling becomes possible.

2.4.2 Measurements Against Samples

The reader may argue that a sample-to-sample distinction of the different behaviors is more natural than a measurement-dependent one (although intuition leads to assume that the two are related). This was indeed our first approach to the problem (it was, in fact, proposed in [Par12a]). However, we found that the approach described in the previous section is preferable, both for practical and conceptual reasons.

On the practical side, a sample-to-sample separation implies that from each sample we get only one data point: For any observable, we limit ourselves to its thermal average. In this case we would need a limitless amount of samples to be able to construct a reasonable $P(\hat{q})$. Moreover, the simulations should last a huge number of autocorrelation times τ if we want to have small enough errors on the thermal averages of each sample. Otherwise, we would introduce a large bias that is not reduced when increasing the number of samples.

On the conceptual side, representing each sample merely with a single number (namely the thermal expectation value), is a severe oversimplification. As we show in Fig. 2.4, even though we are in the paramagnetic phase, the behavior within each sample is far from trivial. For a non-negligible fraction of the samples, the overlap distribution is wide, often with a multi-peak structure. The barriers among peaks can be deep, hence suggesting extremely slow dynamics (which is indeed the case for physical dinamics [BJ14b], or for the parallel tempering dynamics [Huk96, Mar98]).

In summary, we find that using instantaneous measurements to classify the available information is the best solution.

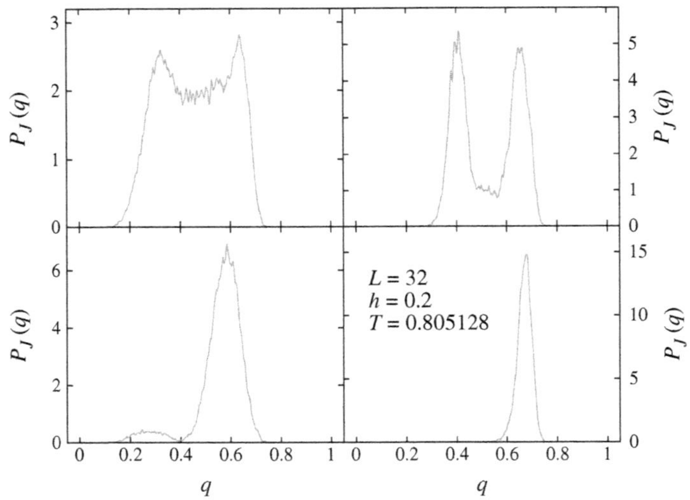

Fig. 2.4 Sample-dependent pdfs $P_J(q)$, for four different samples, each representing a different type of $P_J(q)$ we encountered. As well as the averaged $P(q)$, also the sample-dependent density function can be wide and with a structure. The plotted data comes from samples with $L = 32$, $h = 0.2$ and $T = 0.805128$

2.4.3 *The Selection of the Conditioning Variate*

2.4.3.1 A Quantitative Criterion

In Appendix D we show how to decompose the moments of a generic variable $\mathcal{O}$ as sums of averages conditioned to $\hat{q}$. For the variance we find that

$$\mathrm{var}(\mathcal{O}) = \int_{-1}^{1} d\hat{q}\, P(\hat{q}) \left\{ \mathrm{var}(\mathcal{O}|\hat{q}) + \left[E(\mathcal{O}) - E(\mathcal{O}|\hat{q})\right]^2 \right\}, \tag{2.5}$$

where

$$\mathrm{var}(\mathcal{O}|\hat{q}) = E\left(\left[\mathcal{O} - E(\mathcal{O}|\hat{q})\right]^2 \mid \hat{q} \right). \tag{2.6}$$

A convenient CV is the one that mostly discerns the different behaviors of the model. We can get a quantitative criterion for the selection of a good $\hat{q}$ by rewriting Eq. (2.5) as:

$$\mathrm{var}(\mathcal{O}) = c_1 + c_2, \tag{2.7}$$

where

$$c_1 \equiv \int_{-1}^{1} d\hat{q}\, P(\hat{q}) \mathrm{var}(\mathcal{O}|\hat{q}),$$

$$c_2 \equiv \int_{-1}^{1} d\hat{q}\, P(\hat{q})[E(\mathcal{O}) - E(\mathcal{O}|\hat{q})]^2, \tag{2.8}$$

and studying the relation between the terms c_1 and c_2. Both are positive, and their sum is fixed independently from the used CV.

We will show intuitively that a useful CV has $c_2 \gg c_1$.

If $c_1 = 0$ the fluctuations of $\mathcal{O}$ would be explained solely by the fluctuations of $\hat{q}$. In this case c_2 is large and assumes its largest possible value, meaning that different values of $\mathcal{O}$ are mostly spread apart by $\hat{q}$.

On the other side, $c_2 = 0$ implies $E(\mathcal{O}) = E(\mathcal{O}|\hat{q})$ and signals an insensitive CV, with null correlation between $\mathcal{O}$ and $\hat{q}$.

Equations (2.7) and (2.8) can thus be used to quantify the quality of the CV $\hat{q}$: We look for the highest quotient c_2/c_1.

2.4.3.2 Candidates for $\hat{q}$

To select an appropriate CV we need to chose $\mathcal{O}$ and propose some test definitions for $\hat{q}$. The functions of the observables that one could use as a CV are infinite, but physical intuition lead us to try with simple functions of the overlap and of the link overlap Eq. (1.21). On the other side, a natural choice of $\mathcal{O}$ is the estimator of the replicon susceptibility [see Eq. (1.64)]. This means that

$$\mathcal{O} \longrightarrow \frac{1}{3\mathcal{N}} \sum_{\substack{\text{equiv.wave} \\ \text{vectors}\,\boldsymbol{k}}}^{\mathcal{N}} \left[|\Phi_{\boldsymbol{k}}^{(ab;cd)}|^2 + |\Phi_{\boldsymbol{k}}^{(ac;bd)}|^2 + |\Phi_{\boldsymbol{k}}^{(ad;bc)}|^2 \right], \tag{2.9}$$

where $\mathcal{N}$ is the number of equivalent wave vectors one can construct. This is a 4-replica quantity [see Appendix (B.9)], so six instantaneous overlaps (and six link overlaps) are associated to each instance of the correlators. To define $\hat{q}$ we need to propose a function of the six overlaps in order to get a one-to-one correspondence.

Let us reorder each 6-plet of instantaneous overlaps $\{q^{(ij)}\}$ in the form of six sorted overlaps $\{q_k\}$

$$\left\{q^{(\text{ab})}, q^{(\text{ac})}, q^{(\text{ad})}, q^{(\text{bc})}, q^{(\text{bd})}, q^{(\text{cd})}\right\} \longrightarrow \{q_1 \le q_2 \le q_3 \le q_4 \le q_5 \le q_6\}, \tag{2.10}$$

and do the same thing with the link overlap

$$\left\{q_{\text{link}}^{(\text{ab})}, q_{\text{link}}^{(\text{ac})}, q_{\text{link}}^{(\text{ad})}, q_{\text{link}}^{(\text{bc})}, q_{\text{link}}^{(\text{bd})}, q_{\text{link}}^{(\text{cd})}\right\} \longrightarrow$$

$$\longrightarrow \left\{q_{\text{link},1} \le q_{\text{link},2} \le q_{\text{link},3} \le q_{\text{link},4} \le q_{\text{link},5} \le q_{\text{link},6}\right\}, \tag{2.11}$$

Table 2.2 Criterion for the choice of the CV $\hat{q}$ for $h = 0.1$, $L = 32$, $T = 0.805128$, by looking at the indicators c_1 and c_2 relatively to $\chi_R(\mathbf{0})$ and $\chi_R^{\text{link}}(\mathbf{0})$

$\hat{q}$	$\chi_R^{\text{spin}}: c_1$	$\chi_R^{\text{spin}}: c_2$	c_2/c_1	$\chi_R^{\text{link}}: c_1$	$\chi_R^{\text{link}}: c_2$	c_2/c_1
$q_{\text{min}}^{\text{spin}}$	399000 ± 37000	121000 ± 15000	0.30(6)	8.35 ± 0.47	0.297 ± 0.023	0.36(5)
$q_{\text{max}}^{\text{spin}}$	514000 ± 51000	6230 ± 690	0.012(3)	8.54 ± 0.49	0.1070 ± 0.0073	0.013(2)
$q_{\text{med}}^{\text{spin}}$	162000 ± 10000	358000 ± 45000	2.2(4)	7.35 ± 0.39	1.30 ± 0.11	0.18(2)
$q_{\text{av}}^{\text{spin}}$	328000 ± 26000	192000 ± 28000	0.6(1)	7.51 ± 0.41	1.141 ± 0.094	0.15(2)
$q_{\text{min}}^{\text{link}}$	461000 ± 46000	59300 ± 5800	0.13(3)	8.38 ± 0.48	0.271 ± 0.020	0.032(4)
$q_{\text{max}}^{\text{link}}$	460000 ± 46000	59700 ± 5900	0.13(3)	8.56 ± 0.49	0.0838 ± 0.0067	0.010(1)
$q_{\text{med}}^{\text{link}}$	360000 ± 36000	160000 ± 18000	0.44(9)	7.36 ± 0.38	1.29 ± 0.11	0.17(2)
$q_{\text{av}}^{\text{link}}$	415000 ± 42000	105000 ± 10000	.25(5)	7.72 ± 0.42	0.927 ± 0.073	0.12(2)

We want the $\hat{q}$ to split as much as possible the different measured susceptibilities. This is obtained, see (2.8), when the ratio c_2/c_1 is maximised. From the data we see that this occurs with $\hat{q} = q_{\text{med}}$

The following are natural test CVs:

$$\hat{q} = \begin{cases} q_{\text{min}} &= q_1 & \text{(the minimum)} \\ q_{\text{link,min}} &= q_{\text{link},1} \\ q_{\text{max}} &= q_6 & \text{(the maximum)} \\ q_{\text{link,max}} &= q_{\text{link},6} \\ q_{\text{med}} &= \frac{1}{2}(q_3 + q_4) & \text{(the median)} \\ q_{\text{link,med}} &= \frac{1}{2}(q_{\text{link},3} + q_{\text{link},4}) \\ q_{\text{av}} &= \frac{1}{6}(q_1 + q_2 + q_3 + q_4 + q_5 + q_6) & \text{(the average)} \\ q_{\text{link,av}} &= \frac{1}{6}(q_{\text{link},1} + q_{\text{link},2} + q_{\text{link},3} + q_{\text{link},4} + q_{\text{link},5} + q_{\text{link},6}). \end{cases} \tag{2.12}$$

We checked how each of the CVs sorted the overlap and link susceptibilities $\chi_R(\mathbf{0})$ and $\chi_R^{\text{link}}(\mathbf{0})$. Table 2.2 depicts the c_1 and c_2 terms, and their ratio, for all the CV, for a single triplet (T, L, h) and $\boldsymbol{k} = (0, 0, 0)$.

The best CV is clearly the median overlap, since it has the highest c_2/c_1 ratio. The situation is similar for other choices of (T, L, h).

For a qualitative description of the difference between the diverse CVs, in Fig. 2.5 (top) the reader can appreciate the probability distribution functions for each of the CVs, while in Fig. 2.5 (bottom) we plotted the conditioned susceptibilities. From (2.4) we stress that the integral of the values on the top times the values of the bottom set yields the average susceptibility, which is indicated with a horizontal line on the

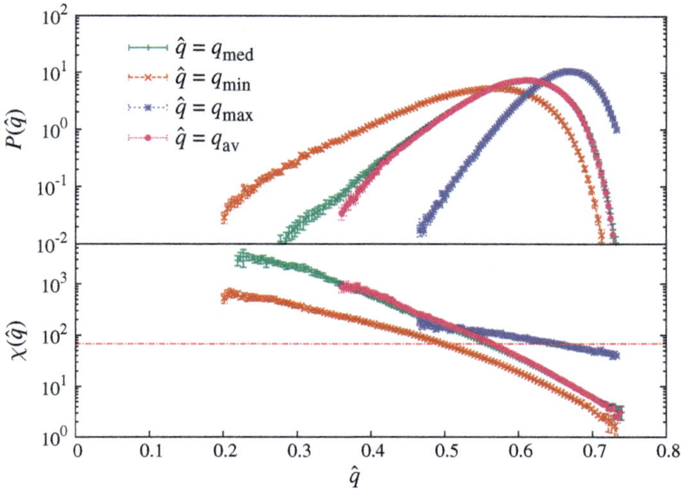

Fig. 2.5 Features of the diverse CVs we proposed for $L = 32, h = 0.2$ and $T = 0.805128$. The *top* figure shows the histograms $P(\hat{q})$ for four candidates of conditioning variate: the minimum overlap q_{min} [of the six we can make with four replicas, recall equations (2.12)], the maximum q_{max}, the median q_{med} and the average q_{av}. The histograms were constructed as explained in Appendix C. The *bottom* figure depicts the size of the susceptibility χ for each value of the CV. The horizontal line marks the value of χ when it is averaged over the full set of measurements. For aesthetic reasons in both figures we have cut the curves at the two end points, where they become extremely noisy due to poor sampling

bottom plot of Fig. 2.5. As it is also reflected by Table 2.2, q_{max} is the worst CV, as its χ does not vary much with the fluctuations of q_{max}. The steepest slope is obtained when the CV is q_{av} or q_{med}, but the latter is smoother and covers a wider range of χ.

Figure 2.5 also displays the large deviations present in the system. In fact one can see that the value of q_{med} at which the $P(q_{med})$ has its maximum is significantly different with respect to the value of q_{med} at which $\chi(q_{med})$ assumes the value of the average.

Let us compare the overlap with the link-overlap signal. Besides the fact that the link overlaps appear to be bad CVs, one can see from Table 2.2 that on one side the fluctuations on $\chi_{link,R}(\mathbf{0})$ are much smaller than $\chi_R(\mathbf{0})$, and on the other none of the CVs seems to separate the behaviors (the ratio c_2/c_1 is much smaller). We can see this better from Fig. 2.6, that depicts the results of a sorting with the median (link-)overlap on $C_R(\boldsymbol{r})$ and $C_{link,R}(\boldsymbol{r})$. The bold line stands for the average behavior, while the thin ones represent a sorting of the data according to the quantile of the distribution of the CV.[5] If the average is in the middle of the thin lines it is a good descriptor of the data, otherwise it is a biased estimator. Very spread thin lines indicate that $c_2 \gg c_1$: the CV separates behaviors properly.

[5]A quantile is the value of $\hat{q}$ that separates a fixed part of the pdf (Sect. 2.5 later on).

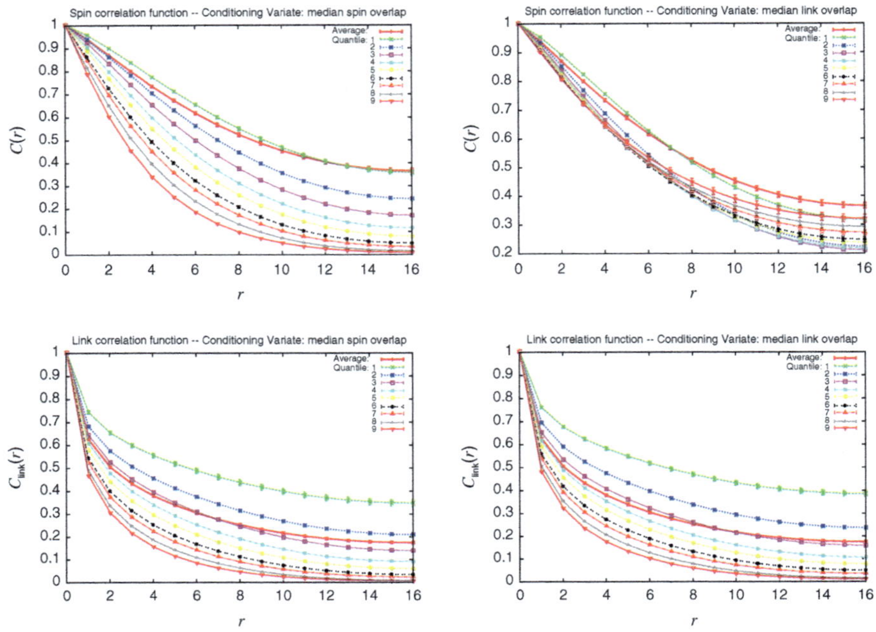

Fig. 2.6 Spin and link plane replicon correlation functions with $h = 0.1$, $L = 32$, $T = 0.805128$. The thin lines indicate different quantiles of the conditioning variate's distribution (see Sect. 2.5), the bold lines indicate the average. *Top* Spin correlation functions, *bottom* link correlation functions. *Left* $\hat{q} = q_{\text{med}}$, *right*: $\hat{q} = q_{\text{link,med}}$. Discussion in the main text

The data illustrates that, while the average spin correlation function is not representative of the majority of outcomes for both the CV, the link correlation function is well-described by its average. This suggests that the link overlap might be a more accurate indicator of the critical behavior of the EA spin glass in a field. Analyses on the link-overlaps will be object of further future studies.

On another side, if we concentrate on the spin correlation function $C(r)$, we see that the link is not a suitable CV, both because it separates less the behaviors, and because the separation has a dependency on the distance r.

2.5 Quantiles and a Modified Finite-Size Scaling Ansatz

We stated in Sect. 2.3 that the set of measurements with low $\hat{q}$ has a very different behavior from the measurements with high $\hat{q}$ (recall Fig. 2.3). From now on, we shall restrict ourselves to $\hat{q} = q_{\text{med}}$, since we evinced that the median is our best CV. Our next goal will be to carry out a finite-size scaling analysis based on the $P(q_{\text{med}})$ that lets us observe different parts of the spectrum of behaviors of the system.

In order to analyse separately these different sets of measures, we divide the $P(q_{\text{med}})$ in 10 sectors, each containing 10 % of the measured q_{med}. We focus our analysis on the values of q_{med} that separate each of these sectors. They are called quantiles (see, e.g., [Hyn96]), and we label them with the subscript $i = 1, \ldots, 9$. If we call $\tilde{q}_i(h, T, L)$ the value of the ith quantile, we can define it in the following implicit way:

$$\int_{-1}^{\tilde{q}_i} \mathrm{d}\hat{q}\, P(\hat{q}) = \frac{i}{10}. \tag{2.13}$$

In Appendix C we explain how $\tilde{q}_i(h, T, L)$ was computed.

We can adapt to the ith quantile the definitions we gave in Sect. 1.2.4:

$$\chi_{\mathrm{R},i}(\boldsymbol{k}) = \frac{1}{N} E\left(|\hat{\Phi}_{\boldsymbol{k}}^{(ab;cd)}|^2 \,\middle|\, \tilde{q}_i \right), \tag{2.14}$$

$$\xi_{L,i} = \frac{1}{2\sin(k_{\min}/2)} \sqrt{\frac{\chi_{\mathrm{R},i}(\mathbf{0})}{\chi_{\mathrm{R},i}(2\pi/L, 0, 0)} - 1}, \tag{2.15}$$

$$R_{12,i} = \frac{\chi_{\mathrm{R},i}(2\pi/L, 0, 0)}{\chi_{\mathrm{R},i}(2\pi/L, \pm 2\pi/L, 0)}. \tag{2.16}$$

This way we can extend the finite-size scaling methodology to the ith quantile:

$$\left.\frac{\xi_L}{L}\right|_{T,h,L,i} = f_{\xi_i}\left(L^{1/\nu}(T - T_{\mathrm{c}})\right) + \cdots, \tag{2.17}$$

$$R_{12}|_{T,h,L,i} = f_{R_i}\left(L^{1/\nu}(T - T_{\mathrm{c}})\right) + \cdots. \tag{2.18}$$

This is a new approach for finite-size scaling. Although it demands a very large amount of data because it is done over a small fraction of the measurements (in Appendix E we explain a method we used to reduce rounding errors), it allows us to perform finite-size scaling on selected sets of measurements.

Let us stress that no a priori knowledge is required on the probability distribution function $P(q_{\text{med}})$: Quantiles are conceived in order to define a scaling that self-adapts when the volume increases.

2.6 Testing the Quantile Approach

We take advantage of our $h = 0$ data from [BJ13] to validate our new FSS ansatz and the quantile description, by showing its behavior in the zero-field case. Two replicas would be enough to construct connected correlators in $h = 0$, and using the 4-replica definitions proposed in Sect. 1.2.4 only adds noise to the results. Yet, we opted for the latter option because the objective of the current section is the validation of the full procedure proposed herein.

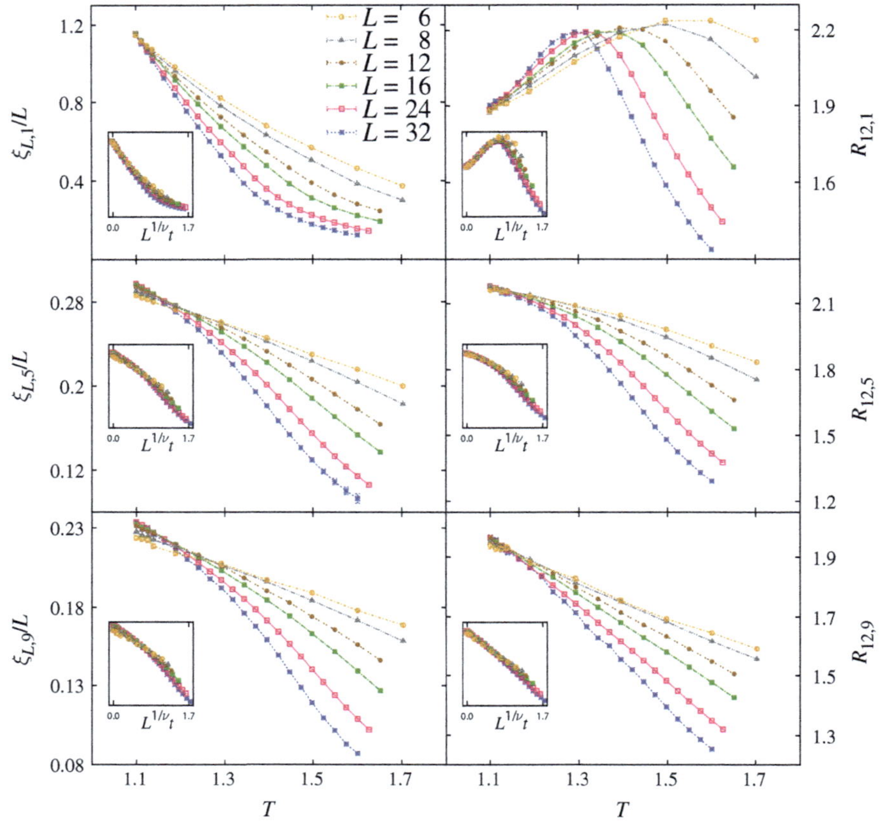

Fig. 2.7 Finite-size indicators of a phase transition, computed for $h = 0.2$. On the *left* side we plot, for quantiles 1 (*top*), 5 (*middle*) and 9 (*bottom*), the correlation length in units of the lattice size ξ_L/L versus the temperature, for all our lattice sizes. The *right* side is equivalent, but for the R_{12}, defined in Eq. (1.46). The curves crossings are compatible with the well-known temperature of the zero-field transition. The data come from [BJ13]. We used 256,000 samples for each lattice size. The *insets* show the same data of the larger sets, but as a function of the scaling variable $L^{1/\nu}t$, where t is the reduced temperature $t = (T - T_c)/T_c$

In the absence of a magnetic field we expect that the curves $\xi/L(T)$ and R_{12} cross no matter the quantile, since the behavior of the system is not dominated by extreme events and crossover fluctuations. Also, in this case the data in our hands arrive down to the critical point, so the crossings ought to be visible.

One can see in fact from Fig. 2.7 that all the quantiles show visible signs of a crossing at T_c both in the case of ξ_L/L and of R_{12}. Furthermore, if we plot the same data as a function of the scaling variable $L^{1/\nu}(T - T_c)/T_c$ the data collapses well for all the quantiles (Fig. 2.7, insets).

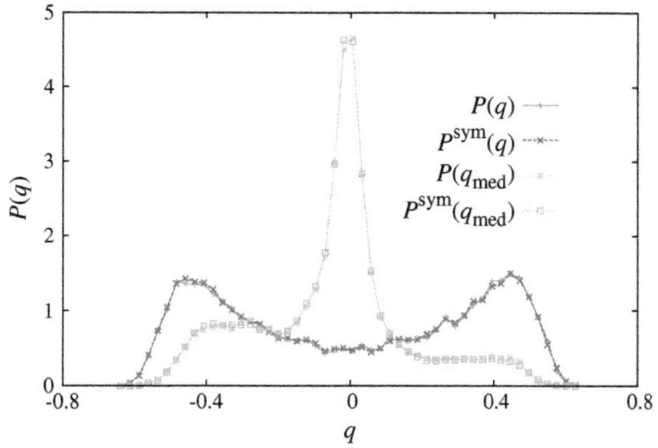

Fig. 2.8 Probability distribution function for $h = 0$, $L = 24$, $T = 1.1$. The data come from 512 samples where we took all the 16^4 combinations of overlaps per sample. We show $P(q)$, that in null field is symmetric, and $P(q_{\text{med}})$, that is not. We also plot the symmetrized histograms $P^{(\text{sym})}(q)$ and $P^{(\text{sym})}(q_{\text{med}})$, that overlap on the respective curves. As more extendedly explained in the main text, the symmetrized overlap is obtained by averaging each q_{med} over the values it would acquire by imposing all the combinations of Z_2 symmetry (flip all the spins) on the configurations on which the q_{med} is calculated

Some reader may be surprised that quantiles 1 and 9 show different behavior, being $P(q)$ symmetrical (Fig. 2.8). The reason is that, although $P(q)$ is symmetrical, $P(q_{\text{med}})$ is not. In fact, given six overlaps q^{ab}, q^{ac}, q^{ad}, q^{bc}, q^{bd}, q^{cd} coming from four configurations $|s^{(\text{a})}\rangle, |s^{(\text{b})}\rangle, |s^{(\text{c})}\rangle, |s^{(\text{d})}\rangle$, each enjoying a Z_2 symmetry, the distribution of their median privileges negative values.[6] We show this in Fig. 2.8, where we give both the $P(q)$ and the $P(q_{\text{med}})$ for $h = 0$, $L = 32$, $T = 1.1$. The first is symmetrical and the second is not. To convince the reader that the starting configurations do enjoy Z_2 symmetry, we also construct the symmetrized functions $P^{(\text{sym})}(q)$ and $P^{(\text{sym})}(q_{\text{med}})$. These two functions are obtained by explicitly imposing the reflection symmetry Z_2: for each measurement we construct the 2^4 overlaps with both $|s\rangle$ and $|-s\rangle$. It is visible from Fig. 2.8 that $P^{(\text{sym})}(q_{\text{med}})$ is asymmetric even though we imposed by hand the Z_2 symmetry on the configurations.

[6]Let us give a simple example. Take 4 Z_2-symmetric single-spin systems that can assume different values $s_1 = \pm 1$, $s_2 = \pm 2$, $s_3 = \pm 3$, $s_4 = \pm 4$. We can construct 6 overlaps $q_{ij}(s_1, s_2, s_3, s_4)$. If we explicitate the Z_2 symmetry, taking all the combinations of our random variables, the histogram of q will be symmetric with zero mean. Yet, if we take the histogram of the median overlap, it will be asymmetric with mean $\langle q_{\text{med}}\rangle = -3$. This can easily be checked by computing all the possible combinations of the signs of the s_i and computing the median in each case: $q_{\text{med}}(+1, +2, +3, +4) = 5$, $q_{\text{med}}(+1, +2, +3, -4) = -1$, $q_{\text{med}}(+1, +2, -3, -4) = -3.5$, and so on.

2.6.1 The $P(q_{\mathrm{med}})$

To our knowledge, the median overlap q_{med}, despite its simplicity, has not been object of previous study. Yet, since we base our analysis on this quantity, it is necessary to dedicate passing attention to its features.

By its definition, the probability distribution $P(q_{\mathrm{med}})$ of the median overlap has narrower tails than $P(q)$ (recall Fig. 2.2), although from Fig. 2.5 (top) it is clear that the strong fluctuations persist also with q_{med}.

The median of $P(q_{\mathrm{med}})$ corresponds to the fifth quantile. We will prefer to call it "5th quantile" rather than "median of the median overlap". Of the nine studied quantiles it is the smoothest and has the least finite-size effects, as one can see from Fig. 2.9 (inset). Further analysis is given in Sect. 2.8.

We remark also that the separation between the different $\tilde{q}_i$'s can be used as order parameter, since its thermodynamic limit should be zero in the paramagnetic phase, and greater than zero in the possible low-temperature phase due to the (would-be) replica symmetry breaking. Figure 2.9 shows the difference between the 8th and the 2nd quantile, i.e., the q_{med}-span of the central 60 % of the data. If we were able to extrapolate a clean $L \to \infty$ limit for this curve, we would be able to answer to whether the transition exists or not. Unfortunately, even for $T > T_c(h = 0) = 1.1019(29)$, where we know that we are in the paramagnetic phase, it is not possible to make good extrapolations since the trend is strongly non-linear. In Sect. 2.8 we will show that extrapolations to the thermodynamic limit were only possible in the trivial case of

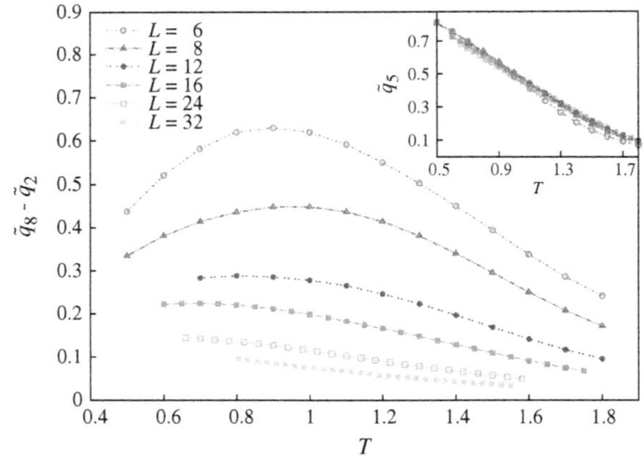

Fig. 2.9 Using q_{med} as CV, we show the temperature dependence of the difference between quantiles $\tilde{q}_8 - \tilde{q}_2$, for all our lattice sizes, in a field of intensity $h = 0.2$. This corresponds to the width of the central 60 % of area of $P(q_{\mathrm{med}})$. This quantity can reveal a phase transition, since in the paramagnetic phase the $P(q_{\mathrm{med}})$ should be a delta function, while in the spin-glass phase it should have a finite support. We show the central 60 % and not a wider range because it is an equivalent indicator of the phase transition, and it is safer from rare events that would vanish in the thermodynamic limit. In the *inset* we show the position of 5th quantile as a function of temperature in all our lattice sizes. It is a very smooth curve with very small finite-size effects

$h = 0.4$ (deep paramagnetic phase), and that between all the quantiles, the median curve is the one that shows less finite-size effects.

2.7 A *caveat* for the Quantile Description

In the absence of an applied field, the overlap probability distribution function $P(q)$ is symmetric, with a single peak centred in $q = 0$. In the presence of a field, instead, we expect the $P(q)$ to be strictly positive, at least in the thermodynamic limit. Similarly, we expect that the probability distribution function $P(q_{\rm med})$ have only one peak at positive $q_{\rm med}$ when a field is applied, and a peak in $q = 0$ if $h = 0$.

If the system sizes are too small, it may occur that the $h = 0$ behavior bias the $P(q_{\rm med})$. This is what happens, for example, when $L = 6$, $h = 0.2$ and the temperature is sufficiently low: a second peak around $q_{\rm med} \simeq 0$ develops upon lowering T (Fig. 2.10, top). This second peak disappears when we increase the lattice size (Fig. 2.10, centre), and the $P(q_{\rm med})$ assumes only positive values when L is large enough (Fig. 2.10, bottom). The lower the field, the easier it is to find multiple peaks, and the greater the system has to be to be able to neglect the $h = 0$ behavior. For $h = 0.05$, even lattices with $L = 12$ show a double peak.

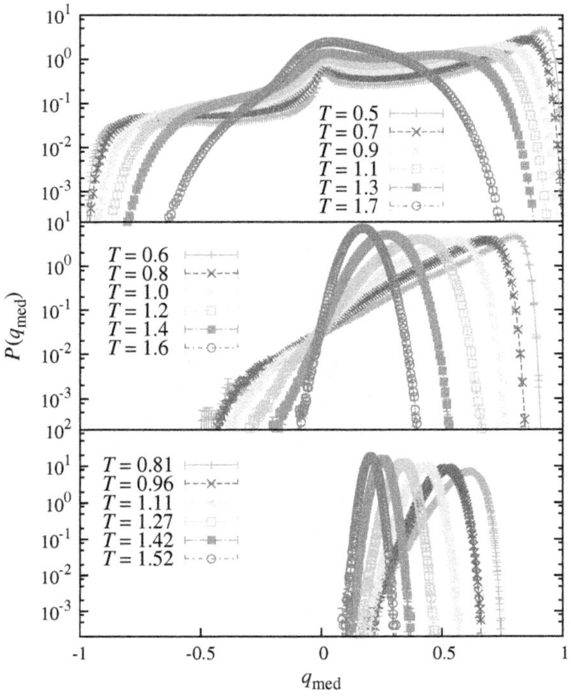

Fig. 2.10 Median overlap probability distribution function $P(q_{\rm med})$ with $h = 0.2$ for different temperatures (the ones from $L = 32$ are an approximation to the second decimal digit). The *top* figure shows the case of $L = 6$, where the lowest temperature curves display a second peak around $q_{\rm med} \simeq 0$, which disappears when T increases. For $L = 16$ (*middle*) the $P(q_{\rm med})$ are single-peaked, but assume also negative values. In the *bottom* curve we have $L = 32$, where the $P(q_{\rm med})$ are single-peaked and defined only on positive $q_{\rm med}$, since we are closest to the asymptotic behavior

A second peak in $P(q_{\rm med})$ is a clear signal that we are observing and echo of $h = 0$. When we make the quantile classification, and have a quantile on a peak, we are seeing *only* non-asymptotic data. Thus, quantile 1 for the smallest lattices gives us no relevant information.

If we plot versus the temperature any observable $\mathcal{O}$ related to the first quantile, the information will be biased for low temperatures, and the bias will gradually disappear as we increase T. The result is that the curve $\mathcal{O}(T)$ will have a strange shape and will be of no use (see, e.g., the $h = 0.05$ data in Fig. 2.11). This is why we did not include the $L = 6$ points in the top set of Fig. 2.13 later on.

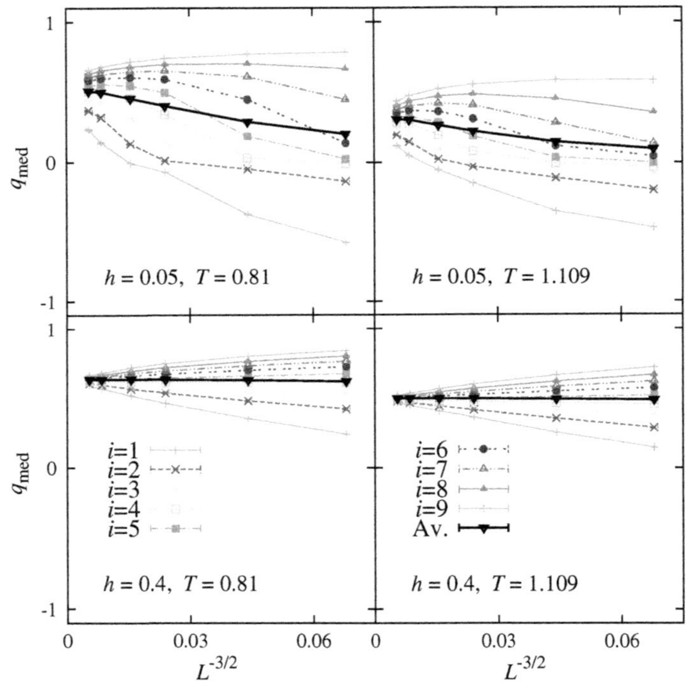

Fig. 2.11 Extrapolations to infinite size of the quantile overlap $\tilde{q}_i$, for $T = 0.81$ (*left*) and $T = 1.109$ (*right*), and fields $h = 0.05$ (*top*) and $h = 0.4$ (*bottom*). We show quantiles $i = 1, \ldots, 9$ (*thin lines*), and the average behavior (*bold line*). The $h = 0.4$ extrapolations to infinite volume were clean ($\chi^2/\mathrm{DOF} < 1$), while for $h = 0.05$ (and all the other fields we simulated), we encountered too strong and nonlinear finite-size effects to get reasonable extrapolations. We choose $1/L^{D/2}$ as scaling variable because in conditions of validity of the central limit theorem, the fluctuations should be of order $1/\sqrt{N}$

2.8 Finding a Privileged q

Since all our simulations are in the paramagnetic phase the thermodynamic limit of the $P(q)$ is a delta function, so all the quantiles should tend to the a common $q = q_{\text{EA}}$ in the $L \to \infty$ limit. We tried to perform these extrapolations at fixed (reasonably low) temperature, to see if we could look at the problem from such a privileged position. In Fig. 2.11 we see this type of extrapolation for $h = 0.4$ and $h = 0.05$, at temperatures $T = 0.81$ and 1.109. The first is the lowest temperature we simulated in all our lattices, while the second is the zero-field critical temperature [BJ13]. Since we are in the paramagnetic phase and we are plotting $\tilde{q}_i$ versus the inverse lattice size, the curves should cross at the intercept. This is indeed what appears to happen, but although in the case of $h = 0.4$, the extrapolations were clean, for all the other simulated fields the finite-size effects were too strong and nonlinear to make solid extrapolations. We remark, yet, that once $L > 8$ the 5th quantile is the one with the least finite-size effects.

2.9 The Silent Majority

As already emphasized, the behavior of the system is characterized by very strong fluctuations, and a wide and asymmetric $P(q)$. As a result, the average and median behavior are very different. In Fig. 2.12, we show the replicon susceptibility: its average χ on the left plot, and its fifth quantile χ_5. Motivated by the arguments in Sect. 2.4 all the quantiles we show in this section use the CV $\hat{q} = q_{\text{med}}$.

Visibly, not only is the average susceptibility much larger than the 5th quantile, but also the two have peaks at different temperatures. Also, finite-size effects are much stronger in the case of χ_5 (yet, recall the inset in Fig. 2.9, finite-size effects on $\tilde{q}_5$ are tiny).[7]

We show in Fig. 2.13 how sorting the data with the quantiles revealed the presence of different types of behavior, by plotting the ξ_L/L and the R_{12} for quantiles 1, 5 and 9 at $h = 0.2$. There are two vertical lines in each figure. The one on the left represents the upper bound $T^{\text{up}}(h)$ for the phase transition (meaning that no phase transition can occur for $T > T^{\text{up}}(h)$) given in [BJ14b], while the one on the right indicates the zero field critical temperature $T_c = 1.1019(29)$ [BJ13].

We can see that the 1st quantile has the same qualitative behavior of the average (Fig. 2.1), but lower values, since the main contribution to the average comes from data whose q_{med} is even lower than $\tilde{q}_1$. Moreover, one can notice that in Fig. 2.1 the indicators ξ_L/L and R_{12} show a different qualitative behavior when the lattices are small (R_{12} shows a crossing). This discrepancy vanishes when we look only at the

[7] We made power law extrapolations to $L \to \infty$ of the maxima of the susceptibility, but they were not satisfactory (too large χ^2/DOF). Only for $h = 0.2, 0.4$ were we able to fit the maxima's heights and obtained $\eta(h = 0.2) \approx 0.6$ and $\eta(h = 0.4) \approx 0.9$..

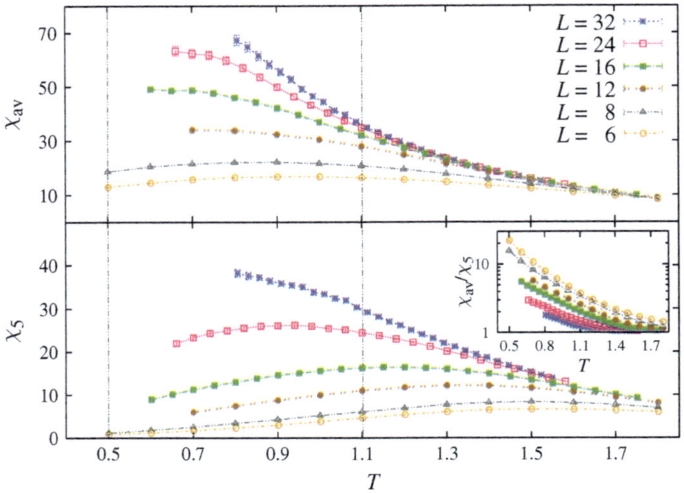

Fig. 2.12 The replicon susceptibility χ as a function of the temperature, for all the simulated lattice sizes and the field $h = 0.2$. We represent its average χ (*top*), and the 5th quantile χ_5 with $\hat{q} = q_{\text{med}}$ (*bottom*). In both plots, the two vertical lines represent the upper bound of the possible phase transition $T^{\text{up}}(h = 0.2) = 0.5$ given in [BJ14b], and the zero-field critical temperature $T_c(h = 0) = 1.109(29)$ [BJ13]. The amplitudes and the positions of the peaks of χ are strikingly different (mind the different scales in the y axes). The *inset* shows the ratio between the two, which we expect to tend to an order one constant in the thermodynamic limit. This is actually what we see at high temperatures

first quantile: Separating different behaviors enhances the consistency between ξ_L/L and R_{12}.

The behavior of the 5th quantile is quite different, since now it appears reasonable that the curves cross at some $T \lesssim T^{\text{up}}(h)$. The crossings become even more evident when we consider the highest quantile.

All this is consistent with the arguments of Sect. 2.3, where we showed how the correlation function is dominated by a little portion of data, near the first quantile (Fig. 2.3), while the behavior of the majority of the samples is hidden.

Unfortunately, the high non-linearity of the curves impedes an extrapolation of the crossing points, but they are apparently compatible with the upper bound T^{up}, and their heights apparently do not depend on the intensity of the applied field h (see also Fig. 2.14).

The careful reader might have noticed that the upper bound $T^{\text{up}}(h)$ for the possible phase transition given in [BJ14b] is higher when the field is lower: $T^{\text{up}}(0.1) = 0.8 > T^{\text{up}}(0.2) = 0.5$. It is then justified to ask oneself how do the quantile plots look like for $h = 0.1$. We show them in Fig. 2.14. Since the field is lower, the effects on the double peak on the first quantile (Sect. 2.7) extend to larger lattices than for $h = 0.2$. Thus, we show only the non-biased sizes, i.e., $L > 12$.

Although the 9th quantile shows signs of scale invariance at $T = T^{\text{up}}(0.1)$, the behavior of the 5th quantile suggests a scale invariance around $T = 0.5$. We believe

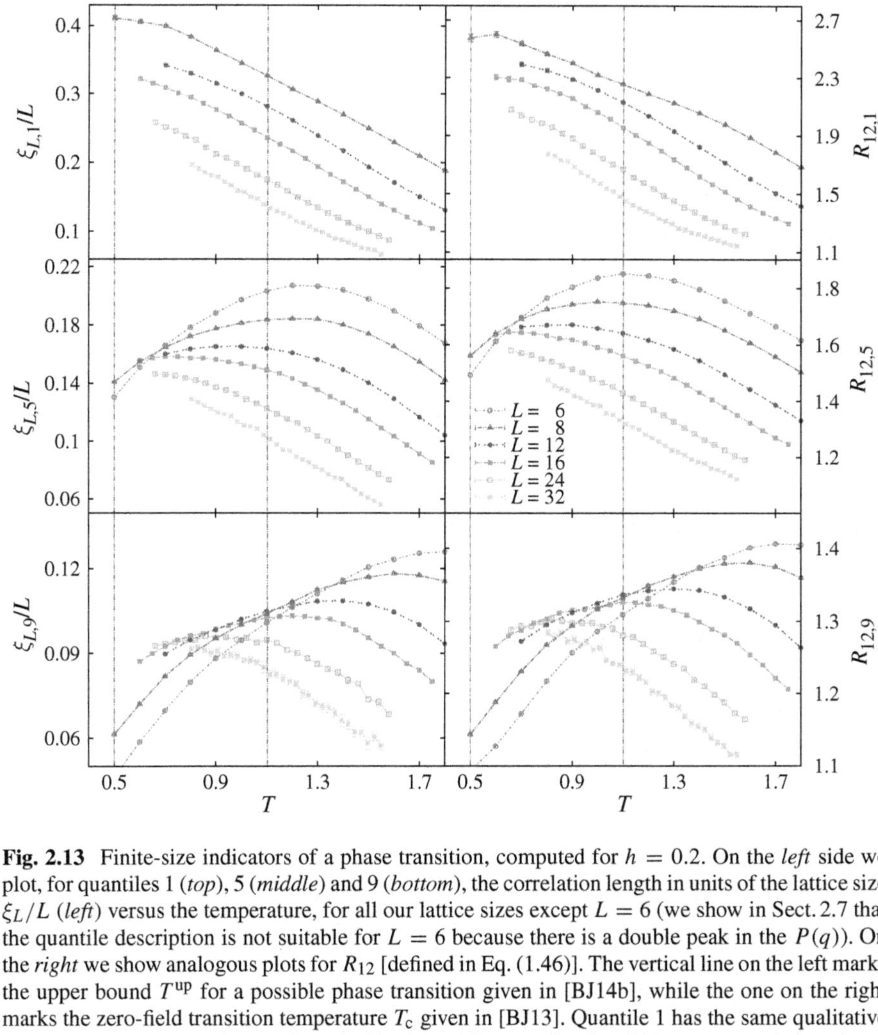

Fig. 2.13 Finite-size indicators of a phase transition, computed for $h = 0.2$. On the *left* side we plot, for quantiles 1 (*top*), 5 (*middle*) and 9 (*bottom*), the correlation length in units of the lattice size ξ_L/L (*left*) versus the temperature, for all our lattice sizes except $L = 6$ (we show in Sect. 2.7 that the quantile description is not suitable for $L = 6$ because there is a double peak in the $P(q)$). On the *right* we show analogous plots for R_{12} [defined in Eq. (1.46)]. The vertical line on the left marks the upper bound T^{up} for a possible phase transition given in [BJ14b], while the one on the right marks the zero-field transition temperature T_c given in [BJ13]. Quantile 1 has the same qualitative behavior of the average ξ_L/L, shown in Fig. 2.1, while quantiles 5 and 9 suggest a scale invariance at some temperature $T_h < T^{\text{up}}$

that the 5th quantile is a better indicator, since the position of the fifth quantile $\tilde{q}_5$ has less finite-size effects (it practically has none, Fig. 2.9–inset) than $\tilde{q}_9$.

It is interesting to focus on the height of the crossings of each quantile from Fig. 2.13, and compare them with $h = 0.2$ (Fig. 2.13). This is expected to be a universal quantity, and in the hypothesis of a phase transition it should be the same for both fields. Although it is not possible to assign error bars to the these values, it is possible to see that both for $h = 0.1$ and $h = 0.2$ the heights are similar ($\xi_{L,5}/L \approx 0.15$, $\xi_{L,9}/L \approx 0.09$, $R_{12,5} \approx 1.6$, $R_{12,9} \approx 1.3$).

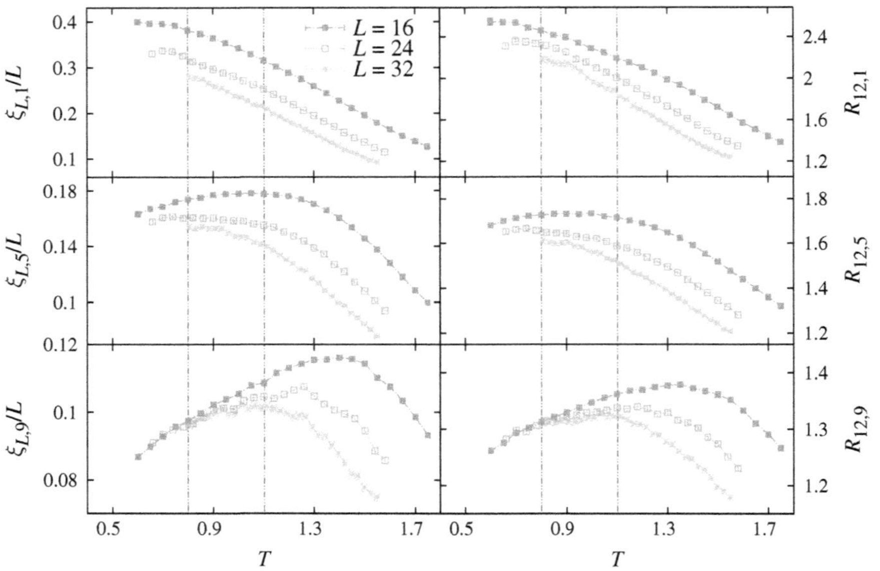

Fig. 2.14 Same as Fig. 2.13, but for $h = 0.1$. This time the effects of the zero-temperature transition are stronger, so we removed from the plot sizes $L = 6, 8, 12$. In Sect. 2.7 we show that the quantile description is not suitable for smaller lattices due to crossover effects from the zero-field behavior.

2.10 This is not an Echo of the $h = 0$ Transition

The crossing suggested by the quantiles 5 and 9 in Fig. 2.13 is unlikely to be caused by the zero-field transition, since it appears at $T < T_c$, and shifts towards lower temperatures as the lattice size increases. Also, the value of ξ_L/L (R_{12}) at the possible crossing point of the fifth quantile is upper-bounded to $\xi_L/L \simeq 0.16$ ($R_{12} \simeq 1.65$), while for $h = 0$ it is considerably larger ($\xi_L(T_c)/L \simeq 0.28$ [$R_{12}(T_c) \simeq 2.15$]), recall Sect. 2.6. In this section, we will advance more arguments contradicting the notion that what was seen resulted from the effects of zero-field transition.

2.10.1 An Escaping Transition

As pointed out in Sect. 2.3, there is a controversy because we observe a wide $P(q)$, just like in the mean-field model, but the curves $\xi_L/L(T)$ and $R_{12}(T)$ do not show any sign of a crossing. If we were in the presence of a phase transition, a straightforward explanation could reside in an anomalous exponent η close to 2 [BJ14d], since at the critical temperature the replicon susceptibility scales as $\chi_R(L) \sim L^{2-\eta}$ (1.91). It is possible to calculate η with the quotients' method [Nig75, Bal96], by comparing the susceptibility χ_L of different lattice sizes at the critical point T_c:

$$\frac{\chi_{2L}(T_c)}{\chi_L(T_c)} = 2^{2-\eta} + \cdots, \tag{2.19}$$

where the dots stand for subleading terms. This definition only makes sense at criticality, but we can extend it in an effective manner to a generic temperature. This way we can delineate an effective exponent

$$\eta_{\text{eff}}(T; L, 2L) = 2 - \log_2 \frac{\chi_{2L}(T)}{\chi_L(T)}. \tag{2.20}$$

In case there were a phase transition at a finite temperature T_h, we would have $\eta_{\text{eff}}(T_h) = \eta$. We should have $\eta_{\text{eff}} = 2$ in the paramagnetic phase, $\eta_{\text{eff}} = -1$ in the deep spin-glass phase[8] and signs of a crossing at $\eta_{\text{eff}} = \eta(h = 0) = -0.3900(36)$ [BJ13] in the limit of a complete domination by the $h = 0$ transition.

In Fig. 2.15 we show $\eta_{\text{eff}}(T)$ for $h = 0.4$, $h = 0.1$, and $h = 0$ (the $h = 0$ data come from the simulations we performed in [BJ13]).[9] If a phase transition were present, but hidden by heavy finite-size effects, we would expect at least that the L-trend of η_{eff} be decreasing. Contrarily, the larger our lattices, the wider the temperature range in which $\eta_{\text{eff}} = 2$. The apparent phase transition shifts towards lower temperature when we suppress finite-size effects. The data in our possession is not enough to state whether this shift will converge to a positive temperature. In any case, this is compatible with the upper bounds to a possible transition given in [BJ14b].

On the other side, η_{eff} stays positive for all our simulated lattices (except $h = 0.05$, $L = 6$), and that even for $T < T_c(h = 0)$ it tends to some value around 0.5, so it is unlikely that the null field transition is dominating the system's behavior.

2.10.2 Scaling at $T = T_c(h = 0)$

From the scaling with the lattice linear size of ξ_L/L at $T_c = T_c(h = 0)$, we can get another element to discard the hypothesis that the $h = 0$ transition is biasing significantly our measures. Assuming that there is no critical line for $h > 0$, a very large correlation length could be due to an echo of the zero-field transition or a low-temperature effect. In a theory that predicts that system is critical only at $h = 0$, $T = T_c$, the effects of this echo on the $h > 0$ behavior should be maximal near $T = T_c$. So, if we find a ξ that is large compared to our lattice sizes for $T < T_c$, a primary check is to monitor the scaling of the coherence length at T_c. Figure 2.16 shows the scaling of ξ_L/L at T_c with $h = 0.2$. We plot the average, the first, the fifth and the highest quantile. All of them show a clear decrease of ξ_L/L when increasing the lattice size, so our lattice sizes are large enough to state that the divergence at

[8] See Appendix B.21, keeping in mind that $\eta_{\text{eff}} = -1$ is somewhat trivial in the limit $h \to 0$, where χ reduces to $\chi = VE(q^2)$.

[9] For each jackknife block we calculated $\eta_{\text{eff}}(T)$ and made a cubic spline temperature interpolation.

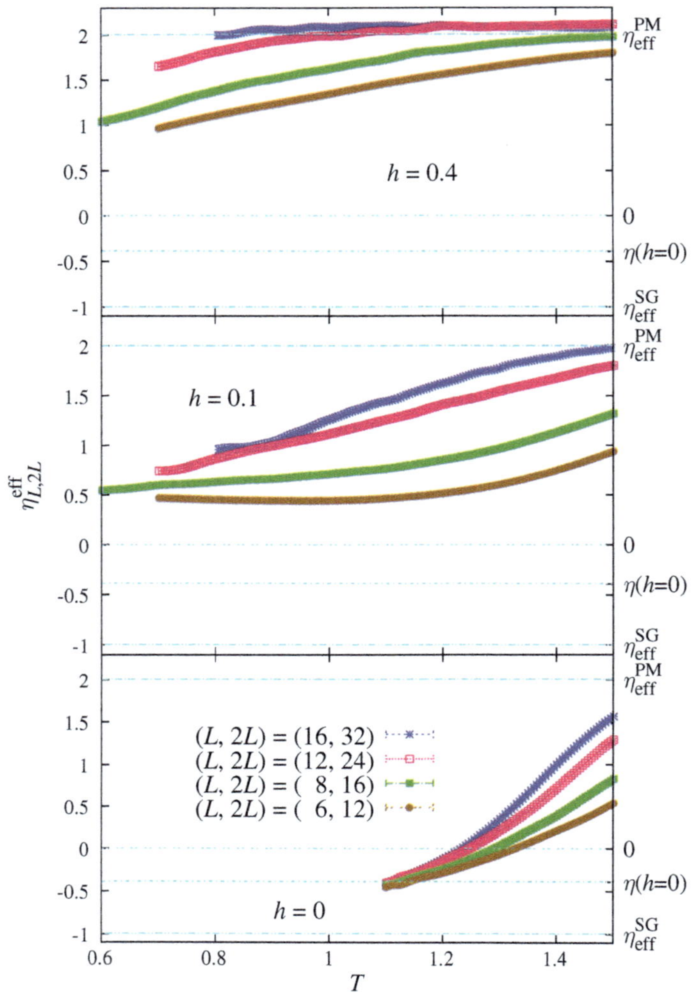

Fig. 2.15 We plot $\eta_{\text{eff}}(T)$, defined in (2.20), for all the pairs $(L, 2L)$ we could form. The magnetic fields are $h = 0.4$ (*top*), $h = 0.1$ (*center*) and $h = 0$ (*bottom*). The $h = 0$ data comes from [BJ13]. In each plot we uses horizontal lines to underline meaningful limits, and we label them with a tic on the right axis. From up to down, we depict the limit $\eta^{\text{PM}}_{\text{eff}} = 2$ of a system in the paramagnetic phase, the $\eta_{\text{eff}} = 0$ axis, the zero-field value $\eta_{\text{eff}}(h = 0, T_c) = -0.3900(36)$ [BJ13], and its value in a deep spin-glass phase $\eta^{\text{SG}}_{\text{eff}} = -1$. Notice the difference between the case with or without a field. For $h = 0.1$ the curves appear to converge to a positive $\eta_{\text{eff}} \simeq 0.5$, while in the latter all the curves become negative and merge at $\eta_{\text{eff}}(h = 0, T_c)$

$h = 0$ is not dominating ξ_L's behavior. On the other side, we are still far from the thermodynamic limit, since when the lattices are large enough, $\xi_L(T_c)/L$ should decay to zero linearly in $1/L$.

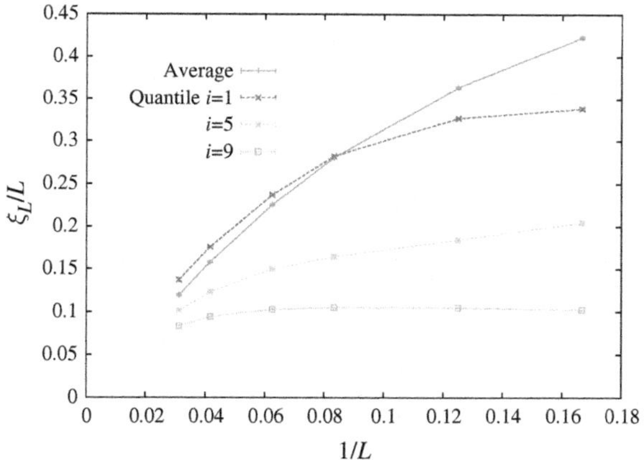

Fig. 2.16 Scaling of ξ_L/L at the null-field critical temperature $T_c = 1.109(29)$ [BJ13], with $h = 0.2$. We show the behavior of the average, and of quantiles 1, 5 and 9. If L is large enough, ξ_L/L should go as $1/L$, while if the system is seeing purely an echo of the divergence of the $h = 0$ transition transition, then ξ_L/L should be constant

2.11 Overview

We have studied the equilibrium behavior of the three-dimensional Ising Edwards–Anderson spin glass in an external magnetic field. Thermalizing the system at sufficiently low temperature was a computationally hard task and required the use of the JANUS dedicated computer to thermalize lattice sizes up to $L = 32$, down to temperatures $T \geq 0.8$.

First of all, we carried out a traditional analysis of our data. We chose observables that would be scale invariant at the critical temperature, and compared them for different lattice sizes, looking for crossings in their temperature curves. With this procedure we found no traces of a phase transition.

Yet, the scenario is more complicated. Despite the absence of crossings, indications that something non-trivial is going on are given by signals such as a growing correlation length (even for our largest lattices), peaks in the susceptibility, and a wide probability distribution function of the overlap.

We noticed a wide variety of behaviors within the same set of simulation parameters. Some measurements presented signs of criticality, while others did not. So, we tried to classify them in a meaningful way. We sorted our observables with the help of a CV, and came up with a quantitative criterion to select the best CV. Between the ones we proposed, the function of the instant overlaps that made the best CV turned out to be the median overlap q_{med}.

As a function of the median overlap, the scenario appeared rather non-trivial. The averages turned out to be dominated by a very small number of measurements. Those with a small q_{med} behaved similarly to the average: long correlation lengths, very large susceptibilities, and no signs of criticality. On the other side, the median behavior was far from the average, and the behavior of most of the measurements was qualitatively different from the average, with smaller correlation lengths and susceptibilities, but non-negligible indications of scale invariance right below the upper bound $T^{\text{up}}(h)$ given in [BJ14b]. Furthermore, separating the different behaviors of the system we obtain mutually consistent indications of criticality from our primary dimensionless magnitudes $\xi_{L,i}/L$ and $R_{12,i}$. The achievement of this consistency is an important step forward with respect to [Bn12a], where the phase transition was revealed only by the R_{12} indicator, but it was invisible to ξ_L/L.

Unfortunately we were not able to make a quantitative prediction on the critical temperatures $T_c(h)$, because the observables as a function of the lattice size and of the temperature were very nonlinear, and the temperatures we reached were not low enough reliably to identify the crossing points of the quantile-dependent $\xi_{L,i}/L$ and $R_{12,i}$.

Overall, the presence of a phase transition appears plausible from our simulations. Perhaps more importantly, now the challenge is well defined: in order to be able to give, numerically, a conclusive answer on the presence of a de Almeida-Thouless line we need push our simulations down to $T \simeq 0.4$ (at $h = 0.2$). We believe that *Janus II*, the next generation of our dedicated computer [BJ14b], will be able to assume this challenge.

References

[AB10a] R. Alvarez Baños, A. Cruz, L.A. Fernandez, J.M. Gil-Narvion, A. Gordillo-Guerrero, M. Guidetti, A. Maiorano, F. Mantovani, E. Marinari, V. Martin-Mayor, J. Monforte-Garcia, A. Muñoz Sudupe, D. Navarro, G. Parisi, S. Perez-Gaviro, J.J. Ruiz-Lorenzo, S.F. Schifano, B. Seoane, A. Tarancon, R. Tripiccione and D. Yllanes (Janus Collaboration): J. Stat. Mech. **2010**, P06026 (2010). doi:10.1088/1742-5468/2010/06/P06026. arXiv:1003.2569

[AB10b] R. Alvarez Baños, A. Cruz, L. A. Fernandez, J.M. Gil-Narvion, A. Gordillo-Guerrero, M. Guidetti, A. Maiorano, F. Mantovani, E. Marinari, V. Martin-Mayor, J. Monforte-Garcia, A. Muñoz Sudupe, D. Navarro, G. Parisi, S. Perez-Gaviro, J.J. Ruiz-Lorenzo, S.F. Schifano, B. Seoane, A. Tarancon, R. Tripiccione, D. Yllanes (Janus Collaboration): Phys. Rev. Lett. **105**, 177202 (2010). doi:10.1103/PhysRevLett.105.177202. arXiv:1003.2943

[BJ12] M. Baity-Jesi, R. A. Baños, A. Cruz, L.A. Fernandez, J.M. Gil-Narvion, A. Gordillo-Guerrero, M. Guidetti, D. Iniguez, A. Maiorano, F. Mantovani, E. Marinari, V. Martin-Mayor, J. Monforte-Garcia, A. Munoz Sudupe, D. Navarro, G. Parisi, M. Pivanti, S. Perez-Gaviro, F. Ricci-Tersenghi, J.J. Ruiz-Lorenzo, S.F. Schifano, B. Seoane, A. Tarancon, P. Tellez, R. Tripiccione, D. Yllanes, Eur. Phys. J. Special Topics **210**, 33 (2012). doi:10.1140/epjst/e2012-01636-9. arXiv:1204.4134

[BJ13] M. Baity-Jesi, R.A. Baños, A. Cruz, L.A. Fernandez, J.M. Gil-Narvion, A. Gordillo-Guerrero, D. Iniguez, A. Maiorano, F. Mantovani, E. Marinari, V. Martin-Mayor, J. Monforte-Garcia, A. Muñoz, Sudupe, D. Navarro, G. Parisi, S. Perez-Gaviro, M. Pivanti, F. Ricci-Tersenghi, J.J. Ruiz-Lorenzo, S.F. Schifano, B. Seoane, A. Tarancon, R. Tripiccione, D. Yllanes (Janus Collaboration). Phys. Rev. B **88**, 224416 (2013). doi:10.1103/PhysRevB.88.224416. arXiv:1310.2910

[BJ14a] M. Baity-Jesi, R.A. Baños, A. Cruz, L.A. Fernandez, J.M. Gil-Narvion, A. Gordillo-Guerrero, D. Iniguez, A. Maiorano, M.F., E. Marinari, V. Martin-Mayor, J. Monforte-Garcia, A. Muñoz Sudupe, D. Navarro, G. Parisi, S. Perez-Gaviro, M. Pivanti, F. Ricci-Tersenghi, J.J. Ruiz-Lorenzo, S.F. Schifano, B. Seoane, A. Tarancon, R. Tripiccione, D. Yllanes, J. Stat. Mech. **2014**, P05014 (2014). doi:10.1088/1742-5468/2014/05/P05014. arXiv:1403.2622

[BJ14b] M. Baity-Jesi, R.A. Baños, A. Cruz, L.A. Fernandez, J.M. Gil-Narvion, A. Gordillo-Guerrero, D. Iniguez, A. Maiorano, M. F., E. Marinari, V. Martin-Mayor, J. Monforte-Garcia, A. Muñoz Sudupe, D. Navarro, G. Parisi, S. Perez-Gaviro, M. Pivanti, F. Ricci-Tersenghi, J.J. Ruiz-Lorenzo, S.F. Schifano, B. Seoane, A. Tarancon, R. Tripiccione, D. Yllanes, Phys. Rev. E **89**, 032140 (2014). doi:10.1103/PhysRevE.89.032140. arXiv:1307.4998

[BJ14d] M. Baity-Jesi, L.A. Fernandez, V. Martin-Mayor, J.M. Sanz, Phys. Rev. **89**, 014202 (2014). doi:10.1103/PhysRevB.89.014202. arXiv:1309.1599

[Bal00] H.G. Ballesteros, A. Cruz, L.A. Fernandez, V. Martin-Mayor, J. Pech, J.J. Ruiz-Lorenzo, A. Tarancon, P. Tellez, C.L. Ullod, C. Ungil, Phys. Rev. B **62**, 14237–14245 (2000). doi:10.1103/PhysRevB.62.14237. arXiv:cond-mat/0006211

[Bal96] H.G. Ballesteros, L.A. Fernandez, V. Martin-Mayor, A. Muñoz, Sudupe. Phys. Lett. B **378**, 207 (1996). doi:10.1016/0370-2693(96)00358-9. arXiv:hep-lat/9511003

[Bn11] R.A. Baños, A. Cruz, L.A. Fernandez, J.M. Gil-Narvion, A. Gordillo-Guerrero, M. Guidetti, D. Iñiguez, A. Maiorano, F. Mantovani, E. Marinari, V. Martin-Mayor, J. Monforte-Garcia, A. Muñoz Sudupe, D. Navarro, G. Parisi, S. Perez-Gaviro, F. Ricci-Tersenghi, J.J. Ruiz-Lorenzo, S.F. Schifano, B. Seoane, A. Tarancón, R. Tripiccione, D. Yllanes, Phys. Rev. B **84**, 174209 (Nov 2011). doi:10.1103/PhysRevB.84.174209. arXiv:1107.5772, http://link.aps.org/doi/10.1103/PhysRevB.84.174209

[Bn12a] R.A. Baños, A. Cruz, L.A. Fernandez, J.M. Gil-Narvion, A. Gordillo-Guerrero, M. Guidetti, D. Iniguez, A. Maiorano, E. Marinari, V. Martin-Mayor, J. Monforte-Garcia, A. Muñoz, Sudupe, D. Navarro, G. Parisi, S. Perez-Gaviro, J.J. Ruiz-Lorenzo, S.F. Schifano, B. Seoane, A. Tarancon, P. Tellez, R. Tripiccione, D. Yllanes, Proc. Natl. Acad. Sci. USA **109**, 6452 (2012). doi:10.1073/pnas.1203295109

[Bel08a] F. Belletti, M. Cotallo, A. Cruz, L.A. Fernandez, A. Gordillo, A. Maiorano, F. Mantovani, E. Marinari, V. Martin-Mayor, A. Muñoz Sudupe, D. Navarro, S. Perez-Gaviro, J.J. Ruiz-Lorenzo, S.F. Schifano, D. Sciretti, A. Tarancon, R. Tripiccione, J.L. Velasco (Janus Collaboration): Comp. Phys. Comm. **178**, 208–216 (2008). doi:10.1016/j.cpc.2007.09.006. arXiv:0704.3573

[Bel09a] F. Belletti, A. Cruz, L.A. Fernandez, A. Gordillo-Guerrero, M. Guidetti, A. Maiorano, F. Mantovani, E. Marinari, V. Martin-Mayor, J. Monforte, A. Muñoz, Sudupe, D. Navarro, G. Parisi, S. Perez-Gaviro, J.J. Ruiz-Lorenzo, S.F. Schifano, D. Sciretti, A. Tarancon, R. Tripiccione, D. Yllanes (Janus Collaboration). J. Stat. Phys. **135**, 1121 (2009). doi:10.1007/s10955-009-9727-z. arXiv:0811.2864

[Bel06] F. Belletti, F. Mantovani, G. Poli, S.F. Schifano, R. Tripiccione, I. Campos, A. Cruz, D. Navarro, S. Perez-Gaviro, D. Sciretti, A. Tarancon, J.L. Velasco, P. Tellez, L.A. Fernandez, V. Martin-Mayor, A. Muñoz, Sudupe, S. Jimenez, A. Maiorano, E. Marinari, J.J. Ruiz-Lorenzo (Janus Collaboration). Comput. Sci. Eng. **8**, 41 (2006)

[Bel09b] F. Belletti, M. Guidetti, A. Maiorano, F. Mantovani, S.F. Schifano, R. Tripiccione, M. Cotallo, S. Perez-Gaviro, D. Sciretti, J.L. Velasco, A. Cruz, D. Navarro, A. Tarancon, L.A. Fernandez, V. Martin-Mayor, A. Muñoz-Sudupe, D. Yllanes, A. Gordillo-Guerrero, J.J. Ruiz-Lorenzo, E. Marinari, G. Parisi, M. Rossi, G. Zanier, Janus collaboration. Comput. Sci. Eng. **11**, 48 (2009). doi:10.1109/MCSE.2009.11

[Bra80b] A.J. Bray, M.A. Moore, J. Phys. C: Solid St. Phys. **13**, 419 (1980). doi:10.1088/0022-3719/13/3/016

[Cas05] T. Castellani, A. Cavagna, J. Stat. Mech. **2005**, P05012 (2005). doi:10.1088/1742-5468/2005/05/P05012

[Cav09] A. Cavagna, Phys. Rep. **476**, 51–124 (2009). arXiv:0903.4264

[Deb97] P.G. Debenedetti, *Metastable Liquids* (Princeton University Press, Princeton, 1997)

[Deb01] P.G. Debenedetti, F.H. Stillinger, Nature **410**, 259–267 (2001)

[Fer09a] L.A. Fernandez, V. Martin-Mayor, Phys. Rev. E **79**, 051109 (2009). doi:10.1103/PhysRevE.79.051109

[Fer13] L.A. Fernandez, V. Martin-Mayor, G. Parisi, B. Seoane, EPL **103**, 67003 (2013). doi:10.1209/0295-5075/103/67003. arXiv:1307.2361

[Fer09b] L.A. Fernandez, V. Martin-Mayor, S. Perez-Gaviro, A. Tarancon, A.P. Young, Phys. Rev. B **80**, 024422 (2009). doi:10.1103/PhysRevB.80.024422

[Ful13] C.J. Fullerton, M.A. Moore (2013). arXiv:1304.4420

[Gun91] K. Gunnarsson, P. Svedlindh, P. Nordblad, L. Lundgren, H. Aruga, A. Ito, Phys. Rev. B **43**, 8199–8203 (1991). doi:10.1103/PhysRevB.43.8199

[Huk96] K. Hukushima, K. Nemoto, J. Phys. Soc. Japan **65**, 1604 (1996). doi:10.1143/JPSJ.65.1604. arXiv:cond-mat/9512035

[Hyn96] R. Hyndman, Y. Fan, Am. Stat. **50**, 361 (1996)

[Jön05] P.E. Jönsson, H. Takayama, H.A. Katori, A. Ito, Phys. Rev. B **71**, 180412(R) (2005). doi:10.1103/PhysRevB.71.180412. arXiv:cond-mat/0411291

[Jör08b] T. Jörg, H.G. Katzgraber, F. Krzakala, Phys. Rev. Lett. **100**, 197202 (2008). doi:10.1103/PhysRevLett.100.197202. arXiv:0712.2009

[Kir87] T.R. Kirkpatrick, D. Thirumalai, Phys. Rev. B **36**, 5388 (1987). doi:10.1103/PhysRevB.36.5388

[Kir89] T.R. Kirkpatrick, D. Thirumalai, P.G. Wolynes, Phys. Rev. A **40**, 1045 (1989)

[Kot83] G. Kotliar, P.W. Anderson, D.L. Stein, Phys. Rev. B **27**, 602 (1983). doi:10.1103/PhysRevB.27.602

[Lar13] D. Larson, H.G. Katzgraber, M.A. Moore, A.P. Young, Phys. Rev. B **87**, 024414 (2013). doi:10.1103/PhysRevB.87.024414. arXiv:1211.7297

[Leu13] L. Leuzzi, G. Parisi, Phys. Rev. B **88**, 224204 (2013). arXiv:1303.6333

[Leu08] L. Leuzzi, G. Parisi, F. Ricci-Tersenghi, J.J. Ruiz-Lorenzo, Phys. Rev. Lett. **101**, 107203 (2008). doi:10.1103/PhysRevLett.101.107203

[Mar98] E. Marinari, *In Advances in Computer Simulation*, ed. by J. Kondor (Springer-Verlag, Kerstész and I, 1998)

[Moo02] M.A. Moore, B. Drossel, Phys. Rev. Lett. **89**, 217202 (2002). doi:10.1103/PhysRevLett.89.217202. arXiv:cond-mat/0201107

[New99] M.E.J. Newman, G.T. Barkema, *Monte Carlo Methods in Statistical Physics* (Clarendon Press, Oxford, 1999)

[Nig75] M.P. Nightingale, Physica A **83**, 561 (1975). doi:10.1016/0378-4371(75)90021-7

[Pal99a] M. Palassini, S. Caracciolo, Phys. Rev. Lett. **82**, 5128–5131 (1999). doi:10.1103/PhysRevLett.82.5128. arXiv:cond-mat/9904246

[Par12a] G. Parisi, F. Ricci-Tersenghi, Phil. Mag. **92**, 341 (2012). doi:10.1080/14786435.2011.634843. arXiv:1108.0759v1

[Par12b] G. Parisi, T. Temesvári, Nucl. Phys. B **858**, 293 (2012). arXiv:1111.3313

[Pet99] D. Petit, L. Fruchter, I.A. Campbell, Phys. Rev. Lett. **83**, 5130 (1999). doi:10.1103/PhysRevLett.83.5130. arXiv:cond-mat/9910353

[Pet02] D. Petit, L. Fruchter, I.A. Campbell, Phys. Rev. Lett. **88**, 207206 (2002). doi:10.1103/PhysRevLett.88.207206. arXiv:cond-mat/011112

[Seo13] B. Seoane: *Spin glasses, the quantum annealing, colloidal glasses and crystals: exploring complex free-energy landscapes*. Ph.D. thesis, Universidad Complutense de Madrid (January 2013)

[Tab10] Y. Tabata, K. Matsuda, S. Kanada, T. Yamazaki, T. Waki, H. Nakamura, K. Sato, K. Kindo, J. Phys. Soc. Japan. **79**, 123704 (2010). arXiv:1009.6115

[Tem08] T. Temesvári, Phys. Rev. B **78**, 220401 (2008)

[Tem02] T. Temesvári, C. De Dominicis, Phys. Rev. Lett. **89**, 097204 (2002). doi:10.1103/PhysRevLett.89.097204. arXiv:cond-mat/0207512

[Yll11] D. Yllanes, *Rugged Free-Energy Landscapes in Disordered Spin Systems*. Ph.D. thesis, Universidad Complutense de Madrid (2011). arXiv:1111.0266

[You04] A.P. Young, H.G. Katzgraber, Phys. Rev. Lett. **93**, 207203 (2004). doi:10.1103/PhysRevLett.93.207203. arXiv:cond-mat/0407031

Chapter 3
Heisenberg Spin Glass with Random Exchange Anisotropy

In the current chapter we expose the physical results of a work that required the use of unusual computing resources that revealed crucial for our results [BJ14d]. We carried out a massive campaign of MC simulations, exploiting the GPU clusters *Minotauro*,[1] in Barcelona, and *Tianhe-1a* in Tianjin, China,[2] and developing parallel codes in C, CUDA C and MPI to run our programs on one or multiple GPUs. This chapter will be dedicated to the physical results, while useful information on the computational aspects of our campaign is supplied in Appendix A.

3.1 The Kawamura Scenario

Already in the late '80s–early '90s there was general agreement on that experimental SGs undergo a phase transition at sufficiently low temperature [Bou86, Lév88, Gun91].

On the other hand, theoretical work in three dimensions was less advanced, even though one works with extremely simple models. For the Ising SG there were arguments supporting the existence of a phase transition [Fra94], that were later confirmed numerically [Pal99b, Bal00]. In the Heisenberg case, instead, all the attempts carried out during the '80s and '90s failed in finding a phase transition at a finite temperature $T_{SG} > 0$ [McM85, Oli86, Mor86, Mat91]. In fact, Matsubara et al. showed in 1991 that once a small anisotropic term is added to the Heisenberg Hamiltonian the phase transition becomes visible [Mat91]. This was in agreement with a later domain-wall computation [Gin93]. The accepted picture at the time was that the lower critical dimension (i.e. the spatial dimension below which there is no phase transition) lie somewhere between $3d$ and $4d$ [Col95].

[1] Barcelona Supercomputing Center, Barcelona, Spain, http://www.bsc.es.

[2] National Super-Computing Center, Tianjin, China, http://www.nscc-tj.gov.cn/en/.

M. Baity Jesi, *Spin Glasses*, Springer Theses, DOI 10.1007/978-3-319-41231-3_3

However, the story was slightly more complicated. Villain and coworkers made a provocative suggestion hypothesizing that, although maybe there was no spin glass transition, a different order parameter called chirality (or vorticity) could be critical [Mau90]. Chirality is a scalar observable that describes vorticity and alignment between neighboring spins [recall its definition (1.34) in Sect. 1.2] with the idea of mapping XY and Heisenberg to Ising SGs [Vil77, Vil78].

Villain's idea was elaborated by Kawamura in his *spin-chirality decoupling scenario* [Kaw92, Kaw98]. In the ideal case of a purely isotropic system the spin and chiral glass order parameters would be decoupled, the CG order parameter (1.36) would be critical whereas the SG overlap (1.33) would not display any phase transition. The introduction of any small anisotropy would couple the two. Since real samples always have some degree of anisotropy (see the following Sect. 3.2) the SG channel, coupled to the chiral one, would appear critical.

Kawamura's scenario was apparently consistent with all the observations until 2003, when Lee and Young employed more efficient simulation algorithms and finite-size scaling techniques to show that the spin glass channel is critical also in the fully isotropic model (i.e. the Heisenberg limit) [Lee03].

Both order parameters seemed to become positive at the same temperature. Further simulations confirmed the existence of a SG phase transition, although uncertainty remains on whether the transition is unique [Cam06, Fer09b] or chiralities order at a slightly higher temperature T_{CG} [Vie09].

A parallel issue is measuring the chiral order parameter in experiments. Kawamura proposed in 2003 that the extraordinary Hall resistivity is a simple function of the linear and non-linear CG susceptibilities [Kaw07]. Experiments based on this proposal observed the chiral transition and measured, for instance, the critical exponent δ [Tan07].

Interestingly enough, the value of δ turned out to be in between spin and chiral glass prediction. Nonetheless, it was impossible to identify a universality class despite the critical exponents of these systems had been extensively measured (at least in the SG sector) [Bou86, Lév88, Pet02]: the impression was that they change in a continuous way from the Heisenberg to the Ising limit [Cam10], as we increased the anisotropy.

However, analogy with ferromagnetic materials suggests a different interpretation. Anisotropy would be a relevant parameter in the sense of the renormalization group [Ami05]. There should be a new dominant FP, and symmetry considerations lead to think it should belong to the Ising-Edwards-Anderson (IEA) universality class. Yet, when we add a relevant parameter to the Hamiltonian, there should be some *crossover* effects (recall Sect. 1.3.5.1). In other words, one expects that while the correlation length ξ is small, the critical exponents are closer to the Heisenberg-Edwards-Anderson universality class, and that only for large enough ξ the universality class reveals its nature.

Notwithstanding, it is very hard, both numerically and experimentally, to prepare a SG with a large correlation length, since one should wait very long times (it has been argued that the waiting time t_{w} required to reach a certain coherence length is proportional to almost its seventh power, see e.g. [Bel08b, Bel09a, Joh99]). Probably this explains why the largest measured correlation lengths are of the order of only one hundred lattice spacings [Joh99, Ber04b]. That is a rather small distance to reveal the true universality class, so it is plausible that experiments will find critical exponents between the two Universality classes.

In fact, materials are classified according to the degree of anisotropy in their interactions [Pet02], which turns out to be relevant in their non-equilibrium magnetic response [Ber04b]. On one end of the materials' spectrum we find the extremely anisotropic $Fe_{0.5}Mn_{0.5}TiO_3$, which is maybe the best realization of the ideal limit of an Ising SG (Ising SGs correspond to the idealization of uniaxial spins). On the other end, we have very isotropic alloys such as AgMn or CuMn (whose modelization is notoriously difficult [Pei09], due to the presence of short range spin-density wave ordering [Cab82, Cab84, Lam95]).

To further complicate things, in experiments one has to take in account at least two relevant crossovers. The first is the competition, that we just pointed out, between the isotropic and the anisotropic fixed points. It is the one we treat in this chapter. The second crossover, that we will not address, is about short versus long range interactions. In fact, the Hamiltonian we treat is short range, but some often neglected interactions, such as the Dzyaloshinskii-Moriya interaction (see following Sect. 3.2) have been shown to be quasi-long-range, in the sense that the interactions are long range, but only until a cut-off distance of the order of some tens of atomic spacings [Bra82].[3]

Recent numerical work on the Heisenberg SG with *weak* random exchange anisotropies [MM11], as they would appear in nature, found a foggy scenario over the critical properties of the model. It was observed that:

- The CG critical temperature T_{CG} was significantly higher than T_{SG}, in disagreement with experiments and expectations.
- Apparently, the chiral susceptibility was *not* divergent at T_{CG}. This is surprising and, apparently, in contrast with experiments [Tan07]. Technically, this lacking divergence appeared as a very large anomalous dimension $\eta_{CG} \sim 2$.[4]
- Introducing very weak anisotropies changed dramatically T_{SG}. For example, the T_{SG} found by comparing systems of size $L = 6, 12$ was about twice its equivalent on the fully isotropic model. This is surprising, since one expects that the critical temperature would change very little from the isotropic case when D is as small as in [MM11].

To the light of this stumble, we decided to face again the problem of the phase transition in a model with random anisotropic exchange, but we increased drastically two factors, the degree of anisotropy and the size of the systems, in order to collect data closer to the attractive FP (recall Sect. 1.3.5.1),[5] that we suspected to be in the Ising universality class for symmetry reasons that will be discussed in Sect. 3.3.

[3]For further discussion of the crossover between long and short range interactions see [Ami05], Sect. 1.3.1, and [Car96], Sect. 4.3.

[4]Recall that $\gamma_{CG} = \nu(2 - \eta_{CG})$ where γ_{CG} is the critical index for the CG susceptibility, while ν is the correlation-length exponent. .

[5]The underlying assumption is that the whole critical line is dominated by the same FP.

In this chapter we will focus on the uniqueness of the phase transition and on the Universality class, proposing that there is a unique transition, belonging to the IEA Universality class [Edw75]. We will also give an interpretation to the results of [MM11], showing that the apparent inconsistencies are due to scaling corrections, that we will try to characterize, since we believe them to be fundamental both in the interpretation of numerical simulations and of experiments.

3.2 Anisotropy in Spin Systems

Experimentally, anisotropies affect significantly the glassy response to external magnetic fields and the behavior under cooling protocols [Ber04b], and as we have mentioned in the previous section the anisotropy is the driving element of Kawamura's spin-chirality decoupling scenario.

We quickly review here three of the principal mechanisms that lead to an anisotropy in the Hamiltonian [Myd93]. One one side the single-ion, and on the other the dipolar, and the Dzyaloshinskii-Moriya anisotropies. While the first one is site-dependent and does not depend on how the spins are coupled, the latter two are exchange anisotropies and involve the interactions between spins.

3.2.1 Single-Ion Anisotropy

Single-ion anisotropy is produced by the local crystalline electric fields of the solid. It depends on the spin and orbit angular moment of the modelled magnet and on the morphology of the crystalline structure, for example if the material is made in layers or in chains. Certain orientations of the spins will be preferred and others will be suppressed.

The simplest form of anisotropy we can think of is a strong uniaxial anisotropy that forces the spins to point along a single direction, that we usually identify with the z axis. This is the case of the Ising spins. Also, the system could be forced to lie on a $2d$ plane, in that case we would talk of XY spins. One can think Ising and XY systems as Heisenberg systems with an additional term that strongly inhibits certain components,[6]

$$\mathcal{H}_{\rm Ising} = -\frac{1}{2}\sum_{|x-y|=1} J_{x,y}\vec{s}_x\cdot\vec{s}_y + D_{\rm Ising}\sum_x\left((s_x\cdot\hat{e}_x)^2 + (s_x\cdot\hat{e}_y)^2\right), \quad D_{\rm Ising}\gg 1, \tag{3.1}$$

[6]Note that the anisotropy terms in the two following Hamiltonians are equivalent, $D_{\rm Ising} = -D_{\rm XY}$, just as in Sect. 1.3.5.1.

$$\mathcal{H}_{\mathrm{XY}} = -\frac{1}{2} \sum_{|x-y|=1} J_{x,y} \vec{s}_x \cdot \vec{s}_y + D_{\mathrm{XY}} \sum_x (s_x \cdot \hat{e}_z)^2, \quad D_{\mathrm{XY}} \gg 1. \tag{3.2}$$

For ferromagnetic systems (not SGs), the addition of perturbations of this type to the Hamiltonian changes its universality class [Car96]. Notice that for infinite anisotropy these Hamiltonians become the usual Ising and XY Hamiltonians.

In an amorphous material this anisotropy can be random, meaning that the preferred axis along which the spins want to align varies locally. One way to represent this effect is to choose a preferred axis, but assigning randomly how each spin couples to this axis, through a random term D_x chosen from an appropriate pdf. The resulting Hamiltonian is

$$\mathcal{H}_{\mathrm{r}_1} = -\frac{1}{2} \sum_{|x-y|=1} J_{x,y} \vec{s}_x \cdot \vec{s}_y - \sum_x D_x (\vec{s}_x \cdot \hat{e}_z)^2. \tag{3.3}$$

More in general also the direction of the "easy" axis can vary, so

$$\mathcal{H}_{\mathrm{r}_2} = -\frac{1}{2} \sum_{|x-y|=1} J_{x,y} \vec{s}_x \cdot \vec{s}_y - \sum_x D_x (\vec{s}_x \cdot \hat{n}_x)^2, \tag{3.4}$$

where $\hat{n}_x$ are random vectors on the sphere of radius 1.

3.2.2 Dzyaloshinskii-Moriya Anisotropy

The Dzyaloshinskii-Moriya (DM) [Dzy58, Mor60] interaction between two spins $\vec{s}_x$ and $\vec{s}_y$ describes the scattering of a conduction electron by $\vec{s}_x$. The electron then interacts with a non-magnetic scatterer with large spin-orbit coupling, and ends up scattering on spin $\vec{s}_y$.

This mechanism can be described with a term

$$\mathcal{H}_{x,y}^{\mathrm{DM}} = -\vec{B} \cdot (\vec{s}_x \times \vec{s}_y), \tag{3.5}$$

where $\vec{B} = \vec{r}_x \times \vec{r}_y$, and $\vec{r}_x$ is the position of $\vec{s}_x$. If we write the DM term in the form $-\vec{s}_x \cdot \mathbf{D}_{x,y}^{\mathrm{DM}} \vec{s}_y$, then

$$\mathbf{D}_{x,y}^{\mathrm{DM}} = \begin{pmatrix} 0 & B^z & -B^y \\ -B^z & 0 & B^x \\ B^y & -B^x & 0 \end{pmatrix}. \tag{3.6}$$

This antisymmetric matrix has $\det \mathbf{D}_{\mathrm{DM}} = 0$, $\mathrm{Tr} \mathbf{D}_{\mathrm{DM}} = 0$, and has rank 2 (so one null eigenvalue).

3.2.3 Dipolar Anisotropy

The dipolar anisotropy is a weak term: it is never the dominant term of the Hamiltonian. Yet, this type of anisotropy is always present in any kind of spin system, due to the fact that there always is a dipolar interaction between spins. This makes it a perfect candidate for the justification of the Kawamura scenario.

The dipolar interaction takes the form

$$\mathcal{H}^{\text{dip}}_{x,y} = \frac{1}{r^3_{xy}}\left[\vec{s}_x \cdot \vec{s}_y - 3(\vec{s}_x \cdot \hat{r}_{xy})(\vec{s}_y \cdot \hat{r}_{xy})\right], \tag{3.7}$$

where $\vec{r}_{xy} = \vec{r}_x - \vec{r}_y$, and $\hat{r}_{xy} = \vec{r}_{xy}/|r_{xy}|$. We can see how the configuration that minimizes the energy actually depends on the mutual orientation of the two dipoles. So for example, if $\vec{s}_x$ and $\vec{s}_y$ are parallel to $\hat{r}_{xy}$, the two spins will align parallel (the energy of the coupling is $-2/r^3_{xy}$ if they are parallel, $+2/r^3_{xy}$ if they are antiparallel), while if they are initially perpendicular to $\hat{r}_{xy}$ they will prefer to be antiparallel (the energy is $1/r^3_{xy}$ if they are parallel, $-1/r^3_{xy}$ if they are antiparallel). Notice that also the energy of the preferred energy minimum is different.

If we express $\mathcal{H}^{\text{dip}}_{x,y}$ in the form $\vec{s}_x \cdot \mathbf{D}^{\text{dip}}_{x,y}\vec{s}_y$, we get $D^{\alpha\beta} = \delta^{\alpha\beta} - 3r^\alpha r^\beta$. Therefore

$$\mathbf{D}_{\text{dip}} = \begin{pmatrix} 1 - 3r^x r^x & r^x r^y & r^x r^z \\ r^y r^x & 1 - 3r^y r^y & r^y r^z \\ r^z r^x & r^z r^y & 1 - 3r^z r^z \end{pmatrix} \tag{3.8}$$

is a symmetric matrix with a non-zero diagonal.

3.3 The Model and Its Symmetries

We study the model introduced by Matsubara et al. [Mat91], which is particularly convenient because of its simplicity. We consider $N = L^3$ three-dimensional unitary vectors $\vec{s}_x = (s^1_x, s^2_x, s^3_x)$ on a cubic lattice of linear size L, with periodic boundary conditions. The Hamiltonian is

$$\mathcal{H}_{\text{ANI}} = -\sum_{<x,y>}(J_{xy}\vec{s}_x \cdot \vec{s}_y + \sum_{\alpha\beta} s^\alpha_x D^{\alpha\beta}_{xy} s^\beta_y), \tag{3.9}$$

where the indexes α, β indicate the component of the spins. J_{xy} is the isotropic coupling between sites $\boldsymbol{x}$ and $\boldsymbol{y}$. D_{xy} is the anisotropy operator: a 3×3 symmetric matrix, where the six matrix elements $D^{\alpha\beta}_{xy}$, $\alpha \geq \beta$, are independent random variables, so it can be a fair descriptor of a dipolar anisotropy.

There is quenched disorder, this means that the time scales of the couplings $\{J_{xy}, D_{xy}\}$ are infinitely larger than those of our dynamic variables, so we represent them as constant in time random variables, with $\overline{J_{xy}} = \overline{D^{\alpha\beta}_{xy}} = 0$, $\overline{J^2_{xy}} = 1$ and $\overline{(D^{\alpha\beta}_{xy})^2} = D^2$.

We stress that if all the matrix elements $D^{\alpha\beta}_{xy}$ are zero we recover the fully isotropic Heisenberg model, with $O(3)$ symmetry. However, if the $D^{\alpha\beta}_{xy}$ are non-vanishing, the only remaining symmetry is time-reversal: $\vec{s}_x \longrightarrow -\vec{s}_x$ for all the spins in the lattice. Time reversal is an instance of the Z_2 symmetry. This is the symmetry group of the IEA model [Edw75]. Hence, we expect that the Z_2 symmetry will be spontaneously broken in a unique phase transition belonging to the IEA Universality class (see e.g. [Gin93]). Of course, underlying this expectation is the assumption that the anisotropic coupling is a relevant perturbation in the RG sense (as it is the case in ferromagnets [Ami05]). In fact, the infinite-anisotropy limit can be explicitly worked out for a problem with *site* anisotropy [rather than link anisotropy as in Eq. (3.9)]: one finds an IEA-like behavior [PT06, Lie07].

As we argued in Sect. 1.3.5.2 it is widely accepted that the universality class does not change with the probability distribution of the couplings. We take advantage of this, and choose a bimodal distribution for J_{xy} and $D^{\alpha\beta}_{xy}$, $J_{xy} = \pm 1$ and $D^{\alpha\beta}_{xy} = \pm D$. These couplings can be stored in a single bit, which is important because we are using GPUs, special hardware devices where memory read/write should be minimized (Appendix A).

We chose the two different values $D = 0.5, 1$. We want to compare our results with those in [MM11], where simulations were done on samples with weak random anisotropies. In that work the $D^{\alpha\beta}_{xy}$ did not follow a bimodal distribution, but were uniformly distributed between -0.05 and 0.05. To make proper comparisons we consider the standard deviation of the distribution. For bimodal distributions it is exactly D, in [MM11] it is $\overline{(D^2)}^{1/2} = 1/\sqrt{1200} \simeq 0.03$.

3.4 Simulation Details and Equilibration

We simulated on the largest lattices to present (up to $L = 64$), over a wide temperature range.[7] This has been possible thanks to an intense use of graphic accelerators (GPUs) for the computations. We made use of the *Tianhe-1A* GPU cluster in Tianjin, China, and of the *Minotauro* GPU cluster in Barcelona.

We used MC dynamics throughout all the work, mixing three different Monte Carlo algorithms, HB, OR and PT as explained in Appendix A, since both HB and OR are directly generalized to the anisotropic exchange case, where the local field is $\vec{h}_x = \partial\mathcal{H}_{\mathrm{ANI}}/\partial\vec{s}_x = \sum_{\|x-y\|=1}[J_{xy}\vec{s}_y + D_{xy}\vec{s}_y]$.

[7] Of course the limiting factor is in the wide range of *relaxation times,* rather than temperatures. However, relaxation times depend on a variety of implementation-dependent factors (such as the temperature spacing in the parallel tempering, or the number of overrelaxation (OR) sweeps). Hence, comparison with other work will be easier in terms of temperatures.

Table 3.1 Details of the simulations

D	L	N_{samples}	$N_{\text{MCS}}^{\text{min}}$	N_{T}	T_{min}	T_{max}
0.5	8	377	2.048×10^4	10	0.588	0.8
0.5	16	377	4.096×10^4	28	0.588	0.8
0.5	32	377	3.28×10^5	45	0.583	0.8
0.5	64	185	4×10^5	45	0.621	0.709
1	8	1024	2.048×10^4	10	0.877	1.28
1	12	716	1.68×10^5	20	0.893	1.28
1	16	1024	4.096×10^4	28	0.877	1.28
1	24	716	1.68×10^5	40	0.900	1.28
1	32	1024	3.28×10^5	45	0.917	1.28
1	64	54	3.44×10^5	45	1.0	1.16009

We show the simulation parameters for each anisotropy D, and lattice size L. N_{samples} is the number of simulated samples. N_{T} is the number of temperatures that were used in parallel tempering. The temperatures followed a geometric sequence between T_{min} and T_{max}, and N_{T} was chosen so that the PT's acceptance was around 15 %. $N_{\text{MCS}}^{\text{min}}$ is the minimum number of EMCS for each simulation. The simulation for $L = 64$, $D = 1$ was intended only to locate T_{CG}

All the simulations were run on NVIDIA Tesla GPUs. Except $L = 64$, $D = 0.5$, where we ran on 45 parallel GPUs, each sample was simulated on a single GPU. The interested reader can find in appendix A details on how they were performed.

Table 3.1 depicts the relevant simulation parameters. For given L and D, the simulations were all equally long, except for $L = 64$, $D = 0.5$, where we extended the simulation of the samples with the longest relaxation times.

To ensure thermalization we made a *logarithmic data binning*. Each bin had twice the length of the previous, i.e. it contained two times more EMCS, and had twice the measures. More explicitly, let us call i_{f} the last bin: i_{f} contains the last half of the MC time series, $i_{\text{f}} - 1$ the second quarter, $i_{\text{f}} - 2$ the second octave, and so on. This allowed us to create a sequence of values $\langle \mathcal{O}_n(i) \rangle$, for every observable $\mathcal{O}$, where n indicates the sample, and i identifies the bin, that has length 2^i EMCS. A set of samples was considered thermalized if $\overline{\langle \mathcal{O}_n(i) \rangle - \langle \mathcal{O}_n(i_{\text{f}}) \rangle}$ converged to zero. This test is stricter than merely requesting the convergence of the sequence of $\overline{\langle \mathcal{O}_n(i) \rangle}$, because neighboring blocks are statistically correlated, so the fluctuation of their difference is smaller [Fer08]. Physical results were taken only from the last block.

Since the $L = 64$, $D = 0.5$ samples were the most GPU-consuming, we were more strict with them. To ensure and monitor thermalization, beyond the previous criteria, we measured the integrated autocorrelation time (mixing time) of the random walk in temperatures of each sample [Fer09b, Yll11]. In a thermalized sample, all the replicas stay a significant amount of time at each temperature. We made sure that all the simulations were longer than 10 times this autocorrelation time. The sample-to-sample fluctuations were not extreme, and the autocorrelation times τ spanned between 10000 EMCS and 50000 EMCS, depending on the sample. Finally, we decided to take measures only over the last 64000 EMCS of each simulation.

3.5 Interpolations, Extrapolations and Errors

We have been able to estimate the critical temperature from the crossing of the curves ξ/L at L and $2L$, and the exponents ν and η with the method of the quotients, as described in Sect. 1.3.4.

To identify the crossing point between the pairs of curves (Figs. 3.1 and 3.2), we used low-order polynomial fits: for each lattice size, we took the four temperatures

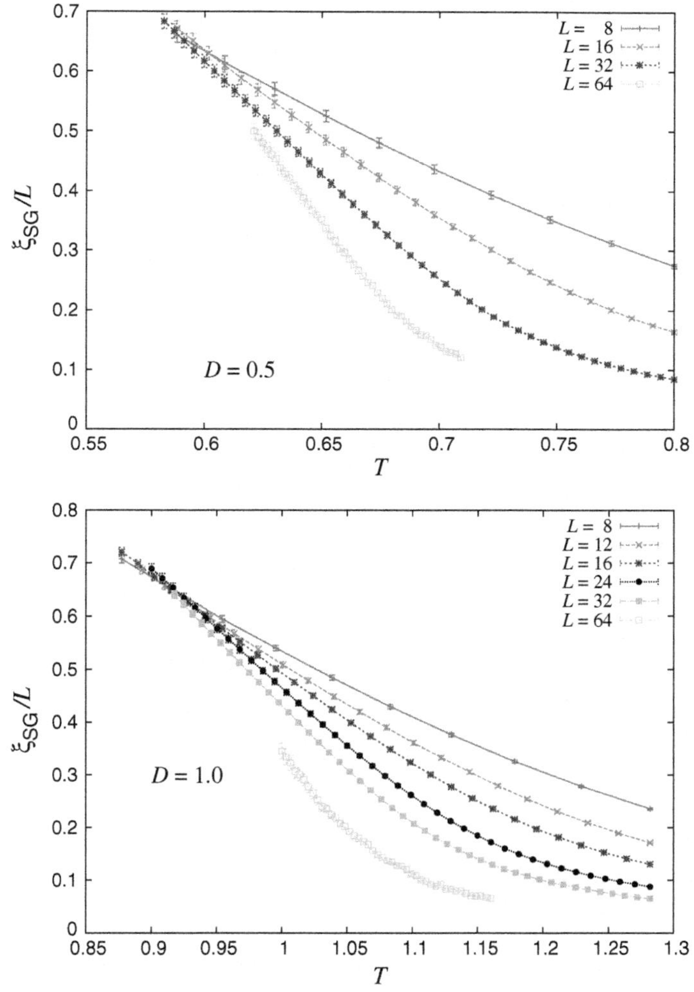

Fig. 3.1 Spin glass correlation length in units of the linear lattice size L for $D = 0.5$ (*top*) and $D = 1$ (*bottom*). All the curves cross at about the same temperature for both anisotropies (see Eq. (1.88)). The data for $D = 1$, $L = 64$, shown here for the sake of completeness, were only used for the chiral sector

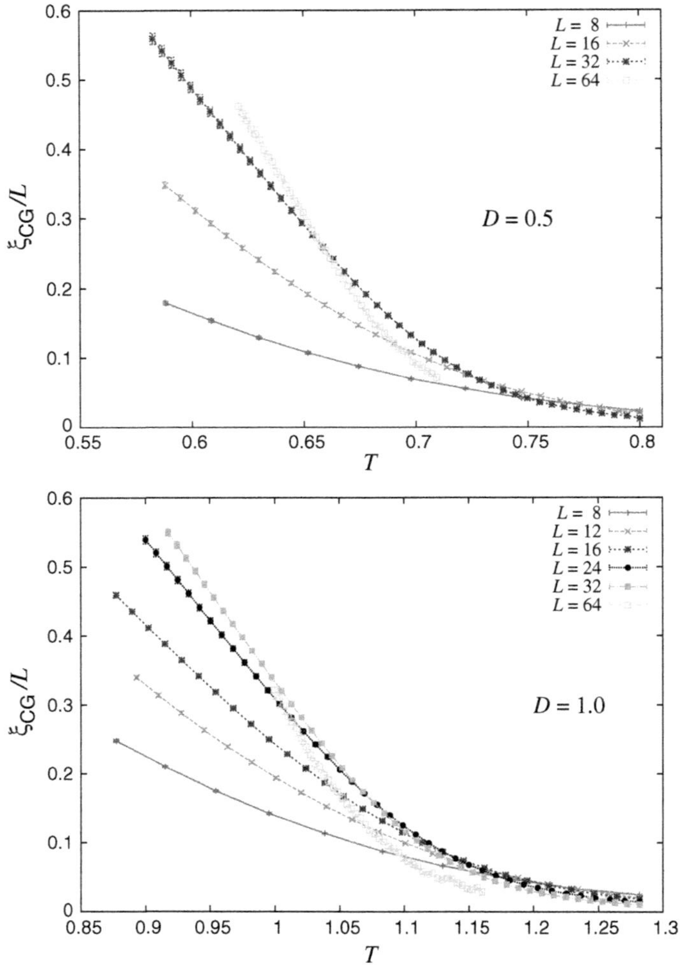

Fig. 3.2 Chiral Glass correlation length in units of the lattice size for $D = 0.5$ (*top*) and $D = 1$ (*bottom*). When L grows, the crossing temperature shifts significantly towards *left*

in the parallel tempering nearest to the crossing point. We fitted these four data points to a linear or quadratic function of the temperature. The obtained results were compatible within one standard deviation (the values reported in this work come from the linear interpolation). In order to calculate ν we needed the derivative of the correlation length at the crossing point. We extracted it by taking the derivative of the polynomial interpolations.

However, there is a difficulty in the calculation of statistical errors: the fits we had to perform came from strongly correlated data (because of the parallel-tempering temperature swap). Therefore, to get a proper estimate of the error, we made jackknife blocks, fitted separately each block, and calculated the jackknife error [Ami05].

The whole mentioned procedure was fluid while $T_{\mathrm{SG}}^{L,2L}$ fell in our simulated temperature span. Yet, since $T_{\mathrm{SG}}^{L,2L}$ was fairly lower than $T_{\mathrm{CG}}^{L,2L}$, it occurred in four cases that we did not reach low enough temperatures in our simulations to be able to interpolate the crossing, and we had to recur to extrapolations. This happened with $D = 1$, $T_{\mathrm{SG}}^{32,64}$ and $T_{\mathrm{CG}}^{32,64}$, and in the lower anisotropy $D = 0.5$, with $T_{\mathrm{SG}}^{16,32}$ and $T_{\mathrm{SG}}^{32,64}$.

The case of $T_{\mathrm{SG}}^{32,64}(D = 1)$ and $T_{\mathrm{SG}}^{16,32}(D = 0.5)$ was not a great issue, because the crossing point was very near to the lowest simulated temperature, so we treated these crossings just like the others.

In the case of $T_{\mathrm{SG}}^{32,64}(D = 0.5)$, instead, we had to extrapolate at a long distance (see Fig. 3.1–top, in the next section). Again, we performed the extrapolation through linear in temperature fits. To make the fit of $L = 64$ more stable, we took in account a progressive number of points (i.e. we fitted to the n lowest temperatures). We increased the number of temperatures, while the crossing temperature was constant. Note that increasing the number of temperatures in the fit results in a smaller statistical error for the crossing-temperature. However, $\xi_L(T)/L$ is not a linear function at high T (see Fig. 3.1). Therefore a tradeoff is needed because, when too high temperatures were included in the fit, the crossing temperature started to change, and we knew that curvature effects were biasing it. Our final extrapolation was obtained from a fit performed on the 10 lowest-temperature points. Unfortunately, this approach was not feasible for the SG susceptibility due to its strongly non-linear behavior. Hence, in the next section we will not give an estimate for $\eta_{\mathrm{SG}}(L = 64)$.

In the case of $T_{\mathrm{SG}}^{32,64}(D = 1)$, the simulation was not devised to reach that crossing point, and we did not extrapolate data.

3.6 Spin Glass Transition

Figure 3.1 show the crossings of $\xi_{\mathrm{SG}}(T)/L$ for $D = 0.5, 1$. Table 3.2 contains the principal results on the SG sector, providing a quantitative description of those figures. As explained in Sect. 3.3, we expect that the transition belongs to the IEA Universality class. This conjecture is supported by the fact that the critical exponents ν_{SG} and η_{SG}, and the height at which the $\xi_{\mathrm{SG}}(T)/L$ cross, are compatible with those of the IEA spin glass, indicated in the last line of Table 3.2. Hence, it is reasonable to extrapolate our results to $L \to \infty$ by assuming the IEA universality class. We took $\omega_{\mathrm{IEA}} = 1.0(1)$ from [Has08], and fitted to Eqs. (1.86), (1.90) and (1.92).[8] In those

[8] At the time these calculations were done and [BJ14d] was submitted, the most precise estimation of the critical parameters of the IEA model was done in [Has08]. At the moment of the drafting of this thesis, a more recent yet article from the JANUS collaboration [BJ13] gives a more precise determination of the critical exponents. The two estimations are compatible and using one or the other does not change qualitatively nor quantitatively our results and conclusions. In fact, the statistical errors on the extrapolations are much larger than those deriving from the uncertainty on ω (see Table 3.2).

Table 3.2 For each anisotropy D, and each pair of lattices $(L, 2L)$, we obtain effective size-dependent estimates for T_{SG}, and the universal quantities ν_{SG}, η_{SG} and $\xi_L(T_{SG})/L$

Determination of the critical quantities for the SG sector

D	$(L, 2L)$	T_{SG}	ν_{SG}	η_{SG}	$\xi_{SG}(T_{SG})/L$
0.5	(8, 16)	0.602(18)	1.91(27)	−0.388(27)	0.629(48)
0.5	(16, 32)	0.577(22)	2.70(63)	−0.449(67)	0.705(76)
0.5	(32, 64)	0.596(14)	2.18(45)	–	0.631(56)
0.5	∞	0.591(16)[0]	2.71(82)[3]	–	0.637(87)[1]
χ^2/d.o.f.		0.55/1	0.47/1	–	0.56/1
1.0	(8, 16)	0.910(21)	2.38(25)	−0.410(44)	0.660(34)
1.0	(12, 24)	0.927(19)	2.32(28)	−0.370(53)	0.629(36)
1.0	(16, 32)	0.910(16)	2.37(28)	−0.400(19)	0.660(35)
1.0	∞	0.917(32)[0]	2.33(67)[0]	−0.391(71)[1]	0.662(83)[0]
χ^2/d.o.f.		0.66/1	0.030/1	0.37/1	0.55/1
IEA	∞		2.45(15)	−0.375(10)	0.645(15)

The thermodynamic limit, indicated with $L = \infty$, is obtained by means of fits to Eqs. (1.88), (1.90), (1.92) and (1.86). Exponent ω was not a fitting parameter (we took $\omega_{IEA} = 1.0(1)$ from [Has08]). The line immediately after the extrapolations displays the estimator of the χ^2 figure of merit of each one. $D =$ IEA represents the critical values of the IEA Universality class, taken from [Has08]. The numbers in square brackets express the systematic error due to the uncertainty of ω_{IEA}. Notice that this systematic error is small compared to the statistical error

fits we took in account both the anticorrelation in the data,[9] and the bias arising from the uncertainty of the exponent ω_{IEA}. Notice, from Table 3.2, that the dependence on L of the data is so weak, that this bias is practically negligible. This situation is different from the one encountered in [MM11], where the anisotropy fields were

[9]Some of the points we used for those extrapolations in this chapter shared some of the data, so the measurements could not be treated as independent. For example, the crossing of ξ_L/L for $L = 8, 16$, had in common the points from size $L = 16$ with the pair $L = 16, 32$. This means that for the estimation of quantities deriving from the crossings, for example the thermal exponent ν (Eq. (1.90)), we need to take in account the non-diagonal part of the covariance matrix that gives a measure of the anticorrelation between measurements that share data.

For the described case, the typical jackknife (JK) statistical error (see Appendix E.1) coming from the diagonal part of the covariance matrix is

$$\sigma^2_{(8,16;8,16)} = (n-1)\sum_{j=0}^{n-1} \frac{(\nu_j^{(8,16)} - \tilde{E}(\nu^{(8,16)}))^2}{n}, \tag{3.10}$$

where n is the number of JK blocks blocks and $\tilde{E}(\ldots)$ is the estimator of the average. The new term we need to take in account in this example is the one coupling the couple (8, 16) to the couple (16, 32)

$$\sigma^2_{(16,32;16,32)} = (n-1)\sum_{j=0}^{n-1} \frac{(\nu_j^{(8,16)} - \tilde{E}(\nu^{(8,16)}))(\nu_j^{(16,32)} - \tilde{E}(\nu^{(16,32)}))}{n}. \tag{3.11}$$

Table 3.3 Same as Table 3.2, but for chirality

Determination of the critical quantities for the CG sector

D	$(L, 2L)$	T_{CG}	ν_{CG}	η_{CG}	$\xi_{CG}(T_{CG})/L$
0.5	(8, 16)	0.7762(43)	1.45(22)	1.9778(23)	0.0321(22)
0.5	(16, 32)	0.7255(29)	1.78(14)	1.8416(98)	0.0735(41)
0.5	(32, 64)	0.659(47)	2.40(47)	0.823(68)	0.258(18)
1.0	(8, 16)	1.2031(33)	1.205(71)	1.9507(27)	0.0418(12)
1.0	(12, 24)	1.1472(40)	1.72(11)	1.8664(51)	0.0691(25)
1.0	(16, 32)	1.1046(38)	2.18(10)	1.6995(75)	0.1098(42)
1.0	(32, 64)	0.987(22)	2.48(84)	0.53(19)	0.368(58)

In this case the corrections to scaling are significant

extremely small ($D \simeq 0.03$, see Sect. 3.3). There, the finite-size effects in the SG sector were huge.

Overall, the strong consistency of our extrapolations to large L with the IEA exponents shows a posteriori that our assumption was proper.

3.7 Chiral Glass Transition

In the CG channel (Fig. 3.2 and Table 3.3) the interpretation is slightly more controversial, since finite-size effects are heavy. For the smaller lattice sizes, T_{CG} is consistently larger than T_{SG}, and ν_{CG} is incompatible with the IEA limit. On the other side, when L is larger, T_{CG} approaches noticeably its SG counterpart, and so does ν_{CG}. We notice that η_{CG} marks the distinction between these two regimes. In fact, when L is small, it is very close to 2. This means that the divergence of χ_{CG} is extremely slow ($\chi \sim L^{2-\eta}$),[10] revealing we are still far from the asymptotic limit. When L is larger, η_{CG} is consistently smaller, the divergence of χ_{CG} is less suppressed, and we can assume the asymptotic behavior is starting to show up. Consistently with this observation, the value of ξ_{CG}/L at the crossing temperature becomes sizeable [indeed, the second-moment correlation length (1.67) is well defined only if $\eta < 2$, see e.g. [Ami05].

3.7.1 *Uniqueness of the Transition*

Although the SG and CG transitions do not coincide yet with our values of L and D, the critical temperatures, as well as ν, become more and more similar as the linear

[10]Recall that $\gamma_{CG} = \nu(2 - \eta_{CG})$ where γ_{CG} is the critical index for the CG susceptibility, while ν is the correlation-length exponent.

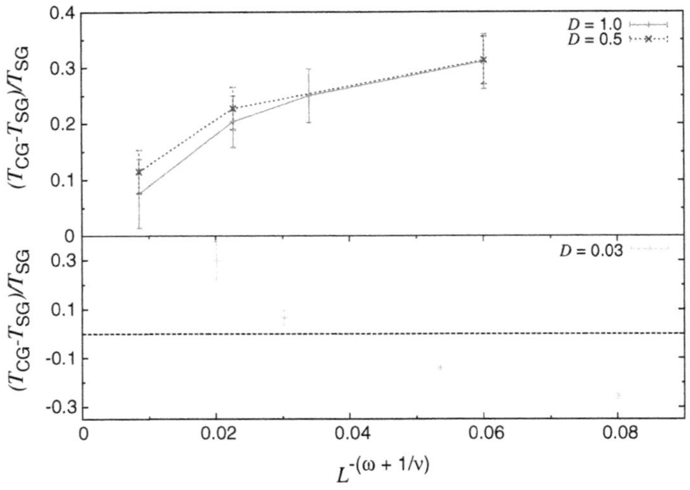

Fig. 3.3 Difference between the chiral crossing T_{CG} and the spin glass transition temperature T_{SG}^{∞}, in units of T_{SG}^{∞} (see Table 3.2 for the extrapolations of T_{SG}^{∞}). The exponents ω_{IEA} and ν_{IEA} are taken from [Has08]. In the *upper plot* we represent our data, for $D = 0.5, 1$. The two transitions get closer when we increase L, and the approach appears faster when the lattice size increases. Notice that a linear interpolation between the two largest lattice sizes intercepts the y axis compatibly with a coupling between the two transitions (i.e. $T_{SG} = T_{CG}$). On the *bottom* plot we show data from [MM11], where much lower anisotropies were considered. Here the scenario is completely different, since the critical temperatures drift apart for large enough L. The *horizontal dashed line* corresponds to $T_{CG} - T_{SG} = 0$

size of the system increases. Moreover, the decrease of η_{CG} as a function of L has not yet stabilized, so it is likely that the chiral quantities will keep changing with bigger lattice sizes.

As explained in Sect. 3.3, we expect that the transition should belong to the IEA Universality class. To confirm this expectation, we make the ansatz of a unique transition, of the IEA Universality class, to seek if the two critical temperatures join for $L \to \infty$. Figure 3.3 (upper half) shows the difference between the critical temperatures as a function of the natural scale for first order corrections to scaling, $L^{-(\omega_{IEA}+1/\nu_{IEA})}$ [Eq. (1.88)]. Again, ω_{IEA} and ν_{IEA} are taken from [Has08]. Not only Fig. 3.3 (top) reveals a marked increase of the speed of the convergence for $L = 64$ (to which corresponds the smallest anomalous exponent η_{CG}), but also, a linear interpolation to infinite volume, taking that point and the previous, extrapolates $T_{SG} = T_{CG}$ within the error.

Figure 3.4 shows how the SG and CG critical temperatures approach each other with L. Again, T_{CG} gets closer to T_{SG}, and the speed of the approach increases with the lattice size. The points in the intercept represent extrapolations to the thermodynamic limit of the T_{SG}. Since the observations are compatible with the ansatz of a unique

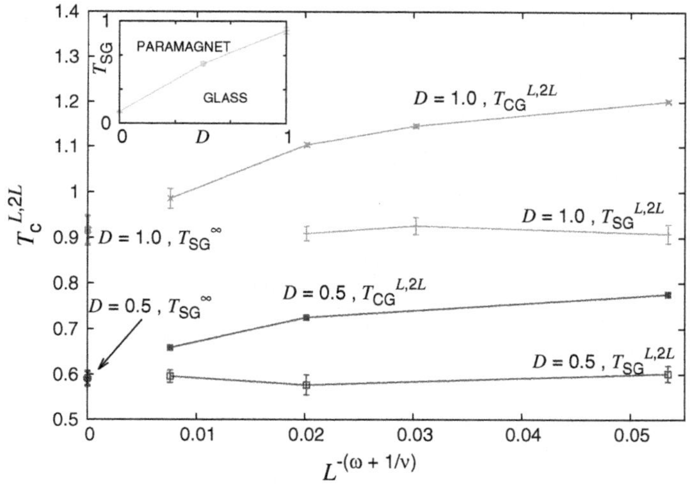

Fig. 3.4 Crossing temperatures as a function of $L^{-(\omega_{\text{IEA}}+1/\nu_{\text{IEA}})}$ (**large** plot). The points on the intercept are the $L \to \infty$ extrapolations from Table 3.2. The *inset* shows the phase diagram of the model with these same points, as the most economic interpretation of our data is that in the thermodynamic limit $T_{\text{SG}} = T_{\text{CG}}$. The $D = 0$ point is borrowed from [Fer09c]

phase transition, belonging to the IEA universality class, we used the infinite-size limit of T_{SG} to plot the model's phase diagram (Fig. 3.4, inset).[11]

3.8 Comparing with Weak Anisotropies

Both plots of Fig. 3.3 show the same observable, for different anisotropies. The top plot depicts our data, in the case of strong anisotropies $D = 0.5, 1$. The bottom one represents the case of weak anisotropies ($D \simeq 0.03$), coming from [MM11]. The behavior is very different between the two cases. For strong anisotropies, the critical temperatures tend to meet as we increase L. That is qualitatively very different from the weak anisotropy case, where their distance increases. We can ask ourselves where this qualitative difference of behavior comes from.

If we compare same system sizes and different D in Table 3.3, we notice that finite-size effects are larger (and η closer to two) the smaller the anisotropy. These differences in the finite-size effects are appreciable with a factor 2 change in the anisotropy (from $D = 1$ to $D = 0.5$), so it is reasonable that suppressing the anisotropy by a factor 17 or 35 will increase drastically the finite-size effects.

[11] In the phase diagram we show, the $D = 0$ point comes from [Fer09b], where chiral and spin glass transition are assumed to be coupled. There is disagreement on whether $T_{\text{SG}} = T_{\text{CG}}$ also in the isotropic case. Yet, we do plot it as a single transition because although T_{SG} might be lower than T_{CG}, their best estimates are compatible (and not distinguishable in the plot).

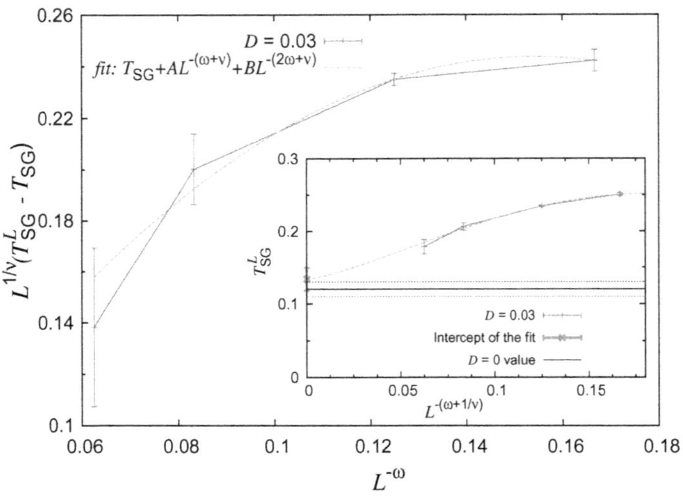

Fig. 3.5 Data from [MM11], corresponding to $D \simeq 0.03$, with extrapolations to the thermodynamic limit assuming the Ising-Edwards-Anderson Universality class. The data is the same in both plots. The *dashed line* is a fit of the scaling in L, considering corrections up to the second order [Eq. (1.88)]. The *large* figure displays the trend of the scaling variable $L^{1/\nu}(T - T_{\text{SG}})$ as a function of $L^{-\omega}$. The *inset* shows the same data set, plotting $T_{\text{SG}}^{L,2L}$ as a function of $L^{-\omega-1/\nu}$, see Eq. (1.88). The extrapolation to large-L (the point in the intercept) is compared with T_{SG} of $D = 0$ from [Fer09b]. The *full horizontal line* is the central value of $T_{\text{SG}}^{D=0}$, and the *dashed lines* define the error

The most economic explanation is then that there is a non-asymptotic effect that disappears with much larger systems or, as we have seen, with larger anisotropies. In other words there is a $L^*(D)$ after which T_{SG} and T_{CG} start joining. For $D \simeq 0.03$, L^* is so large that we observe a growing $T_{\text{CG}} - T_{\text{SG}}$, while for $D \geq 0.5$ we find $L^* < 8$.

Another peculiarity out-coming from [MM11] arises from the SG transition alone. It had been observed that a very weak perturbation on the symmetry of the isotropic system implied huge changes in the critical temperature, while one would expect that the transition line is smooth.

To solve this dilemma, we take advantage of having strong evidence for the Universality class of the transition. So, we take the data from [MM11], and use once again the exponents ν_{IEA} and ω_{IEA} in [Has08] to extrapolate the infinite volume limit with second order corrections to scaling [Eq. (1.88)]. The fit is good ($\chi^2/\text{d.o.f.} = 0.70/1$), and, as we show in Fig. 3.5, its $L \to \infty$ extrapolation for the critical temperature is compatible with $T_{\text{SG}}(D = 0)$ within one standard deviation. Thus, taming the finite-size effects was enough to make the scenario consistent, and the issue reduces to the fact that finite-size effects are extremely strong when the anisotropy is smaller.

3.9 An *ex post* Interpretation

We can reinterpret the results on the Heisenberg model with random anisotropic exchange interaction from an RG perspective. It was already established that in the isotropic $D = 0$ limit there is a phase transition at $T_{\rm SG}^{D=0}$. There are controversies on whether $T_{\rm SG}^{D=0} = T_{\rm CG}^{D=0}$, but this is unimportant to us, because it is generally accepted that $T_{\rm SG}^{D>0} = T_{\rm CG}^{D>0}$, though it was not verified until [BJ14d]. Therefore in the present section we mention the critical temperature as $T_{\rm SG}$.

One of our main questions was whether the universality class changes when $D > 0$. Since in nature anisotropies are always present, though weak, the problem was initially tackled by studying low random anisotropies in [MM11]. To the light of the remarks of Sect. 1.3.5.1, it was expectable that the numerical results be of hard interpretation. In fact when starting the RG flow from a small anisotropy, the system will initially feel strong effects from the $D = 0$ FP. Furthermore, if the flow does not start close to $T_{\rm SG}$, the numerical simulations will only feel at first the effects of the $D = 0$ FP, and then those of the $T = 0$ or $T = \infty$ FP (recall Fig. 1.7 and discussion).

Of the three options that in Sect. 1.3.5.1 are suggested to get away from this hard regime, we are able to adopt two, increasing drastically both the anisotropy and the lattice sizes, and finished obtaining also a better estimate of the critical temperature. The result is depicted in Fig. 3.6. Starting the flow from a large anisotropy leads

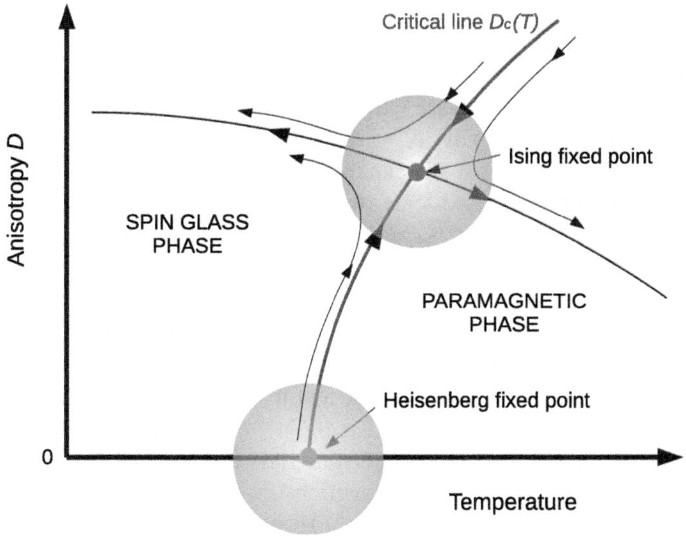

Fig. 3.6 RG flow in the Heisenberg SG with random anisotropies. The *orange zone* represents the zone of the phase diagram where the echoes of the Heisenberg FP are strong (even though it is not an attractive FP). The *blue area* is equivalent, but for the Ising FP. The Ising FP is attractive along the critical line $D_{\rm c}(T)$, but it is not in the rest of the phase diagram, so to approach the *blue* from the orange zone one must follow a flow that starts very close to $D_{\rm c}(T)$. Further discussions in the main text

the system far from the zone where echoes of the $D = 0$ transition are strong, and simulating on larger lattices is equivalent to taking more RG steps, toward the Ising FP. Furthermore, large lattices gave us a better estimate of the critical temperature, so our movement in the phase diagram sped towards the Ising FP in an effective way. In terms of Fig. 3.6 we moved from the outer part of the Heisenberg fixed point influence (drawn in orange, smaller lattices), to the zone where the Ising behavior is strong, blue zone ($L = 64$), so we were able to measure an Ising behavior.

3.10 Overview

We performed a numerical study of the critical behavior of Heisenberg spin glasses with strong bimodal random anisotropies. Our aim was to clarify the role of scaling-corrections, as well as the crossover effects between the Heisenberg and Ising Universality classes, to be expected when the anisotropic interactions are present. In fact, we show that anisotropic interactions are a relevant perturbation in the RG sense: no matter how small the anisotropy, the asymptotic critical exponents are those of the Ising-Edwards-Anderson model. However, a fairly large correlation length maybe needed to reach the asymptotic regime. This observation is relevant for the interpretation of both numerical simulations [MM11], and experiments [Pet02].

It is then clear that large system sizes are needed to make progress, something that calls for extraordinary simulation methods. Therefore, we performed single-GPU and multi-GPU simulations to thermalize lattices up to $L = 64$ at low temperatures. As side benefit, our work provides a proof-of-concept for GPU and multi-GPU massive simulation of spin-glasses with continuous degrees of freedom. This topic is elaborated further in Appendix A.

We performed a finite-size scaling analysis based on phenomenological renormalization (Sect. 1.3.4). We imposed scale-invariance on the second-moment correlation length in units of the system size, ξ_L/L. We followed this approach for both the chiral and spin glass order parameters.

Our results for the spin-glass sector were crystal clear: all the indicators of the Universality class were compatible with their counterparts in the Ising-Edwards-Anderson model. On the other hand, in the chiral sector scaling-corrections were annoyingly large, despite they decrease upon increasing the magnitude of the anisotropic interactions.

Regarding the coupling of chiral and spin glass transition, our numerical results seem to indicate that the two phase-transitions take place at the same temperature (i.e. $T_{\text{CG}} = T_{\text{SG}}$). However, it is important to stress that we need our very largest lattices to observe this trend. Nevertheless, what we see is in agreement with both Kawamura's prediction and experiments, where the phase transitions are apparently coupled, and the chiral glass susceptibility is divergent [Tan07].

Moreover, we were able to rationalize the numerical results in [MM11] with corrections to scaling, by assuming the Ising-Edwards-Anderson Universality class.

We remark that there are strong analogies between the interpretation of numerical and experimental data. In both cases, there is a relevant length scale (the correlation length for experiments, the system size for simulations). If that length is large enough, the asymptotic Ising-Edwards-Anderson Universality class should be observed. Otherwise, intermediate results between Heisenberg and Ising are to be expected, and indeed appear [Pet02].

The difficulty in reaching the asymptotic regime lies on time: the time growth of the correlation length is remarkably slow ($\xi(t_\mathrm{w}) \sim t_\mathrm{w}^{1/z}$ with $z \approx 7$ [Bel08b, Bel09a, Joh99], where t_w is the waiting time). Indeed, the current experimental record is around $\xi \sim 100$ lattice spacings [Joh99, Ber04b], pretty far from the thermodynamic limit.[12] Hence attention should shift to the study of the intermediate crossover regime. An intriguing possibility appears: one could envisage an experimental study of the crossover effects as a function of the *waiting time*. In fact, t_w varies some four orders of magnitude in current experiments [Rod13], which should result in a factor 4 variation of $\xi(t_\mathrm{w})$.

References

[Ami05] D.J. Amit, V. Martin-Mayor, *Field Theory, The Renormalization Group and Critical Phenomena*, 3rd edn. (World Scientific, Singapore, 2005). doi:10.1142/9789812775313_bmatter. http://www.worldscientific.com/worldscibooks/10.1142/5715

[BJ13] M. Baity-Jesi, R.A. Baños, A. Cruz, L.A. Fernandez, J.M. Gil-Narvion, A. Gordillo-Guerrero, D. Iniguez, A. Maiorano, F. Mantovani, E. Marinari, V. Martin-Mayor, J. Monforte-Garcia, A. Muñoz, Sudupe, D. Navarro, G. Parisi, S. Perez-Gaviro, M. Pivanti, F. Ricci-Tersenghi, J.J. Ruiz-Lorenzo, S.F. Schifano, B. Seoane, A. Tarancon, R. Tripiccione, D. Yllanes (Janus Collaboration). Phys. Rev. B **88**, 224416 (2013). doi:10.1103/PhysRevB.88.224416; arXiv:1310.2910

[BJ14d] M. Baity-Jesi, L.A. Fernandez, V. Martin-Mayor, J.M. Sanz, Phys. Rev. **89**, 014202 (2014). doi:10.1103/PhysRevB.89.014202. arXiv:1309.1599

[Bal00] H.G. Ballesteros, A. Cruz, L.A. Fernandez, V. Martin-Mayor, J. Pech, J.J. Ruiz-Lorenzo, A. Tarancon, P. Tellez, C.L. Ullod, C. Ungil, Phys. Rev. B **62**, 14237–14245 (2000). doi:10.1103/PhysRevB.62.14237. arXiv:cond-mat/0006211

[Bel08b] F. Belletti, M. Cotallo, A. Cruz, L.A. Fernandez, A. Gordillo-Guerrero, M. Guidetti, A. Maiorano, F. Mantovani, E. Marinari, V. Martin-Mayor, A.M. Sudupe, D. Navarro, G. Parisi, S. Perez-Gaviro, J.J. Ruiz-Lorenzo, S.F. Schifano, D. Sciretti, A. Tarancon, R. Tripiccione, J.L. Velasco, D. Yllanes, Janus Collaboration, Phys. Rev. Lett. **101**, 157201 (2008). doi:10.1103/PhysRevLett.101.157201. arXiv:0804.1471

[Bel09a] F. Belletti, A. Cruz, L.A. Fernandez, A. Gordillo-Guerrero, M. Guidetti, A. Maiorano, F. Mantovani, E. Marinari, V. Martin-Mayor, J. Monforte, A. Muñoz, Sudupe, D. Navarro, G. Parisi, S. Perez-Gaviro, J.J. Ruiz-Lorenzo, S.F. Schifano, D. Sciretti, A. Tarancon, R. Tripiccione, D. Yllanes (Janus Collaboration). J. Stat. Phys. **135**, 1121 (2009). doi:10.1007/s10955-009-9727-z. arXiv:0811.2864

[Ber04b] F. Bert, V. Dupuis, E. Vincent, J. Hammann, J.-P. Bouchaud, Phys. Rev. Lett. **92**, 167203 (2004). doi:10.1103/PhysRevLett.92.167203

[Bou86] H. Bouchiat, J. Phys. France **47**, 71–88 (1986). doi:10.1051/jphys:0198600470107100

[Bra82] A.J. Bray, M.A. Moore, J. Phys. C: Solid State. Phys. **15**, 3897 (1982)

[12]In a typical system $N = L^3 \sim N_\mathrm{A} \approx 6 \cdot 10^{23} \Rightarrow L \simeq 10^8$.

[Cab82] J. Cable, S. Werner, G. Felcher, N. Wakabayashi, Phys. Rev. Lett. **49**, 829 (1982)
[Cab84] J. Cable, S. Werner, G. Felcher, N. Wakabayashi, Phys. Rev. B **29**, 1268 (1984)
[Cam10] I.A. Campbell, D.C.M.C. Petit, J. Phys. Soc. Jpn. **79**, 011006 (2010). doi:10.1143/JPSJ.79.011006. arXiv:0907.5333
[Cam06] I. Campos, M. Cotallo-Aban, V. Martin-Mayor, S. Perez-Gaviro, A. Tarancon, Phys. Rev. Lett. **97**, 217204 (2006). doi:10.1103/PhysRevLett.97.217204
[Car96] J. Cardy, *Scaling and Renormalization in Statistical Field Theory*, vol. 5, eds. by P. Goddard, J. Yeomans. Lecture notes in physics (Cambridge University Press, Cambridge, 1996). ISBN:0521499593
[Col95] B. Coluzzi, J. of Phys. A: Math. Gen. **28**, 747 (1995)
[Dzy58] I. Dzyaloshinsky, J. Phys. Chem. Sol. **4**, 241 (1958)
[Edw75] S.F. Edwards, P.W. Anderson, J. Phys. F: Metal Phys. **5**, 965 (1975). doi:10.1088/0305-4608/5/5/017. http://stacks.iop.org/0305-4608/5/i=5/a=017
[Fer08] L.A. Fernandez, A. Maiorano, E. Marinari, V. Martin-Mayor, D. Navarro, D. Sciretti, A. Tarancon, J.L. Velasco, Phys. Rev. B **77**, 104432 (2008). doi:10.1103/PhysRevB.77.104432. arXiv:0710.4246
[Fer09b] L.A. Fernandez, V. Martin-Mayor, S. Perez-Gaviro, A. Tarancon, A.P. Young, Phys. Rev. B **80**, 024422 (2009). doi:10.1103/PhysRevB.80.024422
[Fer09c] L.A. Fernandez, V. Martin-Mayor, D. Yllanes, Nucl. Phys. B **807**, 424–454 (2009)
[Fra94] S. Franz, G. Parisi, M. Virasoro, J. Phys. (France) **4**, 1657 (1994). doi:10.1051/jp1:1994213
[Gin93] M. Gingras, Phys. Rev. Lett. **71**, 1637 (1993)
[Gun91] K. Gunnarsson, P. Svedlindh, P. Nordblad, L. Lundgren, H. Aruga, A. Ito, Phys. Rev. B **43**, 8199–8203 (1991). doi:10.1103/PhysRevB.43.8199
[Has08] M. Hasenbusch, A. Pelissetto, E. Vicari, J. Stat. Mech. **2008**, L02001 (2008). doi:10.1088/1742-5468/2008/02/L02001
[Joh99] Y.G. Joh, R. Orbach, G.G. Wood, J. Hammann, E. Vincent, Phys. Rev. Lett. **82**, 438 (1999). doi:10.1103/PhysRevLett.82.438
[Kaw92] H. Kawamura, Phys. Rev. Lett. **68**, 3785–3788 (1992). doi:10.1103/PhysRevLett.68.3785. http://link.aps.org/doi/10.1103/PhysRevLett.68.3785
[Kaw98] H. Kawamura, Phys. Rev. Lett. **80**, 5421–5424 (1998). doi:10.1103/PhysRevLett.80.5421. http://link.aps.org/doi/10.1103/PhysRevLett.80.5421
[Kaw07] H. Kawamura, Phys. Rev. Lett. **19**, 145213 (2007)
[Lam95] F. Lamelas, S. Werner, S. Shapiro, J. Mydosh, Phys. Rev. B **51**, 621 (1995)
[Lee03] L.W. Lee, A.P. Young, Phys. Rev. Lett. **90**, 227203 (2003). doi:10.1103/PhysRevLett.90.227203
[Lév88] L.P. Lévy, Phys. Rev. B **38**, 4963–4973 (1988). doi:10.1103/PhysRevB.38.4963. http://link.aps.org/doi/10.1103/PhysRevB.38.4963
[Lie07] F. Liers, J. Lukic, E. Marinari, A. Pelissetto, E. Vicari, Phys. Rev. B **76**, 174423 (2007). doi:10.1103/PhysRevB.76.174423
[MM11] V. Martin-Mayor, S. Perez-Gaviro, Phys. Rev. B **84**, 024419 (2011). doi 10.1103/PhysRevB.84.024419. http://link.aps.org/doi/10.1103/PhysRevB.84.024419
[Mat91] F. Matsubara, T. Iyota, S. Inawashiro, Phys. Rev. Lett. **67**, 1458 (1991)
[Mau90] A. Mauger, J. Villain, Y. Zhou, C. Rigaux, N. Bontemps, J. Férré, Phys. Rev. B **41**, 4587–4592 (1990). doi:10.1103/PhysRevB.41.4587. http://link.aps.org/doi/10.1103/PhysRevB.41.4587
[McM85] W.L. McMillan, Phys. Rev. B **31**, 340–341 (1985). doi:10.1103/PhysRevB.31.340. http://link.aps.org/doi/10.1103/PhysRevB.31.340
[Mor60] T. Moriya, Phys. Rev. Lett. **4**, 5 (1960)
[Mor86] B. Morris, S. Colborne, M. Moore, A. Bray, J. Canisius, J. Phys. C: Solid State Phys. **19**, 1157 (1986). http://iopscience.iop.org/0022-3719/19/8/014/pdf/0022-3719_19_8_014.pdf
[Myd93] J.A. Mydosh, *Spin Glasses: an Experimental Introduction* (Taylor and Francis, London, 1993)

[Oli86] J. Olive, A. Young, D. Sherrington, Phys. Rev. B **34**, 6341 (1986)

[Pal99b] M. Palassini, A.P. Young, Phys. Rev. Lett. **83**, 5126 (1999)

[PT06] F. Parisen Toldin, A. Pelissetto, E. Vicari: J. Stat. Mech.: Theory Exp. **2006**, P06002 (2006). doi:10.1088/1742-5468/2006/06/P06002

[Pei09] O. Peil, A. Ruban, B. Johansson, Phys. Rev. B **79**, 024428 (2009)

[Pet02] D. Petit, L. Fruchter, I.A. Campbell, Phys. Rev. Lett **88**, 207206 (2002). doi:10.1103/PhysRevLett.88.207206. arXiv:cond-mat/011112

[Rod13] G. Rodriguez, G. Kenning, R. Orbach, Phys. Rev. B **88**, 054302 (2013). doi:10.1103/PhysRevB.88.054302. http://link.aps.org/doi/10.1103/PhysRevB.88.054302

[Tan07] T. Taniguchi, J. Phys. Condens. Matter **19**, 145213 (2007)

[Vie09] D. Viet, H. Kawamura, Phys. Rev. Lett. **102**, 027202 (2009)

[Vil77] J. Villain, J. Phys. C: Solid State Phys. **10**, 4793 (1977). doi:10.1088/0022-3719/10/23/013. http://stacks.iop.org/0022-3719/10/i=23/a=013

[Vil78] J. Villain, J. Phys. C: Solid State Phys. **11**, 745 (1978). doi:10.1088/0022-3719/11/4/018. http://stacks.iop.org/0022-3719/11/i=4/a=018

[Yll11] D. Yllanes: *Rugged Free-Energy Landscapes in Disordered Spin Systems*. Ph.D. thesis, Universidad Complutense de Madrid (2011). arXiv:1111.0266

Part III
Energy Landscapes

Chapter 4
Energy Landscape of m-Component Spin Glasses

Although it is established that typical spin glasses [Méz87] order at a critical temperature T_{SG} for $d \geq 3$ [Bal00, Kaw01, Lee03], the nature of the low-temperature phase of spin glasses under the upper critical dimension $d_u = 6$ is still a matter of debate (Sect. 1.1.2).

Already at the dawning of spin glass theory interest had been given to the behavior of SGs as a function of the number of spin components m [Alm78a]. Increasing the number of spin components m reduces the number of metastable states, and recently renewed interest has been shown towards the properties of these models in the $m \to \infty$ limit, and their energy landscape [Has00]. Interesting features have been pointed out in large-m mean field models, such as a Bose-Einstein condensation in which the spins condense from an m-dimensional to an n_0-dimensional subspace, where n_0 scales with the total number of spins N as $n_0 \sim N^{2/5}$ [Asp04].

It has been argued in [Asp04] that the $m = \infty$ limit could be a good starting point for the study of the low-m SGs,[1] via $1/m$ expansions that have been used, for instance, to try to question the presence of a dAT line [Moo12]. However the Hamiltonian of the $m = \infty$ model has a unique local minimum, that can be found easily by steepest descendent (the determination of the ground state is *not* an NP-complete problem).

Explicit computations also indicate that the $m = \infty$ model is substantially different from any finite-m model (for example there is only quasi long-range order under T_{SG}, the upper critical dimension has been shown to be $d_u = 8$, and the lower critical dimension is suspected to be $d_l = 8$ too [Gre82, Via88, Lee05]), and that it is more interesting to study these models for large but finite m, thus reversing the order of the limits $m \to \infty$ and $N \to \infty$ [Lee05].

To better understand the large (but finite) m limit we undertake a numerical study in a three-dimensional cubic lattice. Our aim is to arrive at a quantitative comprehension of the energy landscape of systems with varying m, expecting, for example, to observe growing correlations as m increases [Has00].

[1]For example, in [Bey12] the infinite-m limit is used to derive exact relations in the one-dimensional spin glass with power law interactions.

M. Baity Jesi, *Spin Glasses*, Springer Theses, DOI 10.1007/978-3-319-41231-3_4

We focus on infinite-temperature ISs, i.e. the local energy minima that one reaches by relaxing the system from an infinite-temperature state, that is equivalent to a random configuration. Examining a system from the point of view of the ISs is a very common practice in the study of structural glasses [Cav09]. Only recently the study of quenches[2] from a high to a lower temperature has stimulated interest also in spin systems, both in presence and absence of quenched disorder.[3]

We analyze the properties of the ISs, and we inspect the dynamics of how the system converges to those configurations.

When one performs a quench from $T = \infty$ to $0 < T = T_0 < T_{SG}$, the system is expected to show two types of dynamics, an initial regime where thermal fluctuations are irrelevant, and a later one where they dominate the evolution (see for example the quenches performed in [Ber04c]). We choose $T_0 = 0$, so we can to show that the origin of the second dynamical regime is actually due to thermal effects. We study the quenches as a function of m. While on one side in the Ising limit $m = 1$ the dynamics is trivial, and correlations never become larger than a single lattice spacing, on the other side an increasing m yields a slower convergence, with the arising of low-temperature correlations that we can interpret as interactions between blocks of spins.

4.1 Model and Simulations

4.1.1 Model

The model is defined on a cubic lattice of side L with periodic boundary conditions. Each of the $N = L^3$ vertices $\boldsymbol{x}$ of the lattice hosts an m-dimensional spin $\vec{s}_x = (s_{x,1}, \ldots, s_{x,m})$, with the constraint $\vec{s}_x \cdot \vec{s}_x = 1$. Neighboring spins $\vec{s}_x$ and $\vec{s}_y$ are linked through a coupling constant $J_{x,y}$. The Hamiltonian is

$$\mathcal{H}_{\text{EA}} = -\frac{1}{2} \sum_{|x-y|=1} J_{x,y}\, \vec{s}_x \cdot \vec{s}_y, \tag{4.1}$$

that was already defined in Sect. 1.1.2. The couplings J_{xy} are Gaussian-distributed, with $\overline{J_{xy}} = 0$ and $\overline{J_{xy}^2} = 1$. The local field $\vec{h}_x$ for (4.1) is $\vec{h}_x = \sum_{y:|x-y|=1} J_{xy}\vec{s}_y$.

[2]By quench we mean the minimization of the energy throughout the best possible satisfaction of the local constraints, i.e. a *quench* is a dynamical procedure, as explained in Appendix F.1.1. Be careful not to confuse it with other uses of the same term. For example, those quenches have little to do with the *quenched approximation* used in QCD, or the *quenched disorder*, that is a property of the system.

[3]In addition to [Ber04c, BJ11] cited several times in this chapter, one can e.g. see [Bla14] for systems without quenched disorder, and [Bur07] for spin glasses.

This Hamiltonian is invariant under the simultaneous rotation or reflection of all the spins [that belongs to the $O(m)$ symmetry group], so the energy minimas may be found modulo a global rotation. For this reason we will use the tensorial definitions of the overlap (Sect. 1.2.1) and correlation functions and lengths (both point and plane, Sect. 1.2.3), so that the observables we measure are rotationally invariant too.

When one of the defined quantities is referred to the ISs (i.e. the final configurations of our quenches), we will stress it by putting the subscript $_{\text{IS}}$.

4.1.2 Simulations

We are interested in the ISs from infinite temperature, hence we need to pick random starting configurations, and directly minimize the energy.

The algorithm we choose is a direct quench, that consists in aligning each spin to its local field $\vec{h}_x$ (Appendix F.1.1). This choice was done because it allows us to compare ISs from systems with a different m in a general way. For example, the (SOR) (Appendix F.1.2) yields ISs with different properties, depending on the value of a parameter Λ [BJ11], and the same Λ is not equivalent for two different values of m.

For each sample we simulated two replicas, in order to be able to compute overlaps. We fixed the number of full sweeps of a lattice to $N_{\text{sweeps}} = 10^5$, as it had already been done in [Ber04c] with quenches to finite temperature. As it can be seen in Figs. 4.8, 4.9, 4.10 and 4.11 further on, this amount of steps was enough to guarantee the convergence to an IS in all our simulations. To ensure the convergence we required the last (logarithmically spaced) measurements to be equal within the error for each of the measured observables.

In Table 4.1 we give the parameters of our simulations.

Truncated correlators When the correlation function decays very quickly and the noise becomes larger than the signal, one could measure negative values of $C(r)$, that would be amplified by the factors r^2 and r^4 in the integrals (1.65) and (1.66). This would imply very large errors in ξ, or even the square root of a negative number. To overcome this problem, we truncated the correlation functions when they became less than three times the error, as it was first proposed in [Bel09a]. This procedure introduces a small bias, but reduces drastically the statistical error. Furthermore, the plane correlation function required the truncation much more rarely, therefore we compared the behaviors as a consistency check.

4.2 Features of the Inherent Structures Varying m

We want to analyse how the model's behavior changes with m. Intuitively, the more components a spin has, the easier it is to avoid frustration [Has00], and the simpler is the energy landscape. According to this scenario, when m increases, the number of

Table 4.1 Parameters of our simulations. N_{samples} is the number of simulated samples, N_{sweeps} is the number of quench sweeps of the whole lattice, and $\mathcal{N}_{\text{m}}$ is the number of measures we did during the quench. We chose to follow the same roughly logarithmic progression chosen in [Ber04c], measuring at times 2, 3, 5, 9, 16, 27, 46, 80, 139, 240, 416, 720, 1245, 2154, 3728, 6449, 11,159, 19,307, 33,405, 57,797, 100,000

L	m	N_{samples}	N_{sweeps}	$\mathcal{N}_{\text{m}}$
8	1	10,000	10^5	22
8	2	10,000	10^5	22
8	3	10,000	10^5	22
8	4	5000	10^5	22
8	6	10,000	10^5	22
8	8	10,000	10^5	22
16	1	1000	10^5	22
16	2	1000	10^5	22
16	3	1000	10^5	22
16	4	1000	10^5	22
16	8	1000	10^5	22
16	12	1000	10^5	22
16	16	1000	10^5	22
64	3	160	10^5	22

available ISs decreases down to the limit in which the energy landscape is trivial, and there is only one minimum. This should be reflected in the quantity Q^2/Q^2_{self} (recall Definition (1.31)), that should be small when there are many minima of the energy, and go to 1 when there is only one inherent structure, since all the quenches end in the same configuration. As shown in Fig. 4.1 (top), our expectation is confirmed. With Ising spins ($m = 1$) the energy landscape is so rich that ISs have practically nothing in common. When we increase m the overlaps start to grow until the limit $Q^2 = Q^2_{\text{self}}$. By comparing the data for different L, we can dismiss a difference in the behavior between discrete ($m = 1$) and continuous ($m > 1$) spins, since $m = 1$ for $L = 8$ behaves the same as $m = 2$ for $L = 16$. In Sect. 4.4 we will discuss aspects in which we do encounter differences.

Since the number of available ISs depends on both m and L, we can give an operative definition of a ratio $(m/L)_{\text{SG}}$ under which the number of ISs is exponential (so $Q^2/Q^2_{\text{self}} \simeq 0$), and of a ratio $(m/L)_1$ over which there is only one minimum.

This way, we can characterize finite-size effects effectively: An extremely small system $m/L > (m/L)_1$ is trivial and has only one stable state. Increasing the size we encounter a less trivial behavior, but to find a visible signature of a spin glass phase one has to have $L \geq m(L/m)_{\text{SG}}$. From Fig. 4.1 one can see that for $L = 8$, $m_{\text{SG}} = 1$, and for $L = 16$, $m_{\text{SG}} = 2$. Then, for example, we see that to observe a complex behavior for $m = 3$ spin glasses, one should use $L > 16$.

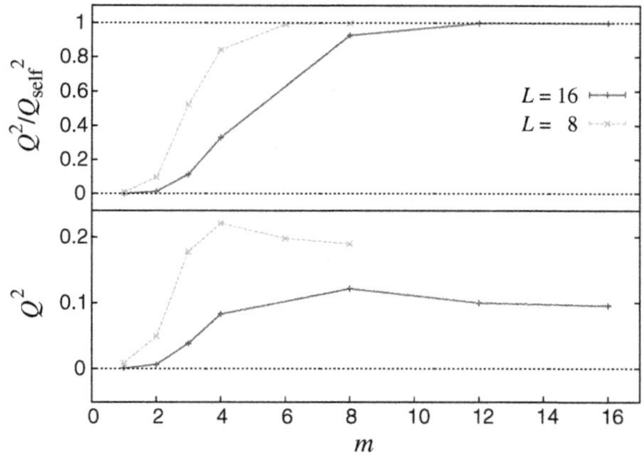

Fig. 4.1 Dependence of the ISs' overlaps from the number of components m of the spins. The *top figure* displays the overlap normalized with the self-overlap, showing that when m is large enough the energy landscape is trivial. In the *bottom* we have the unnormalized overlap Q^2. The *dashed horizontal lines* represent the limits 0 and 1, that bound both observables. Error bars are present though small, so almost not visible

Table 4.2 Properties of the ISs. For each choice of the parameters we show the observables at the end of the quench: The energy e_{IS}, the overlap Q^2_{IS}, the self overlap $Q^2_{\text{self,IS}}$, the point-correlation length $\xi^{\text{point}}_{\text{IS}}$ and the plane correlation length $\xi^{\text{plane}}_{\text{IS}}$

L	m	e_{IS}	Q^2_{IS}	$Q^2_{\text{self,IS}}$	$\xi^{(\text{plane})}_{\text{IS}}$	$\xi^{(\text{point})}_{\text{IS}}$
8	1	−0.4709(1)	0.0095(1)	1	0.68(2)	1.71(1)
8	2	−0.5953(1)	0.0497(3)	0.50297(2)	1.49(1)	2.802(4)
8	3	−0.6151(1)	0.1784(6)	0.33994(4)	2.188(2)	3.2358(7)
8	4	−0.6176(2)	0.2213(5)	0.26229(9)	2.2919(9)	3.2760(5)
8	6	−0.61801(11)	0.1989(1)	0.1997(1)	2.2567(3)	3.2514(2)
8	8	−0.61797(12)	0.1905(1)	0.1905(1)	2.2364(3)	3.2428(2)
16	1	−0.4721(1)	0.00123(6)	1	0.63(2)	1.69(1)
16	2	−0.5965(1)	0.0067(2)	0.500379(8)	1.49(4)	3.20(6)
16	3	−0.6165(1)	0.0382(5)	0.33416(1)	3.37(3)	5.43(1)
16	4	−0.6191(2)	0.0833(6)	0.25144(2)	4.153(7)	6.008(4)
16	8	−0.6200(1)	0.1218(3)	0.13126(5)	4.519(2)	6.187(1)
16	12	−0.6202(1)	0.10031(9)	0.10044(9)	4.3814(8)	6.087(1)
16	16	−0.6197(1)	0.0959(1)	0.0959(1)	4.3412(8)	6.066(1)
64	3	−0.61657(4)	0.00064(2)	0.3333466(4)	3.53(7)	6.74(6)

Moreover, this interpretation gives a straightforward explanation of the finite-size effects one encounters in the energy of an inherent structure (Table 4.2). For example, if we compare $L = 8, 16$ at $m = 8$, we notice two incompatible energies. In fact, there is an intrinsic difference between the two sizes, since $L = 8$ represents

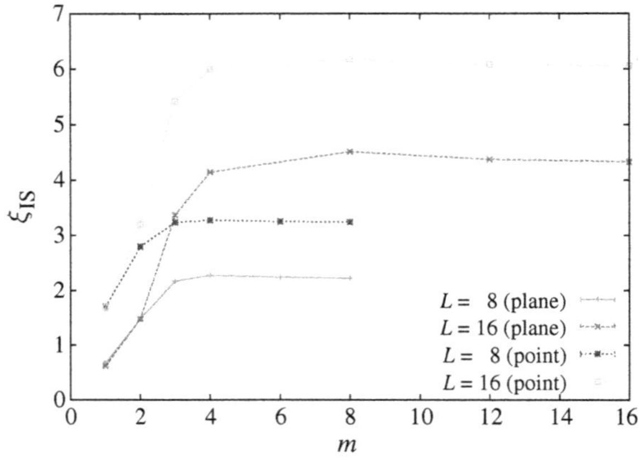

Fig. 4.2 Dependence of the second-moment correlation length ξ_2 on the number of components of the spins m. We show both the plane and the point correlation functions defined in Eqs. (1.57) and (1.58), for $L = 8, 16$

single-basin systems, while $L = 16$ has a variety of inherent structures. On the other side, finite-size effects on lower m are smaller, because we are comparing similar types of behavior.

Notice that, although the ratio $Q^2/Q^2_{\text{self}}(m)$ grows monotonously, this is not true for the pure overlap $Q^2(m)$ (Fig. 4.1, bottom), that has a peak at an intermediate m. Moreover, the position of the peak doubles when we double the lattice linear size, justifying the operational definitions $(m/L)_{\text{SG}}$ and $(m/L)_1$. The same peak at intermediate m is also visible in the energy and in the correlation length (Fig. 4.2), indicating that there is an intrinsic difference in the nature of the reached ISs. In Table 4.2 we give the values of the aforementioned observables at the IS. We see in this behavior the competition between two effects. When m is small, the quench has a vast choice of valleys where to fall. Since, reasonably the attraction basin of the lower-energy ISs is larger, the wide variety of ISs will increase the probability of falling in a minimum with low energy and larger correlations. When m increases, the number of available valleys decreases, so it is more likely that two different replicas fall in the same one. Yet, the *quality* of the reached ISs decreases, since the quench does not have the possibility to choose the lowest-energy minimum.

4.3 Overlap Probability Densities

From these observations it is reasonable to think that overlap and energy of the v are correlated. We looked for these correlations both on the overlap, on the self overlap, and in their ratio, but with a negative result. In Fig. 4.3 we show a scatter-plot of the ratio of the inherent structure's overlaps $Q^2_{\text{IS}}/Q^2_{\text{self,IS}}$ that confirms our statements. An equivalent plot for the link overlap is displayed in the inset.

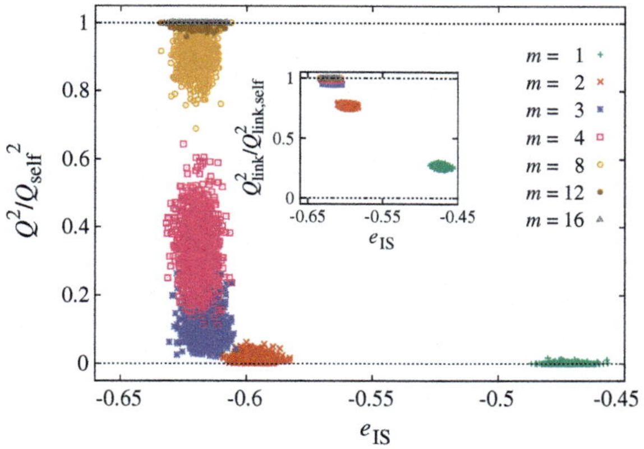

Fig. 4.3 Scatter plots for $L = 16$, at different values of m, of the overlap ratio $Q^2_{\rm IS}/Q^2_{\rm self,IS}$ against mean energy between the two replicas $e_{\rm IS} = (e^{(a)}_{\rm IS} + e^{(b)}_{\rm IS})/2$. Each simulated sample contributes to the plot with a single point. The *inset* displays an analog plot for the link overlap

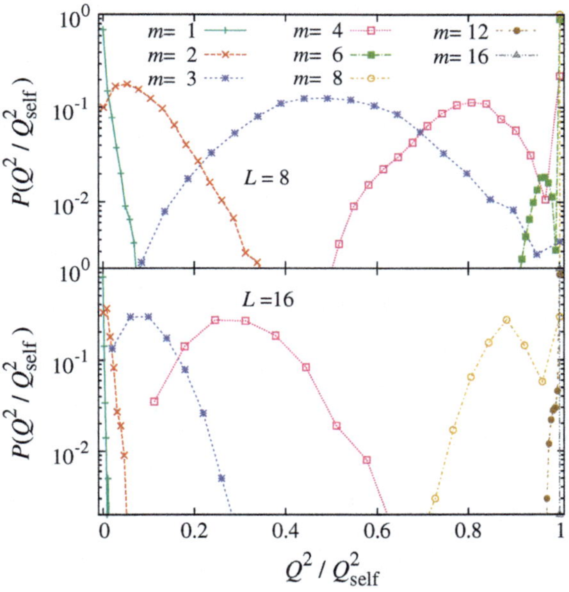

Fig. 4.4 Overlap pdfs of the ISs for different values of m. The *top figure* depicts data for $L = 8$, on the *bottom* we have $L = 16$. The *curves* are normalized to plot all the *curves* together. The actual probability distribution function is obtained by dividing each point by the bin width $\Delta Q/N_{\rm bins}$, where ΔQ is the difference between maximum and minimum Q^2

The cross sections of Fig. 4.3 give an idea of the energy and overlap probability distribution functions. We show explicitly the overlap probability distribution functions (normalized with the bin width) of the ISs in Fig. 4.4. They are qualitatively different from their thermal counterparts (see, e.g., [AB10a]). The ratio $Q^2_{\rm IS}/Q^2_{\rm self,IS}$ is bounded between zero and one. The distributions are extremely wide, and the phenomenology is quite different near the two bounds. In fact, when m is large enough, the limit $Q^2_{\rm IS}/Q^2_{\rm self,IS} = 1$ changes completely the shape of the curves, introducing a second peak (that we could read as an echo of the Bose-Einstein condensation remarked in [Asp04]). Around the lower bound of the $P(Q^2_{\rm IS}/Q^2_{\rm self,IS})$, instead, there is no double peak. We can try to give an interpretation to the presence of this second

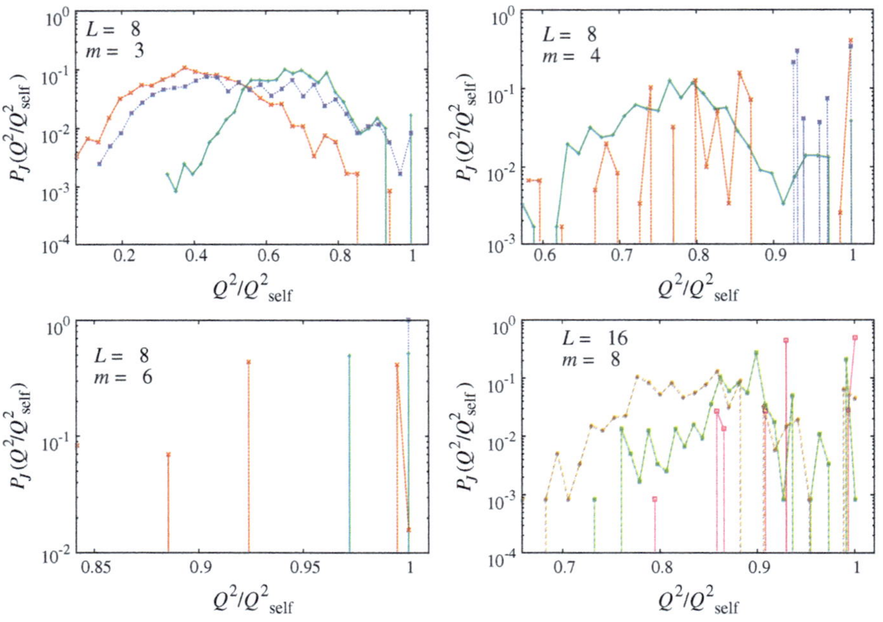

Fig. 4.5 Sample-dependent overlap pdfs $P_J(Q^2_{\text{IS}}/Q^2_{\text{self,IS}})$. Each *curve* depicts data from a separate sample. In each plot we show a selection of three samples with different shapes of the distribution. The choices of the parameters are represented in the key of each plot. We used two different color codes to distinguish the three plots that come from $L = 8$ systems (*top-left* and *right*, and *bottom left*), from the *bottom-right* plot that is for $L = 16$. The *curves* are normalized as in Fig. 4.4

peak by looking at the overlap distribution functions $P_J(Q^2_{\text{IS}}/Q^2_{\text{self,IS}})$ for a given instance of the couplings. In Fig. 4.5 we show that this distribution has relevant sample-to-sample fluctuations. When we increase m, the number of minima of the energy, N_{IS}, gradually becomes smaller. Yet, depending on the specific choice of the couplings, N_{IS} can vary sensibly. For example in Fig. 4.5, top-right, one can see that when $L = 8$ and $m = 4$, N_{IS} can be both large (red curve) or of order one (blue curve). For $L = 8$, $m = 6$ (Fig. 4.5, bottom-left), the situation is similar: for the blue curve $N_{\text{IS}} = 1$, while for others $N_{\text{IS}} > 1$.

As we similarly stated in Sect. 4.2, we notice that the lattice size plays a substantial role on the properties of the reached inherent structure, since when we pass from $L = 8$ to $L = 16$ histograms regarding the same m cover very different ranges of q. We can both see them traditionally as strong finite-size effects, or focus on L as a relevant parameter (as it was suggested, for example, in [BJ14d]), concentrating the interest on finite L.

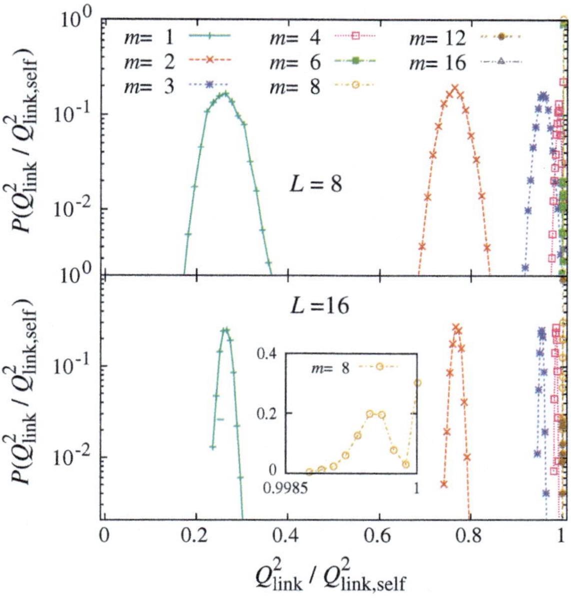

Fig. 4.6 Same as Fig. 4.4, but for the link overlap. The *inset* shows a zoom for the $m = 8$, $L = 16$ data, where we also removed the logarithmic scale on the y axis

4.3.1 Link Overlaps

Since in the past ten years an increasing attention has been devoted to the link overlap $Q^2_{\rm link}$ as an alternative order parameter for the study of the low temperature region of spin glasses [Krz00, Con06, AB10a], in Fig. 4.6 we show also the link-overlap histograms $P(Q^2_{\rm link,IS})$ at the IS. The functions $P(Q^2_{\rm link,IS})$ have much smaller finite-size effects than the $P(Q^2_{\rm IS})$, and are more Gaussian-like (although the Gaussian limit is impossible, since $Q^2_{\rm link}$ is bounded between 0 and 1). The inset shows that the second peak on high overlaps is present also with the link overlap.

We checked also the correlation between spin and link overlaps. At finite temperature there are different predictions between RSB and droplet pictures. According to the RSB picture the conditional expectation value $E(Q^2_{\rm link}|Q^2)$ should to be a linear, strictly increasing function of Q^2, while this should not be true in the Droplet theory (Sect. 1.1.2). When m is small, this correlation is practically invisible, but it becomes extremely strong when we increase the number of components of the spins (Fig. 4.7). Notice how the correlation between spin and link overlap is formidably increased when we normalize the two with the self overlap. The curves in Fig. 4.7 represent $E(Q^2_{\rm link}|Q^2)$. If we exclude the tails, that are dominated by rare non-Gaussian events, the trend is compatible with linearly increasing functions.

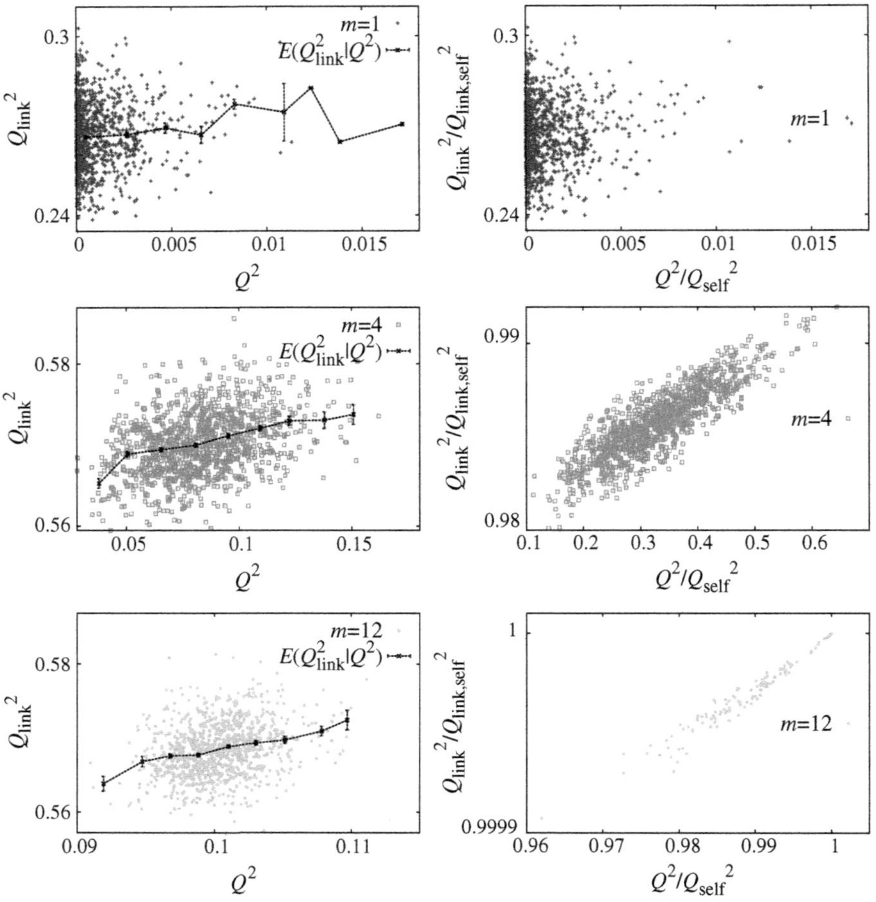

Fig. 4.7 Correlation between the spin and the link overlap of the ISs, for $L = 16$ lattices, with $m = 1$ (*top*), $m = 4$ (*center*) and $m = 12$ (*bottom*). On the *left* we plot the overlaps, while on the *right* they are normalized with the self overlap. Normalizing with the self overlap increases the correlations between the two order parameters. The *two top figures* are the same because the self overlap is one when $m = 1$. The *black lines* on the *left plots* represent $E(Q^2_{link}|Q^2)$, and they show that a correlation exists also without normalization

4.4 Quench Dynamics

Let us get an insight on the dynamics of the quench. For short times, the energy converges towards a minimum with a roughly power law behavior (Fig. 4.8). At longer times there is a cutoff, that grows with the system's size, revealing a change in the dynamics after which the system converges faster to a valley. We stress the great difference in the convergence rate between $m = 1$ and $m > 1$. We can identify two different decrease rates, depending on whether the spins are discrete or continuous.

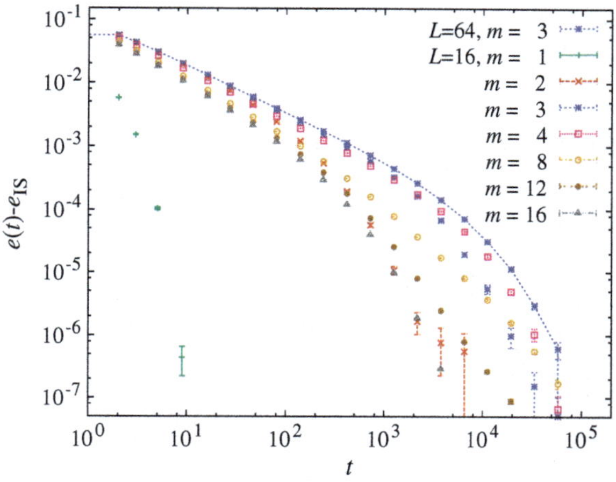

Fig. 4.8 Evolution of the energy during the quench for all the simulated values of m, in $L = 16$ lattices. On the x axis there is the time, measured in full lattice quench sweeps. On the y axis there is the difference between the energy at time t, $e(t) = \left(e^{(a)}(t) + e^{(b)}(t)\right)/2$, and its final value $e_{\mathrm{IS}} = e(t = 10^5)$. The convergence speed is very different between continuous and discrete spins. To stress the finite-size effects we also show points for $L = 64$, $m = 3$ (points connected by segments).

Figure 4.9 shows the evolution of the overlap for $L = 16$, and gives a better understanding of why quantities such as Q^2_{IS} are not monotonous with m. We show both the evolution of Q^2/Q^2_{self} (top), and of Q^2 (bottom). The first one behaves as one would expect when the number of minima is decreasing to one. On the other side, we see from the lower plot *how* the quenches of $m = 8$ reach the highest overlap. A possible interpretation is to ideally separate the quench in two regions. At the beginning there is a search of the valley with a power-law growth of Q^2, and later the convergence inside of the valley. Figure 4.9 shows that the search of the valley stops earlier when $m = 12, 16$, i.e. when their number is of order one.

We remark on a nonlinear trend on the evolution of the self overlap $Q^2_{\mathrm{self}}(t)$. For continuous spins ($m > 1$) it has a different value at infinite and zero temperature (Fig. 4.10). This variation is strikingly visible when m is large, but the same trends are found for $m \leq 3$, though the variations are so small that it is justified that they are usually not found.[4] Moreover $Q^2_{\mathrm{self}}(t)$ is highly nonlinear, and, except for the highest m, it overshoots before having converged.

In Fig. 4.11 we show the evolution of the correlation lengths ξ_2^{plane} during the quenches for $L = 16$ for all our values of m. We see the same variety of behaviors shown by Q^2 (Fig. 4.9), with $\xi_2^{\mathrm{plane}}(m = 12, 16)$ that abruptly stop increasing, while

[4] To our knowledge, the only reference where a non-trivial behavior of the self-overlap was found is in [BJ11]. Yet, in this case it was in the study of ISs from finite temperature, and in the chiral sector (they worked with $m = 3$).

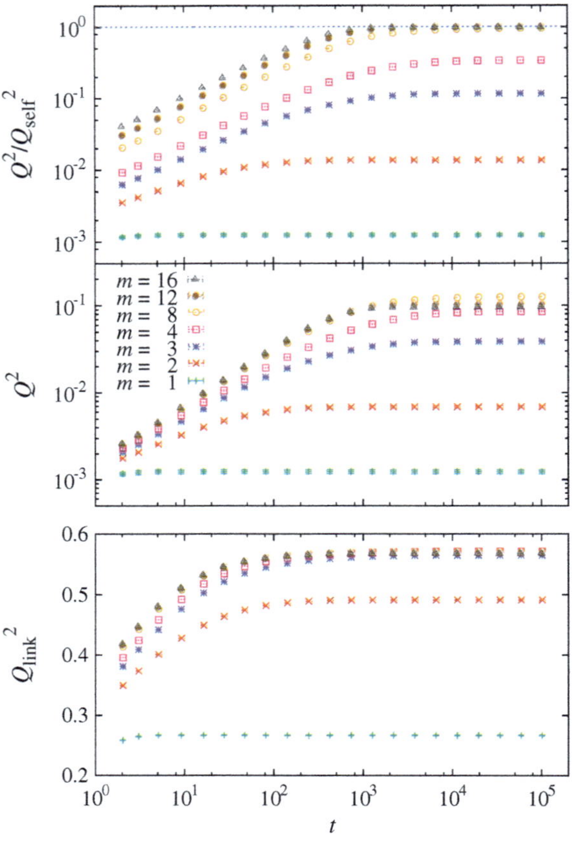

Fig. 4.9 Time evolution of the overlaps in $L = 16$ lattices. In the *top* set we show the overlap Q^2 normalized with the self overlap Q^2_{self}. On the *center* we show Q^2 without normalizing. Notice that differently from the *top case*, in the *center plot* it is the *curve* representing $m = 8$ that reaches the highest values. The *bottom* plot shows that the behavior is analogous with Q^2_{link}

when $m = 8$ the increase is similar but lasts longer and the change of growth is smoother.[5]

We can contrast our results with the ones obtained by Berthier and Young in [Ber04c] for $m = 3$ Heisenberg spin glasses. In that case they measured the evolution of the coherence length in quenches down to positive temperature $T_0 > 0$ ($L = 60$). They remarked two different regimes of growth of the coherence length, and attributed them to the passage from critical to activated dynamics.[6] In that case the slope of the second phase kept being positive and ξ did not appear to converge after 10^5 lattice sweeps. We can make a direct comparison with our quenches to zero-temperature $T_0 = 0$ with $L = 64$ (Fig. 4.9, inset). We obtain a flat second regime after 10^4 sweeps, so we can indeed attribute the growth in the second regime to

[5]The point correlation length ξ_2^{point} behaves analogously.

[6]Note that the definition of the coherence length in [Ber04c] is different from ours.

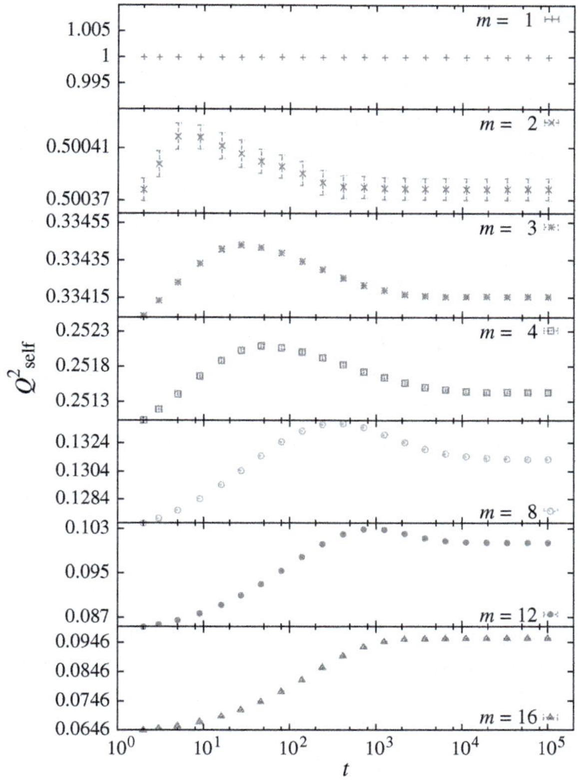

Fig. 4.10 Evolution of the self overlap $Q^2_{\text{self}}(t)$ for lattices of size $L = 16$, for different values of m. Note the differences in the y-scales: For small m the variation of $Q^2_{\text{self}}(t)$ is very small, while for the largest ones it is of the order of the self-overlap

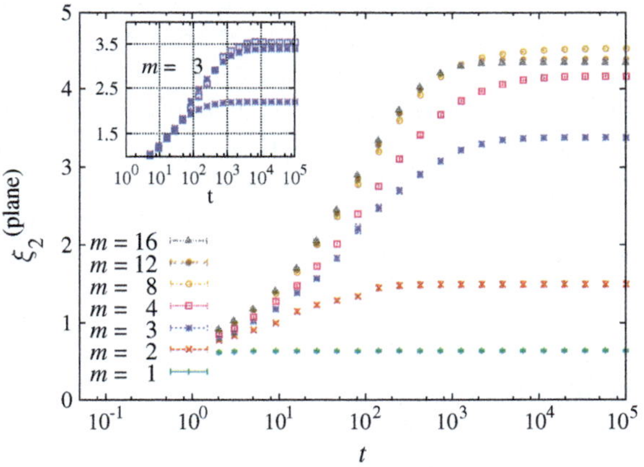

Fig. 4.11 Time evolution of the plane second-moment correlation length ξ_2^{plane}. In the *large figure* we show every simulated m for size L. Notice that the highest correlation length is reached by $m = 8$. The *inset* depicts the sole case of three-dimensional spins ($m = 3$) for sizes $L = 8, 16, 64$

thermal effects. In the inset we compare the coherence length of different lattice sizes to remark that although $\xi_2^{\text{plane}} < 4$, we are clearly far from the thermodynamic limit even for $L = 16$.

4.5 Overview

We performed an extensive study of the energy landscape of three-dimensional vector spin glasses, focusing on their dependence on the number of components m of the spins. We were concerned both with the zero-T dynamics and with the properties of the ISs, remarking various types of finite-size effects.

Increasing m the number of minima in the energy landscape decreases monotonously, down to the limit of a single state. The number of components $m_{\text{SG}}(L)$ after which the number of minima becomes sub-exponential grows with the lattice size. Reversing the relation, we can operatively define $L_{\text{SG}}(m_{\text{SG}})$ as the smallest lattice size needed in order to observe a complex behavior for a given m.

For small m correlations are small and dynamics are trivial, while when m becomes larger correlations increase and the convergence to an inherent structure slows down (for a small enough m/L ratio). We remark on the competition between the $m = 1$ limit, with abundance of ISs, and the large-m limit where at $T = 0$ there is only a single state.

In finite systems neither the overlap, nor the correlation length, nor the energy of the ISs is a monotonous function of m, as one would expect from a decreasing number of available disordered states. They have instead a peak at an intermediate m. We attribute this to the fact that when there are several minima, those of more ordered states have a larger attraction basin, so having many ISs makes it easier to fall into a more ordered state. If one wanted to rule out the non-monotonous behavior it could be useful to redefine the correlations as a function of the normalized overlaps Q^2/Q^2_{self}, as we have seen that the normalized overlaps do exhibit a monotonous trend.

Also, we presented pdfs of the spin and link order parameters Q^2/Q^2_{self} and $Q^2_{\text{link}}/Q^2_{\text{link,self}}$, noticing that the states with $Q^2/Q^2_{\text{self}} = 1$ have a major attraction basin, and create a second peak in the curve. Finite-size effects in the ISs' pdfs were very heavy, as remarked also by looking at other observables, but they were minimal if we considered the link overlap. This can suggests that perhaps the link overlap might be a better descriptor to search a phase transition in a field (Chap. 2).

Finally, we found a non-trivial behavior on the evolution of the self-overlap, that could be used as an indicator of the "quality" of a reached inherent structure.

References

[AB10a] R. Alvarez Baños, A. Cruz, L.A. Fernandez, J.M. Gil-Narvion, A. Gordillo-Guerrero, M. Guidetti, A. Maiorano, F. Mantovani, E. Marinari, V. Martin-Mayor, J. Monforte-Garcia, A. Muñoz Sudupe, D. Navarro, G. Parisi, S. Perez-Gaviro, J.J. Ruiz-Lorenzo, S.F. Schifano, B. Seoane, A. Tarancon, R. Tripiccione, D. Yllanes (Janus Collaboration), J. Stat. Mech. **2010**, P06026 (2010). doi:10.1088/1742-5468/2010/06/P06026. arXiv:1003.2569

[Asp04] T. Aspelmeier, M.A. Moore, Phys. Rev. Lett. **92**, 077201 (2004)

[BJ11] M. Baity-Jesi: *Energy landscape in three-dimensional Heisenberg spin glasses*. Master's thesis, Sapienza, Universitá di Roma, Rome, Italy (January 2011). arXiv:1503.08409

[BJ14d] M. Baity-Jesi, L.A. Fernandez, V. Martin-Mayor, J.M. Sanz, Phys. Rev. **89**, 014202 (2014). doi:10.1103/PhysRevB.89.014202. arXiv:1309.1599

[Bal00] H.G. Ballesteros, A. Cruz, L.A. Fernandez, V. Martin-Mayor, J. Pech, J.J. Ruiz-Lorenzo, A. Tarancon, P. Tellez, C.L. Ullod, C. Ungil, Phys. Rev. B **62**, 14237–14245 (2000). doi:10.1103/PhysRevB.62.14237. arXiv:cond-mat/0006211

[Bel09a] F. Belletti, A. Cruz, L.A. Fernandez, A. Gordillo-Guerrero, M. Guidetti, A. Maiorano, F. Mantovani, E. Marinari, V. Martin-Mayor, J. Monforte, A. Muñoz, Sudupe, D. Navarro, G. Parisi, S. Perez-Gaviro, J.J. Ruiz-Lorenzo, S.F. Schifano, D. Sciretti, A. Tarancon, R. Tripiccione, D. Yllanes (Janus Collaboration), J. Stat. Phys. **135**, 1121 (2009). doi:10.1007/s10955-009-9727-z. arXiv:0811.2864

[Ber04c] L. Berthier, A.P. Young, Phys. Rev. B **69**, 184423 (2004)

[Bey12] F. Beyer, M. Weigel, M. Moore, Phys. Rev. B **86**, 014431 (2012)

[Bla14] T. Blanchard, F. Corberi, L. Cugliandolo, M. Picco, Europhys. Lett. **106**, 66001 (2014)

[Bur07] Z. Burda, A. Krzywicki, O. Martin, Phys. Rev. E **76**, 051107 (2007)

[Cav09] A. Cavagna, Phys. Rep. **476**, 51–124 (2009). arXiv:0903.4264

[Con06] P. Contucci, C. Giardinà, C. Giberti, C. Vernia, Phys. Rev. Lett. **96**, 217204 (2006). doi:10.1103/PhysRevLett.96.217204

[Alm78a] J. de Almeida, R. Jones, J. Kosterlitz, D. Thouless, J. Phys. C: Solid State Phys. **11**, L871 (1978). doi:10.1088/0022-3719/11/21/005. http://stacks.iop.org/0022-3719/11/i=21/a=005

[Gre82] J. Green, A. Bray, M. Moore, J. Phys. A **15**, 2307 (1982)

[Has00] M.B. Hastings, J. Stat. Phys. **99**, 171 (2000)

[Kaw01] H. Kawamura, M. Li, Phys. Rev. Lett. **87**, 18 (2001)

[Krz00] F. Krzakala, O.C. Martin, Phys. Rev. Lett. **85**, 3013 (2000). doi:10.1103/PhysRevLett.85.3013

[Lee05] L.W. Lee, A. Dhar, A.P. Young, Phys. Rev. E **71**, 036146 (2005)

[Lee03] L.W. Lee, A.P. Young, Phys. Rev. Lett. **90**, 227203 (2003). doi:10.1103/PhysRevLett.90.227203

[Méz87] M. Mézard, G. Parisi, M. Virasoro, *Spin-Glass Theory and Beyond* (World Scientific, Singapore, 1987)

[Moo12] M.A. Moore, Phys. Rev. E **86**, 031114 (2012)

[Via88] L. Viana, J. Phys. A **21**, 803 (1988)

Chapter 5
Zero-Temperature Dynamics

In numerous glassy systems, such as electron [Efr75, Dav82, Pan05, Pal12], structural [Wya12, Ler13, Kal14] and spin glasses [Tho77, Dou10, Sha14], it is possible to identify a set of states that exhibit a distribution of soft modes, unrelated to any symmetry, that reaches zero asymptotically. These states with modes infinitely close to zero constitute the manifold that separates stable from unstable states, and are said marginally stable [Mue15].

When we relax an unstable system, it will stabilize the excitations and approach the marginally stable manifold, that we can identify as the region of the space of states where the system becomes stable. When we treat, as we do in this chapter, discrete excitations, the marginal manifold can be attained only in the thermodynamic limit.

Close to null temperature, when marginally stable systems are driven through an external force, the dynamics proceed through discrete changes in some relevant observable. The size of these rearrangements is scale-invariant, and it is usually referred to as crackling noise [Set01].

Often such scale-free bursty dynamics appears for a specific value of the force [Set93, Fis98]. When the crackling noise occurs without the need to tune the external parameters, we talk of self-organized criticality (SOC). When a pseudogap is present in the density of states, and a system displays SOC,[1] then if the stability bounds are saturated the system is marginal [Mue15].

The crackling responses are power-law distributed and span all the system. We study the arisal of crackling and of a pseudogap in the Sherrington-Kirkpatrick spin glass [Eas06, Hor08], that exhibits both marginal stability and SOC. This is done both statically, through stability arguments, and by studying the dynamics of the crackling, that in the SK model appears in form of avalanches of spin flips. At first, we focus on single- and multi-spin stability and scaling arguments. We characterize the pseudogap finding correlations between soft spins and we show that an infinite number of neighbors is needed to have avalanches that span the whole system at

[1] By pseudogap we mean a gap with zero width, i.e. the distribution is zero only in a point.

M. Baity Jesi, *Spin Glasses*, Springer Theses, DOI 10.1007/978-3-319-41231-3_5

$T = 0$, confirming a sensation generated by numerical simulations [And13]. We confirm this impression by stability arguments, indicating that an infinite number of neighbors is needed, and that the presence of the short-range interactions irrelevant: SOC is present in the presence of long-range interactions, and absent in their absence. We then study what happens *during* the avalanches, focusing on their dependency on the type of dynamics, and modelizing them through different types of random walks. The same pseudogap that we find with stability arguments arises spontaneously during the dynamics.

5.1 Self-organized Criticality and Marginal Stability in the SK Model

The SK model, that was introduced in chapter as a SG model for which mean field theory is valid [She75a], treats Ising spins $s_x = \pm 1$ at the vertices of a fully connected graph. We are interested in its hysteresis, so the Hamiltonian includes a magnetic field term,

$$\mathcal{H}_{\mathrm{SK}} = -\frac{1}{2}\sum_{x\neq y} J_{xy}s_x s_y - h\sum_x^N s_x. \tag{5.1}$$

The couplings are Gaussian-distributed, with mean $\overline{J_{xy}} = 0$ [the overline $\overline{(\ldots)}$ indicates an average over the instances of the couplings]. The variance scales as $\overline{J_{xy}^2} = J^2/N$, so the free energy is extensive and the local stability distribution [Eq. (5.4) later on] stays $O(1)$.

We define the local field as

$$h_x \equiv -\frac{\partial\mathcal{H}}{\partial s_x} = \sum_{y\neq x} J_{xy}s_y + h, \tag{5.2}$$

and the local stability of each spin as

$$\lambda_x = h_x s_x. \tag{5.3}$$

If a spin s_x is aligned to its local field, then $\lambda_x > 0$ and that site is stable. If $\lambda_x < 0$ we call it unstable. We will be interested in the distribution of local stabilities

$$\rho(\lambda) = \frac{1}{N}\sum_x^N \delta(\lambda - \lambda_x), \tag{5.4}$$

where $\delta(\ldots)$ is a Dirac delta function. In a stable state, $\rho(\lambda)$ assumes only positive values, whereas if it is non-zero for negative λ the state is unstable.[2] In a marginally stable state the $\rho(\lambda)$ reaches asymptotically zero, creating a pseudogap in the distribution of the local field. For small enough λ we can expect it to scale as [Eas06]

$$\rho(\lambda) \propto \lambda^{\theta}, \tag{5.5}$$

for some θ that we will try to determine.

We work at zero temperature, focusing only on the changes that the variation of field h imposes on the energy landscape. The dynamics are triggered by the variations of h. As soon as the field is strong enough to destabilize a spin, that spin will flip. This flip can both stabilize the system, or destabilize some of its neighbors. When more than one spin is unstable, the most unstable one is flipped (greedy dynamics, [Par03]). This dynamics is not frustrated: the flipping event decreasing the local energy of a spin also lowers the total energy, and thus stable states are achievable after a finite amount of steps.

The magnetization change ΔM between the beginning and the end of the avalanche,[3] and the number of spin flips n, that we call the avalanche size, have distributions $\mathcal{P}(\Delta M)$ and $\mathcal{D}(n)$ that follow a power law

$$\mathcal{P}(\Delta M) \propto \Delta M^{-\tau}\hat{p}(\Delta M/N^{\beta})/\log(N), \tag{5.6}$$

$$\mathcal{D}(n) \propto n^{-\rho}\hat{d}(n/N^{\sigma})/\log(N), \tag{5.7}$$

where $\hat{p}$ and $\hat{d}$ are scaling functions and β and σ are scaling exponents, with $\sigma \simeq 1$ and $0.5 \leq \beta \leq 1$ [Páz99]. The power law exponents are numerically found to be $\tau = \rho = 1$ [Páz99]. The same values of the exponents are found for the ground states (equilibrium avalanches) through replica calculations [LD12]. In Fig. 5.1 we show both distributions $\mathcal{P}(\Delta M)$ and $\mathcal{D}(n)$.

5.2 Stability and Correlations

5.2.1 *Presence of Avalanches*

In order to have avalanches, when a spin is flipped, in average it must trigger at least another spin.[4]

[2] When we say stable we mean that all the local stabilities are positive. In a thermodynamic sense those states are metastable.

[3] The magnetization is $M = \sum_x^N s_x$.

[4] We say at least one, and not one and only one spin, because in principle the average number of triggered spins could be larger than one, and the avalanches stop due to the fluctuations in the number of triggered spins.

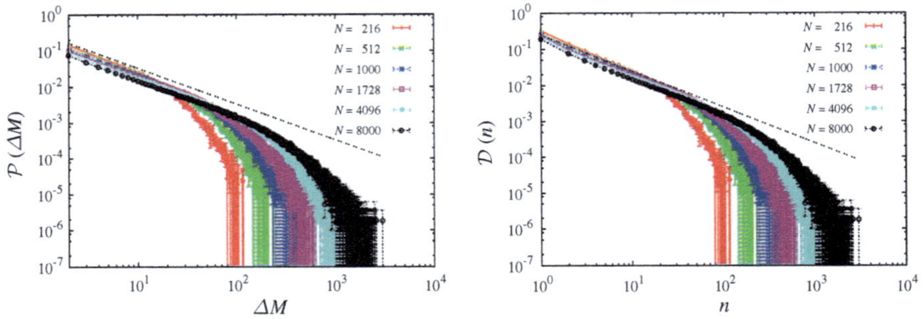

Fig. 5.1 Avalanches in the SK model for several system sizes. *Left* Distribution of the magnetization jumps $\mathcal{P}(\Delta M)$. *Right* Distribution of the avalanche sizes $\mathcal{D}(n)$. The *straight lines* are reference *curves* $\propto \Delta M^{-1}$ and $\propto n^{-1}$

Every spin flip causes a kick K in the local stability of its neighbors, that will be equal to twice the typical coupling J_{typ} between them, so the average kick scales as $K \sim 2J_{\text{typ}} \sim 2/\sqrt{N}$. The probability that spin s_i is triggered by the kick is $P(\lambda_i < K)$, so extending it to the whole system we need

$$(N-1)P(\lambda_i < K) \geq 1. \tag{5.8}$$

Since the kick coming from a single spin is small, we can restrict ourselves to the soft part of the $\rho(\lambda)$, so through Eq. (5.5) we get

$$P(\lambda_i < K) \sim \int_0^{1/\sqrt{N}} \lambda^\theta d\lambda \sim N^{\frac{1-\theta}{2}}, \tag{5.9}$$

that combined with (5.8) implies the stability bound

$$\theta \leq 1. \tag{5.10}$$

If the bound is not satisfied, the avalanches fade off very quickly.

If Eq. (5.8) is satisfied as an equality (we will show that this is the case), it would mean that, in a finite system, in average there is only one element with stability uniformly distributed in $0 < \lambda_i < K$, therefore the $\rho(\lambda)$ displays a kink for small λ and intercepts the y axis at a height $\rho(0) \sim 1/\sqrt{N}$.

Smallest stability We can estimate the scaling of the least stability λ_{MIN} with a similar argument. There has to be a fraction $\frac{1}{N}$ of spins with stability of the order of λ_{MIN} or lower, so

$$\frac{1}{N} \sim \int_0^{\lambda_{\text{MIN}}} \lambda^\theta d\lambda \sim \lambda_{\text{MIN}}^{\theta+1}, \tag{5.11}$$

that implies that the smallest stability scales as

$$\lambda_{\rm MIN} \sim N^{-1/(\theta+1)}. \tag{5.12}$$

This also means that the minimum increase of the external field to trigger an avalanche scales as $h_{\rm MIN} \sim \lambda_{\rm MIN} \sim N^{-1/(\theta+1)}$.

5.2.2 Contained Avalanches

Let us now consider in a stable state, a site $\boldsymbol{x}$, with local stability of the order of $\lambda_{\rm MIN}$, and the site $\boldsymbol{y}$ that, among its neighbors, has the lowest stability. In a finite percentage of cases, the interaction between the two sites will be unfrustrated, meaning that $s_x J_{xy} s_y > 0$. In this situation, the energy cost of the simultaneous flip of both spins will be

$$\Delta E_{xy} = 2(\lambda_x + \lambda_y) - 4|J_{xy}|. \tag{5.13}$$

For stability reasons, ΔE_{xy} should be positive. So, to grant that the second term does not counteract the first two with very large probability, we need $\lambda_{\rm MIN} \geq J_{\rm typ}$, therefore $N^{-1/(\theta+1)} \geq N^{-1/2}$, and

$$\theta \geq 1. \tag{5.14}$$

Constraints (5.8) and (5.14) imply that the two bounds are saturated and the pseudo-gap exponent is $\theta = 1$, confirming numerical simulations [Páz99].

To extend this bound to single-flip stability, one can consider the quantity E, defined as the average number of spins triggered by a flip,

$$E = N \int_0^K \rho(\lambda) d\lambda \sim N^{(1-\theta)/2}. \tag{5.15}$$

If $E \gg 1$ the number of unstable spins grows exponentially, and the avalanche never stops. To avoid this possibility we must have $\theta \geq 1$. Later on we will come back to the participation of E in the dynamics.

5.2.3 Multi-spin Stability

We can also extend the stability criterion to a whole set $\mathcal{F}$ of m spins that are initially stable with respect to a single spin flip. The energy cost of such a change would be

$$\Delta E(\mathcal{F}) = 2 \sum_{x \in \mathcal{F}} \lambda_x - 2 \sum_{x,y \in \mathcal{F}} J_{xy} s_x s_y, \tag{5.16}$$

which is an extension of Eq. (5.13). To study the stability with respect to multi-spin flip excitations, we want to compare the contribution of the two terms in (5.16). This

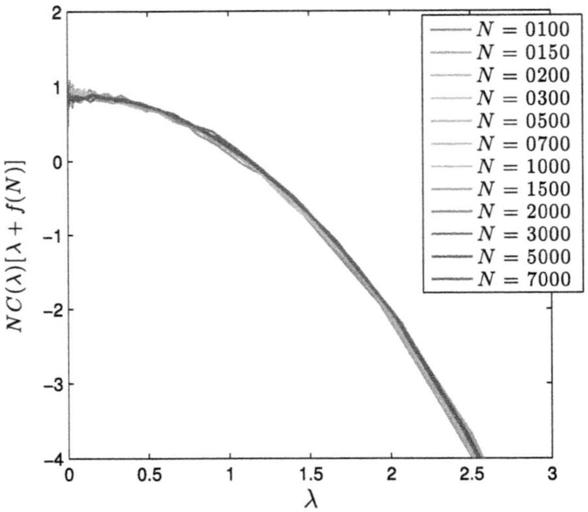

Fig. 5.2 Correlation $C(\lambda) = -2\langle s_x J_{xy} s_y \rangle$ between the least stable spin and the spins with local stability λ, in locally stable states along the hysteresis loop. The function $f(N) = 1.1 \log(N)/N$ is added in order to obtain a collapse. When N goes to infinity $f(N)$ vanishes, and for $\lambda \ll 1$ the *curve* is flat, so in this regime (small λ and thermodynamic limit) we can expect $C(\lambda) \sim 1/\lambda$

had been done by Palmer and Pond by taking in account only the m softest spins [Pal79].

Calling $\lambda(m)$ the mth smallest stability, one has that

$$\frac{m}{N} = \int_0^{\lambda(m)} \rho(\lambda) d\lambda \sim \lambda(m)^{\theta+1}, \tag{5.17}$$

so

$$\lambda(m) \sim \left(\frac{m}{N}\right)^{\frac{1}{1+\theta}}, \tag{5.18}$$

and the first term in the right hand side (r.h.s.) of Eq. (5.16) scales as $m\left(\frac{m}{N}\right)^{\frac{1}{1+\theta}}$. For the second term one has $\sum_{x,y}^{m} J_{xy} s_x s_y \sim \sum_x^m (m/N)^{1/2}$ because of the random signs. The contribution scales then as $m(m/N)^{1/2}$. In [Pal79] it was assumed to be positive, i.e. the softest spins are in average unfrustrated among each other, and from that a stability bound $\theta \geq 1$ was recovered. We can see from Fig. 5.2 that for small λ this hypothesis is not confirmed, so $\Delta E(\mathcal{F})$ is always positive in average.

The nonfulfilling of Palmer and Pond's hypothesis means that for small λ the correlation

$$C(\lambda) = -2\langle s_x J_{xy} s_y \rangle \tag{5.19}$$

between the softest spin and a spin with stability λ is positive in average, and the argument on the average scalings does not imply $\theta \geq 0$. If we postulate a behavior

$$C(\lambda) \sim \lambda^{-\gamma} N^{-\delta}, \tag{5.20}$$

we can predict the scaling

$$\left\langle -\sum_{x,y}^{m} J_{xy} s_x s_y \right\rangle \sim m^2 C\big(\lambda(m)\big) \sim m^{2-\frac{\gamma}{1+\theta}} N^{\frac{\gamma}{1+\theta}-\delta} \tag{5.21}$$

with the help of Eq. (5.18).

5.2.4 Bound Due to the Fluctuations

Even though a multi-spin stability criterion on the averages does not imply a bound $\theta \geq 1$ on the correlation, it is still possible to recover that bound by studying the large fluctuations of the last term of (5.16), that might make $\Delta E(\mathcal{F})$ negative.

Given the set of the m' most unstable spins, let us consider all the subsets $\mathcal{F}$ of $m = m'/2$ spins. We can assume that the ΔE associated with each of the sets $\mathcal{F}$ are independent and Gaussian-distributed, with

$$\langle \Delta E \rangle_{m'} = 2m\, \langle \lambda(m) \rangle_{m'} - 2m^2 \left\langle s_x J_{xy} s_y \right\rangle_{m'} \sim \frac{m^{(2+\theta)/(1+\theta)}}{N^{1/(1+\theta)}} + m^{2-\frac{\gamma}{1+\theta}} N^{\frac{\gamma}{1+\theta}-\delta}, \tag{5.22}$$

$$\mathrm{var}(\Delta E) = \langle \Delta E^2 \rangle_{m'} - \langle \Delta E \rangle_{m'}^2 = 8m^2/N, \tag{5.23}$$

where $\langle \ldots \rangle_{m'}$ is an average over the m' softest sites.[5] We neglected the non-diagonal terms in the variance. So, from Eq. (5.23) it descends that the fluctuations $X = \sum_{x,y}^{m} J_{xy} s_x s_y - \left\langle \sum_{x,y}^{m} J_{xy} s_x s_y \right\rangle$ on $\Delta E(\mathcal{F})$ are of order $m/\sqrt{N}$.[6] As there are 2^{2m} sets $\mathcal{F}$, the number density of having fluctuation X is $\mathcal{N}(X) \sim 2^{2m} \mathrm{e}^{-NX^2/m^2}$ (if ΔE is Gaussian, X has to be Gaussian with zero mean). We can recover the most negative fluctuation by imposing $\mathcal{N}(X_{\mathrm{MIN}}) \sim 1$, that implies straightforwardly $X_{\mathrm{MIN}} \sim -\sqrt{\frac{m^3}{N}}$. Thus, the energy change $\Delta E(\mathcal{F})$ associated with the most negative fluctuation scales as

$$\Delta E(\mathcal{F}_{\mathrm{MIN}}) = m^{(2+\theta)/(1+\theta)} N^{-1/(1+\theta)} + m^{2-\gamma/(1+\theta)} N^{\gamma/(1+\theta)-\delta} - m^{3/2} N^{-1/2}. \tag{5.25}$$

The multi-spin stability condition demands that, for large N and fixed m, the energy change $\Delta E(\mathcal{F}_{\mathrm{MIN}})$ stay positive. This occurs if

---

[5] The second of the two terms on the r.h.s. of Eq. 5.22 comes from Eq. 5.21. To find the first one it is necessary to calculate

$$\langle \lambda(m) \rangle_{m'} = \frac{\int_0^{\lambda(m')} \lambda \rho(\lambda) d\lambda}{\int_0^{\lambda(m')} \rho(\lambda) d\lambda}, \tag{5.24}$$

where the maximum stability of the chosen set, $\lambda(m')$, can be evaluated through Eq. (5.17). Remembering that $m' = 2m$, one obtains $\langle \lambda(m) \rangle_{m'} \sim \left(\frac{m}{N}\right)^{\frac{1}{1+\theta}}$, that multiplied by m gives the term that appears in Eq. (5.22).

[6] We neglect the fluctuations of $\sum_x \lambda_x$, since that sum is always positive and when m is large its fluctuations are small compared to its expectation value.

$$\theta \geq 1, \tag{5.26}$$

or

$$\gamma/(1+\theta) - \delta \geq -1/2, \tag{5.27}$$

depending on which of the two terms in the left hand side (l.h.s.) dominates. Nonetheless, the correlation between spins is bounded by the typical coupling, $C(\lambda) \lesssim N^{-1/2}$, so from Eq. (5.21) we obtain that $\gamma/(1+\theta) - \delta \leq -1/2$. Hence, if (5.26) is not verified,$\theta < 1$, then $\gamma/(1+\theta) - \delta = -1/2$ and (5.27) is saturated.

The scaling with large m of (5.25) also requires $2 - \frac{\gamma}{1+\theta} \geq \frac{3}{2}$, i.e. $\gamma \leq \frac{1+\theta}{2} \leq 1$ and $\delta \leq 1$. In the relevant states all three exponents θ, γ and δ equal 1, and the constraints are satisfied as exact equalities.

5.3 Finite-Size Cutoffs

In finite systems, the avalanches are bounded by cutoffs $n_c(N)$ and $\Delta M_c(N)$. The shape of the avalanche distributions gives a relation between cutoffs and average sizes of the avalanches. In the simplest case $\rho = \tau = \sigma = \beta = 1$, we can incorporate explicitly exponential cutoffs in the distributions of the avalanches, getting

$$\mathcal{D}(n) \propto n^{-1} e^{-\frac{n}{n_c}} \tag{5.28}$$

$$\mathcal{P}(\Delta M) \propto \Delta M^{-1} e^{-\frac{\Delta M}{\Delta M_c}}, \tag{5.29}$$

so if we calculate the mean avalanche size and the mean magnetization jump,[7] they result proportional to their cutoff,

$$\langle n \rangle \propto n_c, \tag{5.30}$$

$$\langle \Delta M \rangle \propto \Delta M_c. \tag{5.31}$$

In the case that the exponents τ and ρ are not equal to unity, $\langle n \rangle$ and $\langle \Delta M \rangle$ can still be used as estimators for the cutoffs, though the relation is not linear anymore.

If the cutoffs diverge as the system size becomes infinite, the system displays SOC, so we can search its presence by looking at $\langle \Delta M \rangle$ and $\langle n \rangle$.

Scaling of $\langle \Delta M \rangle$ Let us consider an ideal driving experiment in which between the beginning and the end we vary the external field of $\Delta h^{(\mathrm{tot})}$. Let the driving be so slow that every time an avalanche is triggered the external field's variation was neglectable, so the field variation is given only by the driving between one avalanche and the next one, h_{MIN}, that as we saw scales like $N^{-1/2}$. Therefore, the number of avalanches in the experiment scales as

[7] In this chapter the averages $\langle \ldots \rangle$ are averages over the avalanches.

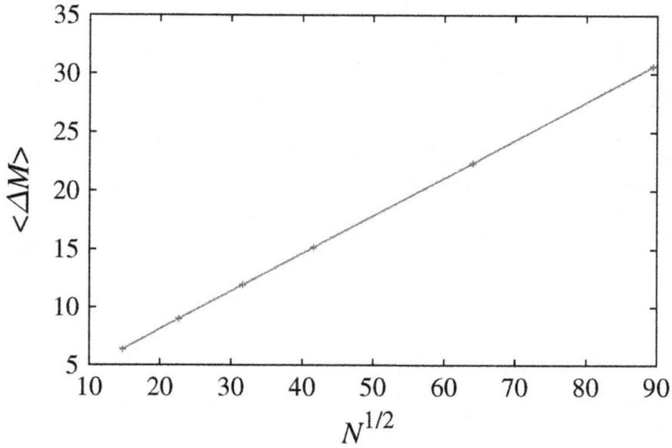

Fig. 5.3 Scaling of the average magnetization jump $\langle \Delta M \rangle$ with the square root of the system size, in the SK model

$$n_{av} = \frac{\Delta h^{(tot)}}{h_{MIN}} \sim \sqrt{N}. \tag{5.32}$$

Also the total magnetization, that will change extensively, $\Delta M^{(tot)} \sim N$, is related to the number of avalanches in the experiment n_{av} by

$$\Delta M^{(tot)} \sim n_{av} \langle \Delta M \rangle , \tag{5.33}$$

implying

$$\langle \Delta M \rangle \sim \sqrt{N}, \tag{5.34}$$

so the cutoff goes to infinity as $N \to \infty$, and the SK model displays SOC, as it is confirmed in Fig. 5.3.

Scaling of $\langle n \rangle$ We can attempt to estimate the scaling of $\langle n \rangle$ by studying the energy, since, differently from the magnetization, its evolution is monotonous in time. In a single avalanche, the energy change is

$$\langle \Delta E_{av} \rangle = \langle n \rangle \left\langle \Delta E_{flip} \right\rangle , \tag{5.35}$$

where $\left\langle \Delta E_{flip} \right\rangle$ is the average energy change per spin flip. Assuming that it is of the order of the typical coupling, $\left\langle \Delta E_{flip} \right\rangle \sim J_{typ} \sim N^{-1/2}$.

For the total energy change during an avalanche, let us consider a full hysteresis loop. Neglecting logarithmic corrections, its area $A = \sum_{i \in \text{drivings}} M dh_i \sim N$ is extensive. [8] The total energy change, $E^{(tot)}$, is zero because the experiment starts and finishes in the same point, but it is also equal to the sum of the contributions of the avalanches and of the field drivings,

[8] With at most logarithmic corrections, that can be neglected in this argument.

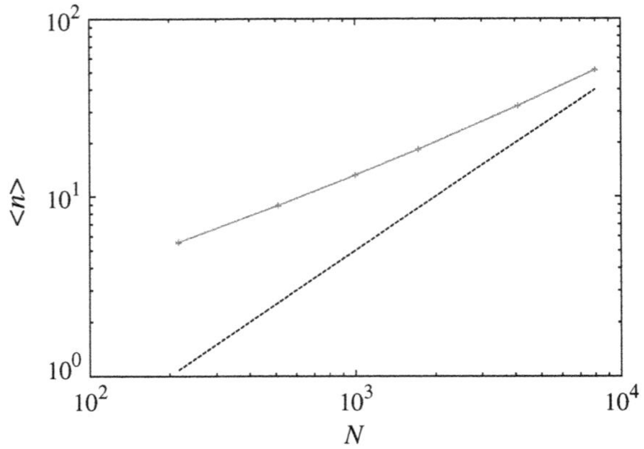

Fig. 5.4 Scaling of the average avalanche size $\langle n \rangle$ with the system size N, in the SK model. The straight line is a reference curve $\propto N$

$$0 \sim E^{(\mathrm{tot})} \sim \sum_{\text{avalanches}} \Delta E_{\mathrm{av}} + \sum_{i \in \text{drivings}} M dh_i \sim \tag{5.36}$$

$$\sim n_{\mathrm{av}} \Delta E_{\mathrm{av}} + A \sim \tag{5.37}$$

$$\sim N + \sqrt{N} \Delta E_{\mathrm{av}}, \tag{5.38}$$

so $\Delta E_{\mathrm{av}} \sim \sqrt{N}$, and as a consequence

$$\langle n \rangle \sim N. \tag{5.39}$$

In Fig. 5.4 we show that numerical data are consistent with an asymptotic behavior $\langle n \rangle \sim N$ (with possibile logarithmic corrections).

Both the cutoffs we recovered go to infinity with the system size, and the SK model displays self-organized criticality.

5.3.1 *Short-Range Models*

Let us consider now models defined on a generic graph where each site has z neighbors. The finite-neighbor (short-range) Hamiltonian is

$$\mathcal{H}_{\mathrm{SR}} = -\frac{1}{2} \sum_{x}^{N} s_x \sum_{y \in \mathcal{V}(x)}^{z} J_{xy} s_y - h \sum_{x}^{N} s_x, \tag{5.40}$$

where $\mathcal{V}(\boldsymbol{x})$ is the set of sites that are connected to $\boldsymbol{x}$ through an edge of the graph. When the interactions are not long-range, i.e. each site has a small connectivity z, it has been observed numerically that θ stays the same, but self-organized criticality vanishes [And13], because the cutoffs of the power law behaviors do not diverge with the system size. Also, the pseudogap disappears, and the intercept of the stability distribution scales as $\rho(0) \sim 1/\sqrt{z}$.

That $\rho(0) \sim 1/\sqrt{z}$ is expectable from the previous argument that in average there is only one element with stability uniformly distributed in $0 < \lambda_i < K$. Since now the kick K is of order $1/\sqrt{z}$, the intercept is at height $\rho(0) \sim 1/\sqrt{z}$, so the distribution of the stabilities becomes

$$\rho(\lambda) \sim \frac{A}{\sqrt{z}} + B\lambda. \tag{5.41}$$

In these conditions the smallest stability is given by

$$\frac{1}{N} \sim P(\lambda < \lambda_{\text{MIN}}) \sim \frac{A'\lambda_{\text{MIN}}}{z} + B'\lambda_{\text{MIN}}^2. \tag{5.42}$$

Since λ_{MIN} is small, we can neglect the quadratic term, so $\lambda_{\text{MIN}} \sim \frac{\sqrt{z}}{N}$. It is straightforward to see that if $z = cN$, for some finite c, the SK limit is recovered.

The cutoff magnetization jump ΔM_c changes consequently

$$\langle \Delta M \rangle = \frac{\Delta M^{(\text{tot})}}{n_{\text{av}}} \sim N\lambda_{\text{MIN}} \sim \sqrt{z}. \tag{5.43}$$

So, if the connectivity z is finite the avalanches have a finite cutoff, while if it diverges we recover the self-organized criticality of the SK model.

This can be seen also through the scaling of n_c, by using the relation $\Delta E_{\text{av}} \sim \langle n \rangle \langle \Delta E_{\text{flip}} \rangle$. The average energy change per flip is of the order of $\Delta E_{\text{flip}} \sim J_{\text{typ}} \sim \frac{1}{\sqrt{z}}$. The hysteresis argument for $\langle \Delta E_{\text{av}} \rangle$ this time yields $n_{\text{av}} \sim \frac{1}{h_{\text{MIN}}} \sim \frac{N}{\sqrt{z}}$. Therefore

$$0 \sim E^{(\text{tot})} \sim n_{\text{av}} \Delta E_{\text{av}} + A \tag{5.44}$$

$$\sim N + \frac{N}{\sqrt{z}} \Delta E_{\text{av}}, \tag{5.45}$$

so $\langle \Delta E_{\text{av}} \rangle \sim \sqrt{z}$ and $\langle n \rangle \sim z$, confirming the absence of self-organized criticality in models with finite connectivity. One could actually expect this by looking at the distributions $\mathcal{P}(\Delta M)$ and $\mathcal{D}(n)$ in Fig. 5.5. For all the sizes, the curves collapse to the same exponential decay, so there cannot be a scaling of the mean values (Fig. 5.6) nor of the cutoffs.

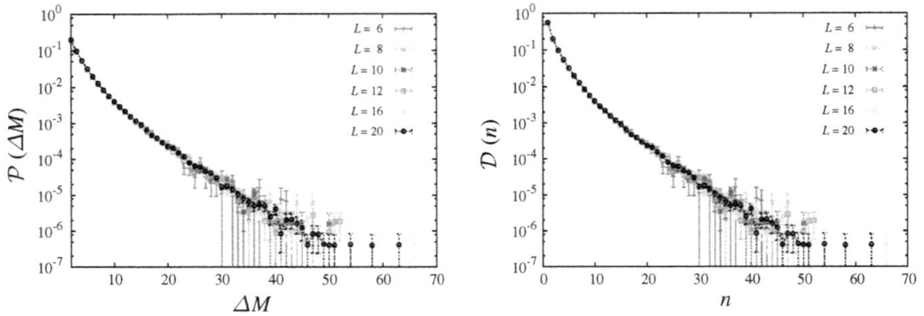

Fig. 5.5 Same as Fig. 5.1, but in the three-dimensional Edwards-Anderson model. The system sizes are the same as Fig. 5.1. In the legend we express them through the linear lattice size L ($L^3 = N$) to stress that the interactions are between nearest neighbors of a cubic lattice

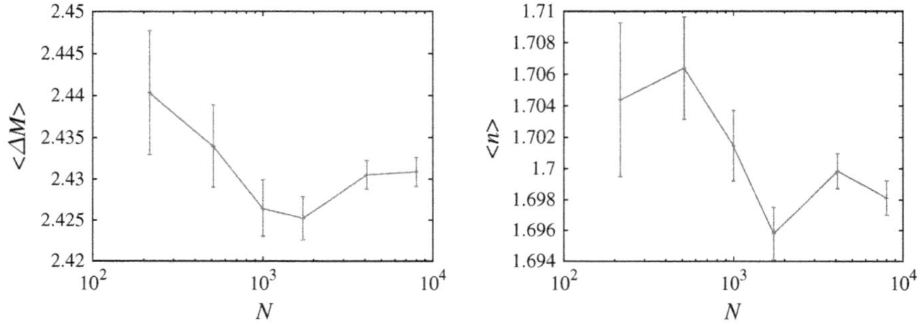

Fig. 5.6 Scaling of the averages $\langle \Delta M \rangle$ (*left*) and $\langle n \rangle$ (*right*) with the system size N, in the EA model

5.3.2 Competition Between Short and Long Range Interactions

Long-range interaction models display SOC, while if the interactions are short-range this is not true. Since the application of the concept of SOC is related to many systems where there might be a coexistence of the two, a question that arises spontaneously is whether it is the presence of long-range interactions that guarantees SOC, the existence of short-range ones that kills it, or it depends on their relative magnitude.

We define thus a model that mixes short and long-range interactions, and try to understand whether or not it displays SOC. A simple way is to get an EA model on a cubic lattice, and add to it an infinite-range interaction term. Let the spacing between nearest neighbors in the lattic be unitary, and L be the side of the full lattice. We impose periodic boundary conditions. Each site $\boldsymbol{x}$ hosts a spin s_x, and interacts with the rest of the spins through a duplex network. One graph follows the geometry of the lattice, and allows only nearest-neighbor interactions, and the other is fully connected.

The Hamiltonian is

$$\mathcal{H}_{SL} = -\sum_{\langle x,y\rangle} J^{(s)}_{xy} s_x s_y - \sum_{x,y} J^{(\ell)}_{xy} s_x s_y - h\sum_x s_x, \tag{5.46}$$

where $J^{(s)}_{xy}$ is the short-range coupling, and $J^{(\ell)}_{xy}$ is the long-range one. Both are gaussian random variables with zero mean $\overline{J^{(s)}_{xy}} = \overline{J^{(\ell)}_{xy}} = 0$, and variances $\overline{J^{(s)2}_{xy}} = J^{(s)}/z$ and $\overline{J^{(s)2}_{xy}} = J^{(l)}/N$. The limit $J^{(s)} = 0$ corresponds to the SK model, while $J^{(l)} = 0$ is the EA model. We work on a cubic lattice, so $z = 2d$.

We impose the stability argument separating the nearest neighbor interactions from the others

$$1 \le (N-z)\int_0^{J^{(\ell)}} \rho(\lambda)d\lambda + \int_0^{\tilde{J}} \rho(\lambda)d\lambda, \tag{5.47}$$

with $\tilde{J}^2 = J^{(\ell)2} + J^{(s)2}$. Taken alone, the first term on the right hand side is always critical, whereas the second one is never.

To verify the presence of both terms, it is convenient to study the limit $J^{(\ell)} \ll J^{(s)}$. Since the typical avalanches do not imply large stability jumps (Fig. 5.12 later on), the kicks on the softest modes will be dictated by $J^{(\ell)}$, and we can assume that the stability distribution be $\rho(\lambda) \propto \alpha\lambda$, where the constant α is to keep track of the competition between the two interactions.

The stability argument becomes then

$$1 \sim (N-z)\int_0^{J^{(\ell)}/\sqrt{N}} \rho(\lambda)d\lambda \sim \tag{5.48}$$

$$\sim \alpha N \int_0^{(\ell)/\sqrt{N}} \lambda d\lambda \sim \tag{5.49}$$

$$\sim \alpha J^{(\ell)2}, \tag{5.50}$$

so $\alpha = 1/J^{(\ell)2}$ and

$$\rho(\lambda) \propto \frac{\lambda}{J^{(\ell)2}}. \tag{5.51}$$

We can use again the argument for the scaling of the magnetization jump, $\langle \Delta M\rangle = \frac{\Delta M^{(tot)}}{\langle n_{\text{av}}\rangle}$, with this $\rho(\lambda)$. The average number of avalanches now scales as $\langle n_{\text{av}}\rangle \sim \frac{1}{h_{\text{MIN}}} \sim \frac{\sqrt{N}}{J^{(\ell)}}$, so

$$\langle \Delta M\rangle \propto J^{(\ell)}\sqrt{N}, \tag{5.52}$$

so even in the presence a the smallest long-range interaction, as the system size grows the average magnetization jump in an avalanche diverges as $\sqrt{N}$, as it is also confirmed numerically in Fig. 5.7, left.

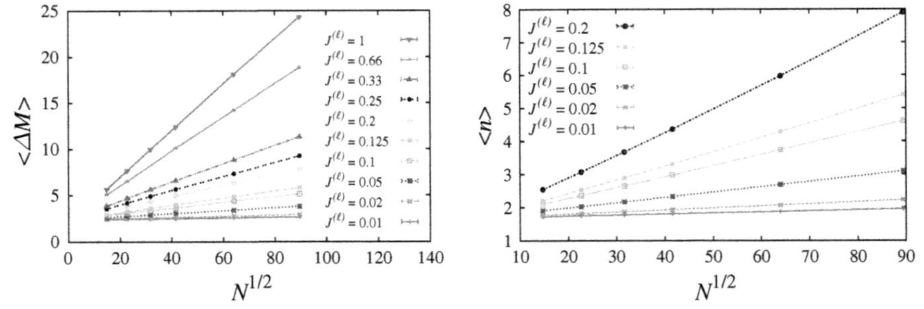

Fig. 5.7 Scaling of the average values in the model that mixes short- and long-range interactions. The short-range coupling is kept fixed to $J^{(s)} = 1$, while the amplitude $J^{(\ell)}$ is tuned in the region $0 < J^{(\ell)} \leq J^{(s)}$. *Left* The average magnetization jump $\langle \Delta M \rangle$ scales clearly as $\sqrt{N}$. *Right* The average avalanche size follows the trend $\langle n \rangle \sim \sqrt{N}$ for small $J^{(\ell)}$

As to the number of spins involved in the avalanche, we also find that it diverges, confirming the self-organized criticality of the model, but this time with a different law than the SK model. In fact $\langle n \rangle \sim \frac{\Delta E_{\rm av}}{\Delta E_{\rm flip}}$. While $\Delta E_{\rm av}$ scales as $J^{(\ell)}\sqrt{N}$, the energy of a flip scales as $\Delta E_{\rm flip} \sim \sqrt{\frac{J^{(\ell)2}}{N} + \frac{J^{(s)2}}{z}}$. The average number of spins taking part in an avalanche then scales like

$$\langle n \rangle = \frac{\Delta E_{\rm av}}{\Delta E_{\rm flip}} \sim \tag{5.53}$$

$$\sim \frac{J^{(\ell)}\sqrt{N}}{\frac{J^{(\ell)}}{\sqrt{N}}\sqrt{1 + \frac{N J^{(s)2}}{z J^{(\ell)2}}}} \sim \tag{5.54}$$

$$\sim \frac{J^{(\ell)}}{J^{(s)}}\sqrt{zN}. \tag{5.55}$$

Numerical simulations, where we tune the amplitude $J^{(\ell)}$ keeping $J^{(s)} = 1$ fixed, confirm this argument (Fig. 5.7, right).

5.4 Dynamics

After having given several conclusions on the self-organized criticality of the SK based on scaling and stability arguments, it is reasonable to ask oneself whether self-organized criticality purely a property of the visited states or the dynamics too play an important role on the crackling. In the following section we try to get some insight from what is happening to the system *during* the avalanches.

5.4.1 A Non-trivial Random Walk

An avalanche starts when a first spin is destabilized, and it finishes when all the local stabilities are positive. With the typical spin update, that we call greedy algorithm, if there is more than one unstable spin, the least stable is updated first. Calling $n_{\text{unst}}(t)$ the number of unstable spins after t spin flips, this reads that the avalanche starts with $n_{\text{unst}}(1) = 1$, it performs a random walk (RW) in the space of n_{unst}, and it end with $n_{\text{unst}}(n) = 0$.

The easiest guess for the dynamics is thus an unbiased RW, where for large avalanches $\mathcal{D}(n)$ would be the return probability of a one-dimensional RW.[9] The return probability of a random walk is $P_{1d} \propto \frac{1}{\sqrt{t}}$ in $1d$ and $P_{2d} \propto \frac{1}{t \log t}$ in $2d$, so the unbiased RW scenario predicts $\rho = 1/2$, that is different from the $\rho = 1$ usually observed (recall Fig. 5.1).

The RW of n_{unst} can be described through two equivalent auxiliary variables $E(n_{\text{unst}}(t))$ and $r(n_{\text{unst}}(t))$, that indicate the likeliness of the avalanche of shrinking or expanding:

$$n_{\text{unst}}(t) = n_{\text{unst}}(t-1)E(t-1), \tag{5.56}$$
$$n_{\text{unst}}(t) = n_{\text{unst}}(t-1) + r(t-1). \tag{5.57}$$

In an unbiased random walk $E(t) = 1\,\forall t$ and $r(t) = 0\,\forall t$. Random walks with constant $E < 1$ ($r < 0$) are attractive, meaning that there cannot be extended avalanches, while if $E \gg 1$ ($r \gg 0$) the system is highly unstable and the avalanches never stop.

Since the number of triggered spins depends exclusively on the links between the flipping spin and its neighbor, which is a static property of the system, it is reasonable to assume—and more in a fully-connected spin glass where it makes no sense to talk of spatial domains—that E and r depend on n_{unst} rather than on how long the avalanche lasted.

In Fig. 5.8 we show E and r for avalanches in the SK and in the $3d$ EA model. Both E and r have a marked dependency on n_{unst}, disclosing non-trivial RWs. In the EA model $E(n_{\text{unst}}) < 1\,\forall n_{\text{unst}}$, meaning that the dynamics is damped and the size of the avalanche can grow only because of fluctuations. Mind that as n_{unst} increases (due to "lucky" fluctuations), $E(n_{\text{unst}})$ approaches 1, reflecting that the connectivity of the unstable domain grows, so it becomes easier to destabilize another spin. In the SK model the situation is more interesting, since the dynamics is critical. Instead of $E(n_{\text{unst}}) = 1\,\forall n_{\text{unst}}$, that could be in principle a good ansatz for a marginal system, the avalanches have a natural tendency to grow up to a size n^*_{unst}. For $n_{\text{unst}} > n^*_{\text{unst}}$, E is slightly smaller than one, meaning that n^*_{unst} is a preferred number of unstable spins. A size-independent n^*_{unst} would entail that the scale invariance is only a low-resolution effect due to the fact that E is smaller than one, but very close to it. From Fig. 5.9 we see that this is not the case: n^*_{unst} grows as $\log(N)$.

[9]It would be exactly the return probability of the random walk if the avalanche started with $n_{\text{unst}} = 0$.

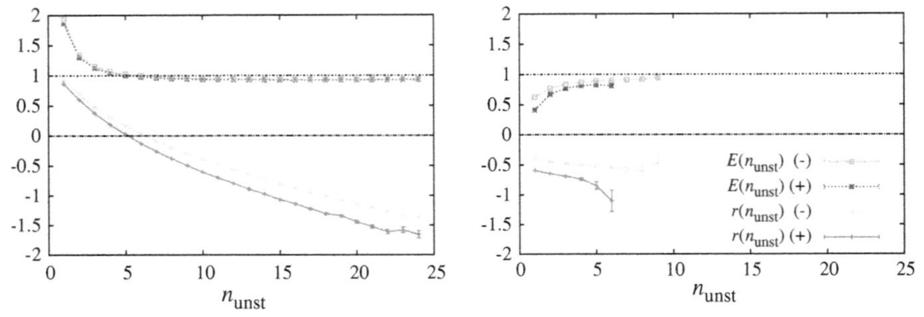

Fig. 5.8 Indicators of the random walk bias E and r as a function of the number of unstable spins n_{unst}. The *curves* tagged with a (+) indicate starting configuration with positive magnetization, those tagged with (−) indicate a negative magnetization. Details on the protocol are given in the main text. The *left plot* shows data from the SK model for $N = 8000$. The *right plot* is from the three-dimensional EA model with $L = 20$ ($N = L^3 = 8000$). The two horizontal lines stress the values of the unbiased RW, $E = 1$ and $r = 0$)

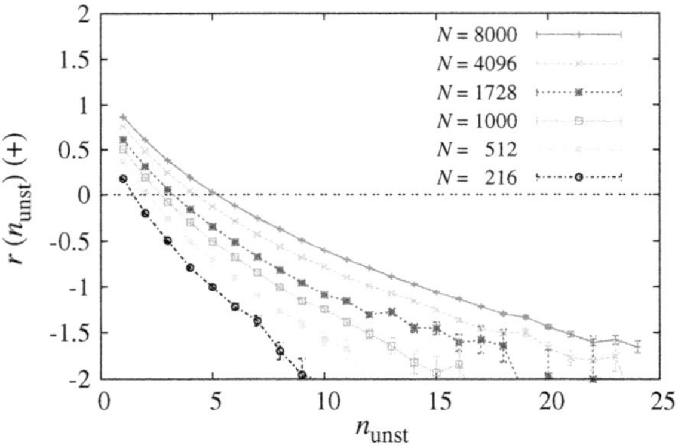

Fig. 5.9 We show $r(n_{unst})$ for different lattice sizes, to stress that the point n^*_{unst} where the *curve* crosses zero grows steadily with the system size. The data is from systems with positive magnetization initial conditions, (+), and is qualitatively equivalent to opposite magnetization starting configurations, (−)

What is clear is that the dynamics of the single spins are far from being independent, and those of the system as a whole are related on the amount of unstable spins. The evolutions and stabilities of the spins are correlated and there is some kind of non-trivial mechanism that keeps the system marginal during the avalanches.

Initial conditions It is legitimate to inquire whether different starting conditions play a pivotal role on the random walk. In Fig. 5.8 we compare two types of initial configuration. We start at zero field with either all spins up (+) or all spins down (−), and we minimize the energy by aligning successively the most unstable spin to its local field until the system becomes stable (greedy algorithm). The two configurations are totally equivalent, except that they have opposite remnant magnetization. In Fig. 5.8 we see that there is an appreciable difference between the two starting conditions, The external field in this numerical experiment varies from 0 to 1.5, that is, the data come from a large number of avalanches, $O(\sqrt{8000})$. If the information on the initial state were lost within the first avalanche, the curves (+) and (−) should differ by the order of 1 %.

5.4.2 Changing the Avalanche Dynamics

A way to understand whether marginality is a property of the static configurations or it depends on the dynamics is to validate it on different types of dynamics. We propose three types of single-spin-flip dynamics. The first is the one used until now, that at each time step updates the most unstable of the spins. We call it *greedy* dynamics (G). The second type of dynamics is inspired from [Par03], and updates the least unstable spin. This is the *reluctant* algorithm (R). It was shown in [Par03] that minimizing the energy with R dynamics leads to inherent structures with much lower energy. The third dynamics we test updates a random spin among those with $\lambda < 0$. We call it *random* dynamics (A).[10]

Avalanche distributions When switching to R and A avalanches, we remark no variation on the $\rho(\lambda)$, that for small λ still grows linearly (only the amplitude changes), but we do see a difference in the exponents of the avalanche distributions. More specifically, for A we see the same exponents $\rho \approx 1$ and $\tau \approx 1$, but with R the avalanches are significantly larger and have $\rho \approx 1.25$, $\tau \approx 1.4$. In figure we show R avalanches. The similarity between G and A can be attributed to the fact that the energy change in a spin flip is of the same order, $\Delta E_{\text{flip}} \sim 1/\sqrt{N}$, while R dynamics implies that the energy dissipated in a spin flip is smaller. Since the $\rho(\lambda)$ is all of order 1, the typical distance between the stabilities is of order $1/N$, so $\Delta E_{\text{flip}} \sim 1/N$.[11]

The data in Fig. 5.10 was obtained by relaxing a totally up configuration, and once the initial IS was found we recorded the data of the avalanches until the overlap with the initial configuration became smaller than $Q = 0.9$. This way we could grant some dependence on the initial IS, and compare avalanches that started with G and R inherent structures. We will use two letters to identify the procedure we refer to: the first one refers to the initial IS, the second to the avalanche dynamics, so for example RG is a greedy avalanche starting from a reluctant IS.

In Fig. 5.10 we compare GR and RR dynamics. Apparently, the exponent does not depend on the initial conditions, but the finite-size effects do visibly. While RR avalanches display a power-law behavior with a finite-size cutoff, in GR one sees that with a probability that decreases with N there can be avalanches with a very large number of spin flips, arriving to $n > N$, that means that in average every spin flips more than once. This suggests that G inherent structures are in some way more unstable with respect to R dynamics than R inherent structures.

---

[10] We use an A, that stands for *aleatory*, because the R of *random* was already picked for the reluctant algorithm.

[11] The arguments of Sect. 5.3 for the scaling of $\langle \Delta M \rangle$ and $\langle n \rangle$ apply also to A and R dynamics. One obtains $\langle \Delta M \rangle \sim \sqrt{N}$ for both the dynamics, $\langle n \rangle \sim N$ for A and $\langle n \rangle \sim N^{3/2}$ for R dynamics. Numerical simulations seem compatible with these trends in the limit of very large systems.

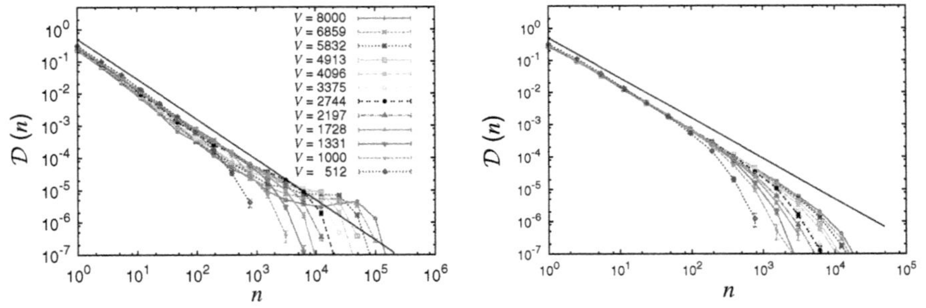

Fig. 5.10 Distribution of sizes $\mathcal{D}(n)$ in avalanches with reluctant dynamics. In the *left* plot the initial IS is obtained with the G algorithm, while on the *right* it is obtained with R dynamics. The straight lines in both plots are reference curves $\propto n^{-1.25}$. Even though the finite-size behaviors are different depending on the starting configuration, the power law is the same, and it is different from G dynamics

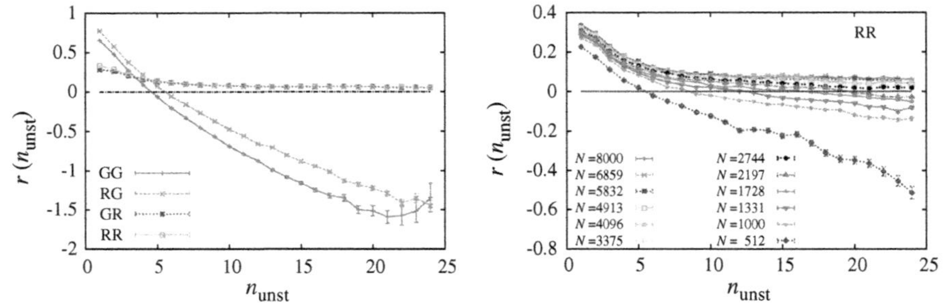

Fig. 5.11 The RW bias indicator r for avalanches that start from the initial inherent structure (i.e. for $Q = 0$). Although our data only extends to $n_{unst} = 24$, the avalanches had also larger numbers of unstable spins. On the *left* we compare greedy and reluctant algorithms in $N = 8000$ systems. In the *right figure* we show RR data for different N. When the system is small, $r(n_{unst})$ crosses zero at a finite n^*_{unst}, that grows with N. For $N \geq 2744$ our data is not able to capture n^*_{unst}, though we still expect it to be large but finite (see also main text). In both plots the horizontal line stresses the unbiased value $r = 0$

Random walks Seeing the avalanche as a RW of the number of unstable spins, we see no remarkable dependency on the initial IS, but we do notice a quite different behavior between G and R avalanche dynamics (Fig. 5.11, left). In the G dynamics $r(n_{unst})$ is initially positive (expansion of the avalanche preferred) becomes negative (shrinking preferred) at a finite n^*_{unst}, justifying avalanches of limited size. Differently, r appears always positive in R avalanches, indicating a tendency towards enlargement. If r is always positive the avalanches can only stop due to large fluctuations or by saturation of the system (we have a trivial bound $n_{unst} < N$), that would mean that the dynamics is unstable. The power law behavior of $\mathcal{D}(n)$ (Fig. 5.10) and the finite-size behavior of $r(n_{unst})$ (Fig. 5.11, right) induce to think that n^*_{unst} is instead finite but large, and that its growth with the system size is significantly quicker than in G avalanches.[12] The different scaling of n^*_{unst} between the G and R could be what leads to different exponents ρ and τ.

5.4.3 Fokker-Planck Description

Coming back to greedy dynamics, we will see now that the same exponents that we obtained through stability constraints arise spontaneously from the dynamics of the avalanches in the SK model. Let us take in account the random walk of each local stability in the space of the local stability space. The random walk starts when a stability becomes negative because of an imposed external magnetic field, and it finishes when all the spins are stable again.

The flipping of the spin s_0 changes its local stability from λ_0 to $\lambda_0' = -\lambda_0$. The stability of all the other spins s_y in the system changes proportionally to their coupling with s_0,

$$\lambda_y \to \lambda_y' = \lambda_y - 2s_0 J_{0y} s_y. \tag{5.58}$$

The stability changes have a random fluctuating part and a non-zero mean value due to the correlations with s_0. As it was similarly done in Refs. [Eas06, Hor08], this dynamics can be modelled with a Fokker-Planck equation for the distribution of stabilities $\rho(\lambda)$,

$$\partial_t \rho(\lambda, t) = -\partial_\lambda \left[v(\lambda, t) - \partial_\lambda D(\lambda, t)\right] \rho(\lambda, t) - \delta(\lambda - \lambda_0(t)) + \delta(\lambda + \lambda_0(t)), \tag{5.59}$$

where now the "time" t is the number of flips per spin that took place during the avalanche and the two delta functions indicate the flipping of s_0. The drift term $v(\lambda, t) \equiv -2N \left\langle s_0 J_{0y} s_y \right\rangle_{\lambda_y = \lambda} = NC(\lambda, t)$ is the average positive kick that a spin with stability λ receives [Eq. (5.58)]. The diffusion constant $D(\lambda, t) \equiv 2N \left\langle J_{0y}^2 \right\rangle = 2$ is the mean square of the kicks. The dynamics have a non-trivial thermodynamic limit only if $v \sim O(1)$, meaning that $\left\langle s_0 J_{0y} s_y \right\rangle \sim 1/N$. This conveys that the exponent δ from Eq. (5.20) must be equal to 1.

As $N \to \infty$, the lowest stability approches zero $\lambda_0(t) \to 0$. We already saw, in fact, that in a driving experiment with a finite field change Δh, the number of avalanches scales as $n_{\text{av}} \sim 1/h_{\text{MIN}} \sim N^{1/(1+\theta)}$. Each avalanche contains on average $\langle n \rangle \sim \int n\mathcal{D}(n)dn \sim N^{(2-\rho)\sigma}$ flip events [recall Eq. (5.6)], so the total number of flips along the hysteresis curve is $n_{\text{av}} \langle n \rangle \sim N^{(2-\rho)\sigma + 1/(1+\theta)}$, that is reasonably larger than N. A diverging number of avalanches implies that the energy dissipation in each avalanche has to be subextensive, ruling out strongly unstable configurations

[12] In G avalanches n^*_{unst} grows logarithmically, $n^*_{\text{unst}} \sim \log(N)$. With R dynamics we have little data because our measurements only go up to $n_{\text{unst}} = 24$. We deduce a roughly linear scaling $n^*_{\text{unst}} \sim N$.

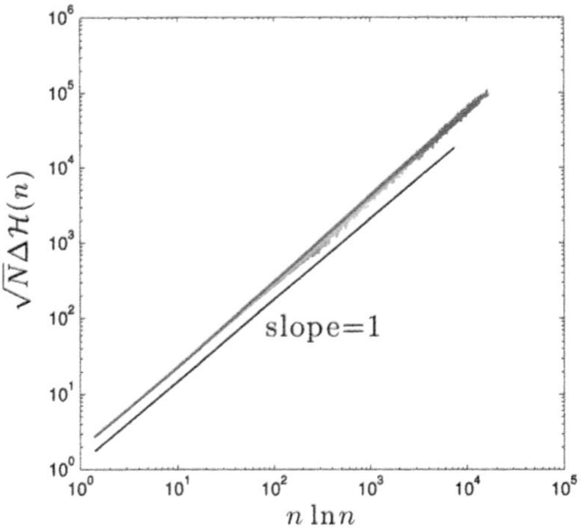

Fig. 5.12 The average dissipated energy $\Delta\mathcal{H}$ in avalanches of size n scales as $\Delta\mathcal{H} \sim n \ln n/\sqrt{N}$. $-\Delta\mathcal{H}/n$ is a measure of the typical value of the stability of most unstable spins, $\lambda_0(n)$. Thus, in the thermodynamic limit, $\lambda_0 \sim \ln n/\sqrt{N} \ll 1$ even for very large avalanches

with an extensive number of spins with negative stability $|\lambda| = O(1)$. So, as we confirm numerically in Fig. 5.12, the smallest local stability must tend to zero in the thermodynamic limit. This observation lets us replace the delta functions in Eq. 5.59 with a reflecting boundary condition at $\lambda = 0$,

$$[v(\lambda, t) - \partial_\lambda D(\lambda, t)]\, \rho(\lambda, t)|_{\lambda=0} = 0. \tag{5.60}$$

Since along the hysteresis loop spins flip a large amount of times, in a finite interval we have a diverging number of time steps. At very large times a steady state must be reached. In such conditions the flux of spins must vanish everywhere, so the steady state drift is

$$v_{ss}(\lambda) = D\partial_\lambda \rho_{ss}(\lambda)/\rho_{ss}(\lambda) \to 2\theta/\lambda, \tag{5.61}$$

where we assumed that the steady-state stability distribution follows (5.5). This implies that $\gamma = 1$ in Eq. (5.20).

5.4.3.1 Arisal of Correlations

We will now argue that the correlations of Eq. (5.20) (with $\gamma = \delta = 1$) arise naturally in the dynamics through the shifts of the local stabilities caused by the spin flips.

Let us define with $C_f(\lambda)$ and $C'_f(\lambda)$ the correlations between the spin s_0 and the spins with local stability λ *before* and *after* the flipping event. After s_0 flips, the stability change is $\lambda'_x = \lambda_x + x_x$, where $x_x = -2s_0 J_{0x} s_x$. The correlation $C'_f(\lambda)$ is an average over all the spins whose stability, after the flip, is λ',

$$C'_f(\lambda) = \frac{1}{\rho'(\lambda)} \int \rho(\lambda - x)(-x) f_{\lambda-x}(x) dx, \tag{5.62}$$

$$\rho'(\lambda) = \int \rho(\lambda - x) f_{\lambda-x}(x) dx. \tag{5.63}$$

$f_\lambda(x)$ is the Gaussian distribution of kicks x given to spins of stability λ: $f_\lambda(x) = \exp\left[-\frac{(x-C_f(\lambda))^2}{4D/N}\right]/\sqrt{4\pi D/N}$. In the integrands we expand $\rho(\lambda - x)$ and $C_f(\lambda - x)$ for small x and keep terms of order $1/N$, which yields

$$C'_f(\lambda) = -C_f(\lambda) + 2\frac{D}{N}\frac{\partial_\lambda \rho(\lambda)}{\rho(\lambda)}, \tag{5.64a}$$

$$\rho'(\lambda) = \rho(\lambda) - \partial_\lambda \left[C_f(\lambda)\rho(\lambda) - \frac{D}{N}\partial_\lambda \rho(\lambda)\right]. \tag{5.64b}$$

Thus, even if correlations are initially absent, $C_f(\lambda) = 0$, they arise spontaneously, $C'_f(\lambda) = 2D\partial_\lambda \rho(\lambda)/N\rho(\lambda)$.

In the steady state, $\rho'_{ss} = \rho_{ss}$, and Eq. (5.64b) implies the vanishing of the spin flux, that is, Eq. (5.61) with $v = NC_f$. Plugged into Eq. (5.64a), we obtain that the correlations are steady, too,

$$C'_f(\lambda) = C_f(\lambda) = \frac{v_{ss}(\lambda)}{N} = \frac{2\theta}{N\lambda}. \tag{5.65}$$

These correlations are expected once the quasi-statically driven dynamics reaches a statistically steady regime, and thus should be present both during avalanches and in the locally stable states reached at their end.

Interestingly, Eq. (5.65) implies that all the bounds of Eqs. (5.26, 5.27) are saturated if the first one is, i.e., if $\theta = 1$. It is intriguing that the present Fokker-Planck description of the dynamics does not pin θ, as according to Eqs. (5.61, 5.65) any value of θ is acceptable for stationary states. However, additional considerations on the applicability of the Fokker-Planck description discard the cases $\theta > 1$ and $\theta < 1$.

Excluding $\theta < 1$ Our Fokker-Planck description only applies beyond the discretization scale of the kicks due to flipping spins, which are of order $J \sim 1/\sqrt{N}$. In particular, from its definition, $C(\lambda)$ must be bounded by $1/\sqrt{N}$. Taking this into account, Eq. (5.61) should be modified to:

$$v_{ss}(\lambda) \approx \min\{D\partial_\lambda \rho_{ss}(\lambda)/\rho_{ss}(\lambda) \sim 1/\lambda, \sqrt{N}\}. \tag{5.66}$$

This modification has no effect when $\theta \geq 1$, since in that case $\lambda_{\min} \sim N^{-1/(1+\theta)} \geq 1/\sqrt{N}$. In contrast, pseudo-gaps with $\theta < 1$ have $\lambda_{\min} \ll 1/\sqrt{N}$. To maintain such a pseudogap in a stationary state, one would require correlations much larger than what the discreteness of the model allows. Pseudogaps with $\theta < 1$ are thus not admissible solutions of Eqs. (5.59, 5.66).

Excluding $\theta > 1$ In this case, $\lambda_{\min} \gg 1/\sqrt{N} \sim J$. Thus when one spin flips, the second least stable spin will not flip in general, and avalanches are typically of size unity [Mue15]. It can easily be shown that in that case the number of flips per spin along the loop would be small (in fact it would even vanish in the thermodynamic limit, which is clearly impossible). In terms of our Fokker-Planck description, the motion of the spin stabilities due to other flips would be small in comparison with the motion of the stabilities inbetween avalanches, due to changes of the magnetic field. Making the crude assumption that the magnetization is random for any λ, the change of external magnetic field leads to an additional diffusion term in the Fokker-Planck equation:

$$\partial_t \rho(\lambda, t) = -\partial_\lambda (v - D\partial_\lambda)\rho(\lambda, t) + D_h \partial_\lambda^2 \rho(\lambda, t), \tag{5.67}$$

where the term D_h is related to the typical field increment $h_{\min} \sim \lambda_{\min}$ required to trigger an avalanche. Indeed $D_h \sim N h_{\min}^2 \sim N^{(\theta-1)/(\theta+1)} \gg D \sim 1$. Under these circumstances, Eq. (5.61) does not hold. The dynamics would be a simple diffusion with reflecting boundary, whose only stationary solution corresponds to $\theta = 0$, violating our hypothesis $\theta > 1$. Thus the last term of Eq. (5.67) provides a restoring force toward dominated dynamics flattens the distribution. As soon as the pseudo-gap is filled up to $\theta = 1$, this diffusion contribution becomes sub-dominant and the dynamics is dominated by the transient dynamics concentrated in the main text. In stationary conditions, a typical pseudo-gap profile must thus converge to $\theta = 1$.

5.5 Overview

The SK model presents presents self-organized criticality (SOC) in its whole hysteresis loop. That is, the external field h triggers power-law distributed avalanches that span the entire system. This SOC is strictly related to marginal stability, since for small λ the distribution of the local stabilities goes as $\rho(\lambda) \propto \lambda^\theta$. Through stability arguments we showed that to have crackling responses $\theta = 1$ is needed. We extended these stability arguments to multiple spins, remarking that the soft spins are in average frustrated with each other (the energy along their links is not minimized): There is a correlation function $C(\lambda)$ that scales inversely with the stability λ.

We then related the averages $\langle \Delta M \rangle$ and $\langle n \rangle$ to the cutoffs of the avalanches. In order to have SOC, the cutoffs need to diverge when $N \to \infty$. With scaling arguments we showed that the SOC of the SK model vanishes when one considers models with a finite number of neighbors, as it is also confirmed by numerical simulations. Through a model that mixes short- and long-range interactions, we showed that fully-connected interactions are a relevant perturbation to the short-range Hamiltonian, so the presence of long-range interactions is strictly necessary to have SOC in the system, independently of the presence or not of short-range interactions, no matter their amplitude. Yet, even though the long-range interactions grant avalanches that extend over all the system, the scaling of the avalanche sizes cutoffs is different depending on the presence of short-range interactions.

We also studied the crackling in the SK model from the point of view of the dynamics. An avalanche can be seen as a discrete random walk (RW) of the number of unstable spins, $n_{\rm unst}$. The end of the avalanche corresponds with the number of time steps that it takes the RW to return to zero. In critical dynamics, these RWs are non-trivial, have a preferred number of unstable spins, $n^*_{\rm unst}$. For $n_{\rm unst} < n^*_{\rm unst}$ the avalanches tendentially grow, for $n_{\rm unst} > n^*_{\rm unst}$ they shrink, suggesting that during the avalanche there is some type of correlation between spins that keeps the system critical. A further extensive study of the relation between $n^*_{\rm unst}$, the correlations $C(\lambda)$ and the size of the avalanches can be a key factor for the understanding of SOC.

To figure out how much of the crackling behavior is related to the type of dynamics one chooses, and how much is more universal, we analyzed different kinds of single-spin flip algorithm. We identified a variation in the exponents of the avalanche distributions, but more fundamental features as the pseudogap exponent θ stay the same.

Finally, through a modelization with a RW in the space of the spin stabilities λ, we found that it is the dynamics itself that, because of a strong correlation among the softest spins, leads the system to a marginal state with a pseudogap. With a Fokker-Planck description of the dynamics we explained the appearance of both the pseudogap and the singular correlation $C(\lambda)$.

References

[And13] J.C. Andresen, Z. Zhu, R.S. Andrist, H.G. Katzgraber, V. Dobrosavljević, G.T. Zimanyi, Phys. Rev. Lett. **111**, 097203 (2013). doi:10.1103/PhysRevLett.111.097203; arXiv:1210.3796; http://link.aps.org/doi/10.1103/PhysRevLett.111.097203

[Dav82] J.H. Davies, P.A. Lee, T.M. Rice, Phys. Rev. Lett. **49**, 758–761 (1982). doi:10.1103/PhysRevLett.49.758; http://link.aps.org/doi/10.1103/PhysRevLett.49.758

[Dou10] P.L. Doussal, M. Müller, K.J. Wiese, EPL (Europhys. Lett.) **91**, 57004 (2010). doi:10.1209/0295-5075/91/57004; http://stacks.iop.org/0295-5075/91/i=5/a=57004

[Eas06] R. Eastham, R.A. Blythe, A.J. Bray, M.A. Moore, Phys. Rev. B **74**, 020406 (2006). doi:10.1103/PhysRevB.74.020406. arXiv:cond-mat/0601402

[Efr75] A.L. Efros, B.I. Shklovskii, J. Phys. C: Solid State Phys. **8**, L49 (1975). doi:10.1088/0022-3719/8/4/003; http://stacks.iop.org/0022-3719/8/i=4/a=003

[Fis98] D. Fisher, Phys. Rep. **301**, 113–150 (1998). ISSN 0370-1573. doi:10.1016/S0370-1573(98)00008-8; http://www.sciencedirect.com/science/article/pii/S0370157398000088

[Hor08] H. Horner, Eur. Phys. J. B **60**, 413 (2008). doi:10.1140/epjb/e2008-00017-1. arXiv:0707.2714

[Kal14] Y. Kallus, S. Torquato, Phys. Rev. E **90**, 022114 (2014). doi:10.1103/PhysRevE.90.022114; http://link.aps.org/doi/10.1103/PhysRevE.90.022114

[LD12] P. Le Doussal, M. Müller, K.J. Wiese, Phys. Rev. B **85**, 214402 (2012). doi:10.1103/PhysRevB.85.214402; http://link.aps.org/doi/10.1103/PhysRevB.85.214402

[Ler13] E. Lerner, G. During, M. Wyart, Soft Matter **9**, 8252–8263 (2013). doi:10.1039/C3SM50515D; http://dx.doi.org/10.1039/C3SM50515D

[Mue15] M. Mueller, M. Wyart, Annu. Rev. Condens. Matter Phys. **6**, 177–200 (2015). doi:10.1146/annurev-conmatphys-031214-014614; arXiv:1406.7669; http://dx.doi.org/10.1146/annurev-conmatphys-031214-014614

[Pal12] M. Palassini, M. Goethe, J. Phys.: Conf. Ser. **376**, 012009 (2012). doi:10.1088/1742-6596/376/1/012009; http://stacks.iop.org/1742-6596/376/i=1/a=012009

[Pal79] R.G. Palmer, C.M. Pond, J. Phys. F: Metal Phys. **9**, 1451 (1979). doi:10.1088/0305-4608/9/7/024; http://stacks.iop.org/0305-4608/9/i=7/a=024

[Pan05] S. Pankov, V. Dobrosavljević, Phys. Rev. Lett. **94**, 046402 (2005). doi:10.1103/PhysRevLett.94.046402; http://link.aps.org/doi/10.1103/PhysRevLett.94.046402

[Par03] G. Parisi, Fractals **11**, 161 (2003). doi:10.1142/S0218348X03001823; http://www.worldscientific.com/doi/abs/10.1142/S0218348X03001823

[Páz99] F. Pázmándi, G. Zaránd, G. T. Zimányi, Phys. Rev. Lett. **83**, 1034–1037 (1999). doi:10.1103/PhysRevLett.83.1034; arXiv:cond-mat/9902156; http://link.aps.org/doi/10.1103/PhysRevLett.83.1034

[Set93] J.P. Sethna, K. Dahmen, S. Kartha, J.A. Krumhansl, B.W. Roberts, J.D. Shore, Phys. Rev. Lett. **70**, 3347–3350 (1993). doi:10.1103/PhysRevLett.70.3347; http://link.aps.org/doi/10.1103/PhysRevLett.70.3347

[Set01] J. Sethna, K. Dahmen, C. Myers, Nature **410**, 242–250 (2001). doi:10.1038/35065675. March

[Sha14] A. Sharma, A. Andreanov, M. Müller, Phys. Rev. E **90**, 042103 (2014). doi:10.1103/PhysRevE.90.042103; http://link.aps.org/doi/10.1103/PhysRevE.90.042103

[She75a] D. Sherrington, S. Kirkpatrick, Phys. Rev. Lett. **35**, 1792–1796 (1975). doi:10.1103/PhysRevLett.35.1792; http://link.aps.org/doi/10.1103/PhysRevLett.35.1792

[Tho77] D.J. Thouless, P.W. Anderson, R.G. Palmer, Phil. Mag. **35**, 593–601 (1977). doi:10.1080/14786437708235992; http://dx.doi.org/10.1080/14786437708235992

[Wya12] M. Wyart, Phys. Rev. Lett. **109**, 125502 (2012). doi:10.1103/PhysRevLett.109.125502; http://link.aps.org/doi/10.1103/PhysRevLett.109.125502

Chapter 6
Soft Modes and Localization in Spin Glasses

More than 40 years ago, it became clear that supercooled liquids and amorphous solids exhibit an excess of low-energy excitations, compared with their crystalline counterparts [Phi81]. This excess was evinced, for instance, from anomalies in the specific heat at low temperatures (below 10 K). A number of scattering techniques such as Raman, neutron [Buc84] and, more recently, inelastic X-ray scattering [Set98], have shown that these excitations are of vibrational nature, and correspond to wave vectors of a few nm^{-1} and frequencies of few mK (see e.g. [Mon09a] and references therein). The corresponding vibrational density of states $g(\omega)$ displays an excess of modes, respect to the conventional Debye behavior $g(\omega) \propto \omega^{d-1}$ (ω^2 in the three-dimensional case treated herein), called *boson peak*. Despite the shape of the $g(\omega)$ depends on numerous factors, such as the considered material, the temperature, the thermal history, etc., the presence of the boson peak is a universal feature [Buc84, Mal91]. The starting point for an analysis of vibrational excitations is the harmonic approximation around stable or metastable states as, for example, this way many low temperature properties of solids can be calculated analytically [Hua87].

Also in liquid systems one encounters the same phenomenology. The density of states in liquids was extensively studied to describe their dynamics, since for small enough times one can characterize them through independent simple harmonic motions (instantaneous normal modes) [Wu92, Key94, Wan94]. In supercooled liquids the dynamics is so damped that it is dominated by the underlying energy landscape [Cav09], and it becomes natural to focus the attention on the harmonic modes of the inherent structures (ISs), the local minima of the energy that can be obtained by quickly relaxing the system, to zero temperature, obtaining metastable configurations called inherent structures [Sti95, Mon09b]. These metastable states are likely to play an important role both in driving the sluggish dynamics of these glassy systems [Gri03], and in their thermodynamic properties as the temperature vanishes or the system becomes jammed [Xu10].

Two main approaches are used to explain the presence of the boson peak, attributing it to the presence of many metastable states.

On one side, there should be a very large number of localized excitations due to the quantum tunneling between very similar states. The system can bounce from one state

M. Baity Jesi, *Spin Glasses*, Springer Theses, DOI 10.1007/978-3-319-41231-3_6

to the other with very little energy exchange. The couples of states described through this phenomenological approach are called two-level systems [And72, Phi72, Phi87]. Although their precise nature has not been clarified, their presence is experimentally detectable [Lis15].

The second cause of an excess of soft modes is motivated by the presence of marginally stable states, that display infinitely soft modes. This excess of soft modes is highly universal among strongly disordered mean field models [Méz87]. Indeed, by means of replica calculations, it has been recently shown that mean field supercooled liquids exhibit a transition to a full RSB phase at high enough pressure [Cha14]. Full RSB implies a complex energy landscape with a hierarchical structure of states and a large amount of degenerate minima separated by small energy barriers [Méz84, Cha14]. These energy barriers can be infinitely small, along with the smallest harmonic excitations, meaning that the system is marginally stable.

Besides to the shape of the energy landscape, marginal stability is also caused by isostaticity [Wya12], the condition of having as many degrees of freedom as independent constraints, that arises at jamming [O'H03]. The strong universality of those features in continuous constraint satisfaction problems suggests that they are a key ingredient for the understanding of the glass and the jamming transition [Fra15a, Fra15b].

A main difference between the two scenarios is that the two-level system picture requires the presence of strongly localized states, whereas the marginal stability is recovered through calculations in infinite dimensions where localization cannot play a crucial role, but a RSB transition is needed. Furthermore, the two-level system descends from a quantum description and requires taking into account anharmonic effects, whereas the boson peak predicted by RSB theories is classical, and can be identified at the harmonic level. Here, we somehow reconcile the two approaches by identifying two-level systems from a purely classical and harmonic starting point.

Even though many of the tools used to explain the boson peak descend from spin glass theory, the investigation of small harmonic excitations of the metastable states has remained relegated to the field of structural glasses. On one hand because in SGs no "crystal phase" can be reached by cooling the system slow enough, on another, perhaps, because the two most studied SG models are the EA and the SK model, both with Ising spins, that are discrete. In the Ising SG the aforementioned phenomena are difficult to study. When the passage from paramagnetic to SG phase is very quick, while in structural glasses there is a large range of temperatures in the disordered phase, where the dynamics is overdamped. Furthermore, it is not straightforward to study soft excitations in a system where the smallest excitation is bounded by its discrete nature.

Still, as we saw in Chap. 4, many types of SG model with continuous degrees of freedom are easy to define. Among those, the Heisenberg model (1.2), where the spins are unitary vectors with $m = 3$ components, is an epitome of the spin glass, as it is the first proposed SG model. Harmonic modes can be easily studied in this model, though due to the $O(3)$ symmetry of the Hamiltonian, the system exhibits an excess of trivial low-frequency modes (Goldstone modes and spin waves) that make this type of analysis less clear. We can decide, thus, to add a random magnetic field to the Heisenberg Hamiltonian to wipe out the symmetries and the soft modes they carry, keeping only those related to marginal stability. A similar procedure of symmetry removal has been carried through in glass-forming liquids, by pinning a certain fraction of particles [Kob12, Cam13]. In those references it was shown that the glass transition survives the pinning. Hence from the above considerations on marginal stability [Méz87, Fra15b] we expect as well a boson peak in pinned systems.

We propose ourselves to extend these considerations to a finite-dimensional system, the Heisenberg SG in a random magnetic field. This lets us verify the extent of the universality of these phenomena. On one side by checking if the soft modes are present with a similar phenomenology on a different type of system, and on the other by extending the ideas of marginal stability to finite dimensions, in non-isostatic systems. As an additional virtue, the model we study gives us the possibility of making this analysis on unprecedentedly large systems, giving us the chance to observe scalings along several orders of magnitude.

Here, we study the inherent structures and we do find that they are marginally stable states where the distribution of eigenvalues of the Hessian matrix stretches down to zero as a power law. Furthermore, we find that the soft modes are localized. This cannot be revealed by computations in infinite dimensions, though it is still possible to observe correlations in pseudo mean field networks such as the Bethe lattice [Lup15], and it was shown that superuniversality (the independence of the behavior on the space dimensionality) can be recovered by removing local excitations [Cha15].

We broaden our analysis by taking in account the anharmonic effects due to the complexity of the energy landscape. We find that the energy barriers along the softest mode are extremely small, in agreement with the mean field picture, and that they connect very similar states with an strong relationship, that we propose as a classical operational definition of two-level systems.

At the end of the game the scenario is consistent, with mean field theory that does apply, but with the necessary finite-dimension corrections due to the presence and importance of localized states.

6.1 Model and Simulations

The model we study is the three-dimensional Heisenberg spin glass in a RF. The RF breaks all rotational and translational symmetry, so there should be no Goldstone bosons. The dynamic variables are spins $\vec{s}_x$ with $m = 3$ components. They are placed at the vertices $\boldsymbol{x}$ of a cubic lattice of linear size L with unitary spacings. We have therefore $N = L^3$ spins, and $2N$ degrees of freedom (dof) due to the constraint $\vec{s}_x \cdot \vec{s}_x = 1$.

The Hamiltonian is

$$\mathcal{H}_{\rm RF} = -\sum_{|x-y|=1} J_{xy}\vec{s}_x \cdot \vec{s}_y - \sum_x^N \vec{h}_x \cdot \vec{s}_x, \tag{6.1}$$

where the fields $\vec{h}_x$ are random vectors chosen uniformly from the sphere of radius $H_{\rm amp}$. The couplings J_{xy} are fixed, Gaussian distributed, with $\overline{J_{xy}} = 0$ and $\overline{J_{xy}^2} = J^2$.

The lattice sizes we simulated are $L = 12, 24, 48, 96, 192$. We chose always $J = 1$, and we compared it with $H_{\rm amp} = 0, 0.01, 0.05, 0.1, 0.5, 1, 5, 10, 50$. In Table 6.1 we resume how many samples we simulated for each couple (L, H).

6.2 Calculating the Density of States

Our goal is to study the dynamical matrix of the system. The dynamical matrix is the Hessian matrix $\mathcal{M}$ of Hamiltonian (6.1), calculated at the local minima of the energy, that we call inherent structures (ISs) in analogy with structural glasses. Each infinite-temperature starting configuration $|\vec{s}\rangle$ can be associated to an IS $\left|\vec{s}^{(\rm IS)}\right\rangle$ through a deterministic relaxation of the system.[1]

6.2.1 Reaching the Inherent Structure

As energy minimization algorithm we use the successive overrelaxation (SOR) (Appendix F.1.2, that was successfully used in [BJ11] for $3d$ Heisenberg spin glasses. This algorithm depends on a parameter Λ, and the convergence speed is maximal for $\Lambda \approx 300$ [BJ11]. Thus, the seek of ISs was done with $\Lambda = 300$, under the reasonable assumption, reinforced in Appendix F, that a change on Λ does not imply sensible changes in the observables we examine. In fact, the concept of IS is strictly related to the protocol one chooses to relax the system, and on the starting configuration. From [BJ11] our intuition is that despite the ISs' energies do depend on these two elements, this dependency is small and we can neglect it (dependencies on the correlation lengths will be examined in a future work [BJ16]).

We validate these hypotheses in Appendix F.2, where we compare the density of states (DOSs) both between $\Lambda = 300$ and $\Lambda = 1$, and between starting configurations at different temperatures.

For most of the simulated fields, the pdf of the overlap of the reached inherent structures, $P(q_{\rm IS})$, is peaked around a non-zero value that is significantly far from 1 (Fig. 6.1). This means that even though all the inherent structures have a very large amount of spins in similar configurations, it is practically impossible with this approach (at least for $L > 12$ lattices), to find two identical inherent structures.

[1] We will show in Appendix F that the starting temperature does not influence visibly the properties we are studying, at least as long as we stay in the paramagnetic phase.

Table 6.1 Number of samples, N_{samples}, and of replicas, N_{replicas}, of our simulations

H_{amp}	L	N_{samples}	N_{replicas}	n_λ	$A(\lvert\vec{\pi}_0\rangle)$	$A(\lvert\vec{\pi}_{\text{RAND}}\rangle)$
50	192	10 (0)	2	35	–	–
50	96	10 (10)	2	80	1	1
50	48	70 (70)	2	500	1	1
50	24	100 (100)	2	500	1	1
50	12	100 (100)	2	500	1	1
10	192	10 (0)	2	35	–	–
10	96	10 (10)	2	80	0.6	0.72
10	48	70 (70)	2	500	0.6	0.72
10	24	100 (100)	2	500	0.3	0.72
10	12	100 (100)	2	500	0.3	0.72
5	192	10 (0)	2	35	–	–
5	96	10 (10)	2	80	0.014	0.3
5	48	70 (70)	2	500	0.014	0.3
5	24	100 (100)	2	500	0.02	0.3
5	12	100 (100)	2	500	0.024	0.3
1	192	10 (0)	2	35	–	–
1	96	10 (10)	2	80	0.004	0.05
1	48	70 (70)	2	500	0.004	0.05
1	24	100 (100)	2	500	0.0045	0.05
1	12	100 (100)	2	500	0.0045	0.05
0.5	192	10 (0)	2	35	–	–
0.5	96	10 (10)	2	80	0.008	0.022
0.5	48	70 (70)	2	500	0.008	0.02
0.5	24	100 (100)	2	500	0.009	0.022
0.5	12	100 (100)	2	500	0.009	0.022
0.1	192	10 (0)	2	35	–	–
0.1	96	10 (10)	2	80	0.006	0.012
0.1	48	100 (70)	2	500	0.006	0.012
0.1	24	100 (100)	2	500	0.1	0.012
0.1	12	100 (100)	2	500	0.1	0.012
0.05	192	10 (0)	2	25	–	–
0.05	96	10 (10)	2	80	0.06	0.011
0.05	48	100 (70)	2	500	0.06	0.011
0.05	24	100 (100)	2	500	0.42	0.011
0.05	12	100 (100)	2	500	0.36	0.011
0.01	192	7 (0)	2	25	–	–
0.01	96	10 (10)	2	80	0.045	0.016
0.01	48	100 (70)	2	500	0.045	0.016
0.01	24	100 (100)	2	500	0.009	0.004
0.01	12	100 (100)	2	500	0.007	0.001

The number between parentheses is the amount of samples used for the forcings. We indicate with n_λ the number of eigenvalues we calculated from the bottom of the spectrum $\rho(\lambda)$ (see Sect. 6.3). $A(\lvert\vec{\pi}_{\text{RAND}}\rangle)$ and $A(\lvert\vec{\pi}_0\rangle)$ are the forcings' parameters from Eqs. (6.15) and (6.16)

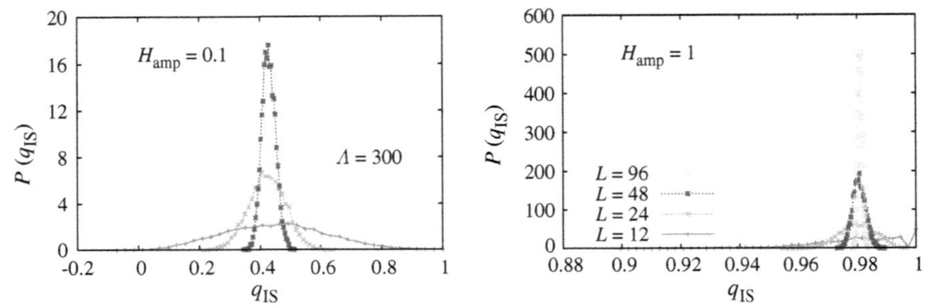

Fig. 6.1 Distribution $P(q_{\mathrm{IS}})$ of the overlaps at the inherent structures obtained with $\Lambda = 300$, for $H_{\mathrm{amp}} = 0.1$ (*left*) and $H_{\mathrm{amp}} = 1$ (*right*).

6.2.2 The Local Reference Frame

Once the IS is found, we want to study the properties of the reached IS. From the Hamiltonian at the inherent structure, $\mathcal{H}_{\mathrm{IS}}$, we want to compute the Hessian matrix $\mathcal{M}$ to study the harmonic behavior at the IS. This is not trivial, because it is necessary to take into account the normalization of the spins $\vec{s}_{\boldsymbol{x}}^{\,2} = 1\ \forall \boldsymbol{x}$.

To this scope we define local perturbation vectors $\vec{\pi}_{\boldsymbol{x}}$, and we call them pions in analogy with the nonlinear σ model [GM60]. The distinguishing feature of the pions is that they are orthogonal to the IS, $(\vec{s}_{\boldsymbol{x}} \cdot \vec{\pi}_{\boldsymbol{x}}) = 0$, and that their global norm is unitary, $\langle \vec{\pi}\,|\vec{\pi}\,\rangle = 1$.[2] We can use the pions to parametrize an order ϵ perturbation around the IS as

$$\vec{s}_{\boldsymbol{x}}^{\,\epsilon} = \vec{s}_{\boldsymbol{x}}^{\,(\mathrm{IS})}\sqrt{1-\epsilon^2\vec{\pi}_{\boldsymbol{x}}^{\,2}} + \epsilon\vec{\pi}_{\boldsymbol{x}}, \quad \vec{\pi}_{\boldsymbol{x}}^{\,2} \equiv \vec{\pi}_{\boldsymbol{x}} \cdot \vec{\pi}_{\boldsymbol{x}}, \tag{6.2}$$

so the position of $\vec{s}_{\boldsymbol{x}}^{\,\epsilon}$ is fully determined by $\vec{\pi}_{\boldsymbol{x}}$. As long as ϵ is small enough to grant $\epsilon^2\vec{\pi}_{\boldsymbol{x}}^{\,2} < 1\ \forall\boldsymbol{x}$, the normalization condition is naturally satisfied without the need to impose any external constraint.

We now build a local reference change. For each site $\boldsymbol{x}$ we define a local basis $\mathcal{B} = \left\{\vec{s}_{\boldsymbol{x}}^{\,(\mathrm{IS})}, \hat{e}_{1,\boldsymbol{x}}, \hat{e}_{2,\boldsymbol{x}}\right\}$, where $\hat{e}_{1,\boldsymbol{x}}, \hat{e}_{2,\boldsymbol{x}}$ are any two unitary vectors, orthogonal to each other and to $\vec{s}_{\boldsymbol{x}}^{\,(\mathrm{IS})}$, and well oriented. In our simulations they were generated randomly. In this basis the pions can be rewritten as

$$\vec{\pi}_{\boldsymbol{x}} = (0, a_1, a_2), \tag{6.3}$$

where now they explicitly depend only on two components, with real values a_1 and a_2. We can therefore rewrite the pions as two-component vectors $\tilde{\pi}_{\boldsymbol{x}}$

$$\tilde{\pi}_{\boldsymbol{x}} = (a_1, a_2). \tag{6.4}$$

---

[2] Recall the notation introduced in Sect. 1.2, according to which $\left\langle \vec{a}\,\middle|\vec{b}\right\rangle \equiv \sum_{\boldsymbol{x}} \vec{a}_{\boldsymbol{x}} \cdot \vec{b}_{\boldsymbol{x}}$.

At this point we completely integrated the normalization constraint with the parametrization, and we can obtain the $2N \times 2N$ Hessian matrix $\mathcal{M}$, that acts on $2N$-component vectors $|\tilde{\pi}\rangle$, by a second-order development of $\mathcal{H}_{\text{IS}}$ (the derivation of $\mathcal{M}$ is shown in Appendix F.3). The obtained matrix is sparse, with 13 non-zero elements per line (1 diagonal element, and 6 two-component vectors for the nearest-neighbors).

The matrix element $\mathcal{M}^{\alpha\beta}_{\boldsymbol{xy}}$ is

$$\mathcal{M}^{\alpha\beta}_{\boldsymbol{xy}} = \mathcal{M}_{\boldsymbol{xy}}(\hat{\mathrm{e}}_{\alpha,\boldsymbol{x}} \cdot \hat{\mathrm{e}}_{\beta,\boldsymbol{y}}), \tag{6.5}$$

with

$$\mathcal{M}_{\boldsymbol{xy}} = \delta_{\boldsymbol{xy}}(\vec{h}^{\,(\text{IS})}_{\boldsymbol{y}} \cdot \vec{s}^{\,(\text{IS})}_{\boldsymbol{y}}) - \sum_{\mu=-D}^{D} J_{\boldsymbol{xy}}\delta_{\boldsymbol{x}+\hat{\mu},\boldsymbol{y}}, \tag{6.6}$$

where the bold latin characters as usual indicate the site, and the greek characters indicate the component of the two-dimensional vector of Eq. (6.4).

Once $\mathcal{M}$ is known, from each simulated H_{amp} we calculate the spectrum of the eigenvalues $\rho(\lambda)$ or equivalently, in analogy with plane waves [Hua87], the DOS $g(\omega)$, by defining $\lambda = \omega^2$. We measure the degrees of freedom (dof) both by means of a convolution with a lorentian function with the method of the moments [Chi78, Tur82, Alo01], and by making the explicit brute-force calculation of the lowest eigenvalues with Arpack [Sor08].

6.3 The Spectrum of the Hessian Matrix

We find that, although for large fields there is a gap in the DOS (as one can easily expect by calculating it exactly in the diagonal limit $H_{\text{amp}} \gg J \simeq 0$) when the field is small enough the gap disappears and the DOS goes to zero developing soft modes (Fig. 6.2, left). In the right set of Fig. 6.2 we show the scaling of the lowest eigenvalue of the Hessian. We see that while for very large fields it remains approximately constant, for smaller fields it approaches zero as we increase the lattice size L.

It is interesting to understand the origin of these soft modes, so we focus on the $\rho(\lambda)$ for small λ, or even better in its cumulative function

$$F(\lambda) = \int_0^{\lambda} \rho(\lambda')d\lambda'. \tag{6.7}$$

In the case that there be no gap and for small λ the function $F(\lambda)$ reach zero as a power law, we can define three exponents δ, α and γ, that describe how the functions g, ρ and F go to zero for small λ[3]:

[3]The exponents δ, α and γ have nothing to do with the critical exponents defined in Sect. 1.3.

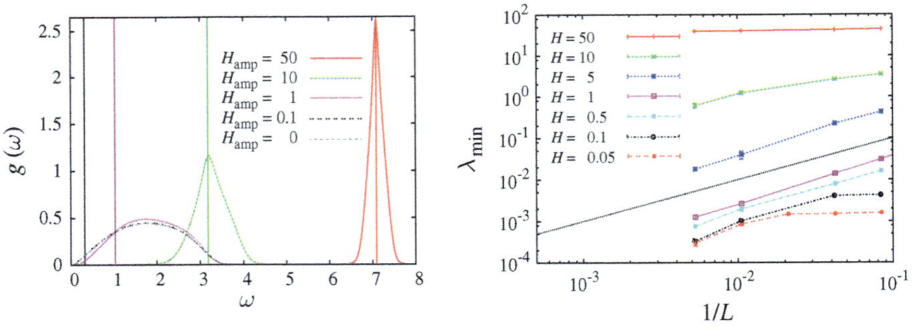

Fig. 6.2 *Left* The DOS $g(\omega)$ calculated with the method of the moments. The *vertical lines* represent its face in the limit of a diagonal hamiltonian, $J = 0$. The $g(\omega)$ corresponding to $H_{\text{amp}} = 0$ and $H_{\text{amp}} = 0.1$ are practically overlapped. *Right* Scaling with $1/L$ of the lowest eigenvalue λ_{min} of the Hessian matrix $\mathcal{M}$ calculated at the IS, for all the simulated fields. The *straight line* is a reference curve $\lambda_{\text{min}} \propto 1/L$

$$g(\omega) \sim \omega^{\delta}, \quad \rho(\lambda) \sim \lambda^{\alpha}, \quad F(\lambda) \sim \lambda^{\gamma}, \tag{6.8}$$

where the exponents are related by $\delta = 2\alpha + 1 = 2\gamma - 1$. In the Debye model, valid for perfect crystals and based on the assumption that all the eigenvectors are plane waves, one has $\delta = d - 1 = 2$ ($\alpha = 0.5$, $\gamma = 1.5$), and this is also what one expects for our model in the absence of a field [Gri11]. In Figs. 6.3 and 6.4 we show the function $F(\lambda)$ for all the fields we simulated. We were able to calculate with Arpack the lowest eigenvalues of the spectrum. The number of calculated eigenvalues n_{λ} is shown in Table 6.1. All the plots are compared with the Debye behavior $\lambda^{1.5}$ and with the power law behavior $\lambda^{2.5}$, because if there is some universality on the exponents γ, our data suggests it has to be around $\gamma = 2.5$ (thus $\delta = 4$ and $\alpha = 1.5$). This is straightforward for $H_{\text{amp}} = 0.1, 0.5, 1, 5$, where when λ is small there is a clear power law behavior, with a power close to 2.5, while it can be excluded for $H_{\text{amp}} = 50$, where the soft modes are suppressed in favor of a gap, as it was also clear from Fig. 6.2. At $H_{\text{amp}} = 10$ we are probably close to where the gap forms. The $F(\lambda)$ goes as a large power law λ^{boh} when λ is large, but at the smallest values of λ, recovered from $L = 192$, there is a slight change of power law towards something that could become 2.5. One could also argue that a $F(\lambda)$ goes to zero as a power law for any finite H_{amp}, as long as one looks at small enough λ. Numerical analysis cannot reply to questions of this type, but still, even if no sharp transition is present, an empirical gap is clearly present for large H_{amp}, since the precision of any experiment (numerical or real) is finite. In the case of the smallest fields $H_{\text{amp}} = 0.01, 0.05$, we suffer from effects from $H_{\text{amp}} = 0$. The spin waves do not hybridize with the bulk of the spectrum, and pseudo-Goldstone modes with a very small eigenvalue appear, making it hard to extract a power law behavior.

Overall, we see good evidence for a γ around 2.5 at several values of H_{amp}, and at other fields the data is not in contradiction with a hypothesis of universality in the exponents (6.8). When the field is small we remark a change of trend from $\gamma \approx 2.5$

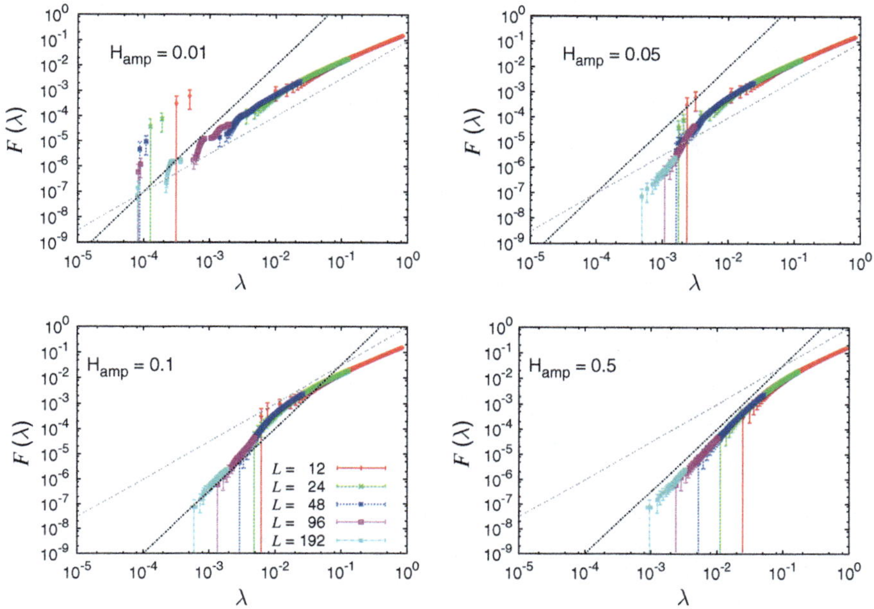

Fig. 6.3 Cumulative distributions $F(\lambda)$ for small random fields $H_{amp} = 0.01, 0.05, 0.1, 0.5$. In each plot we show a *black reference curve* representing the power law $\lambda^{2.5}$, that is our guess for a universal behavior, and a *grey line* indicating the Debye behavior $\lambda^{1.5}$. One could expect a Debye behavior for $\lambda > \lambda^*$, with $\lambda^* \to 0$ as $H_{amp} \to 0$. Instead, we see an excess of eigenvalues even compared to the Debye behavior, indicating a likely boson peak. Further discussions in the main text

to $\gamma < 1.5$ at a value λ^*. The crossover λ^* shifts towards zero as H_{amp} decreases. This probably indicates the presence of a boson peak, an excess of modes at low frequency. Signs of a boson peak in at $H_{amp} = 0$ can be seen in Fig. F.2. In that case the mass of the spectrum is all concentrated at low λ, but there ought to be a Debye behavior, meaning that λ^* is very little.

6.4 Localization

We found that the application of a magnetic field does not induce a gap in the density of states. It goes to zero as a power law even in the presence of a not too large RF, and it develops a gaps when the RF is very large compared with the couplings. What do these soft modes represent? We want to know something more about the $2N$-dimensional eigenvectors $|\tilde{\pi}_\lambda\rangle$ of the matrix $\mathcal{M}$. Similarly as it happens in other types of disordered systems [Xu10, DeG14, Cha15], the soft modes are localized, meaning that the eigenvectors $|\pi_\lambda\rangle$ are dominated by very few components. To observe the localization we can define the inverse participation ratio

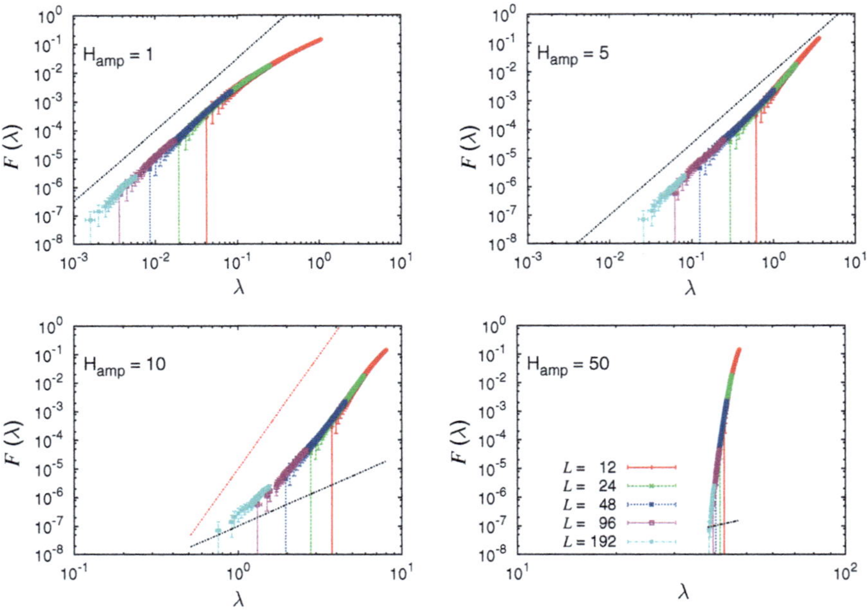

Fig. 6.4 Cumulative distributions $F(\lambda)$ for large random fields $H_{\mathrm{amp}} = 1, 5, 10, 50$. In each plot we show a *reference curve* representing the power law $\lambda^{2.5}$. The *orange line* in the *bottom left* set is proportional to λ^8

$$
Y_\lambda = \frac{\sum_x \left(|\tilde{\pi}_{\lambda,x}|^2\right)^2}{\left(\sum_x |\tilde{\pi}_{\lambda,x}|^2\right)^2} = \frac{\sum_x \left(a_{1,i}^2 + a_{2,x}^2\right)^2}{\left(\sum_x (a_{1,x}^2 + a_{2,x}^2)\right)^2}, \tag{6.9}
$$

where we coupled the two components corresponding to a single site because the local basis vectors have random directions, so there would be no point in trying to distinguish one from the other. If the eigenvector $|\pi_{\lambda,x}\rangle$ is fully localized in one site we will have $Y_\lambda = 1$. On the counterpart, if all its components are the same (fully delocalized) we will have $Y_\lambda = 1/N$. In Fig. 6.5 we show that the softer the eigenvectors the more localized they are.[4] For small random fields (Fig. 6.5, left), we remark sizable finite-size effects, with the passage from localized to delocalized regime that becomes sharper as the lattice size is increased, suggesting the presence of localization threshold that separates a small fixed percentage of localized eigenvectors from the delocalized bulk ones. For larger fields we appreciate no finite-size effects, and it appears that $\sim$1 % of the eigenvectors is localized.

[4]Only in $H_{\mathrm{amp}} = 0.01$ this was not clear, but we attribute it to strong echoes of the $H_{\mathrm{amp}} = 0$ behavior. Due to this interference, we will basically exclude the case of a very small field from our analysis.

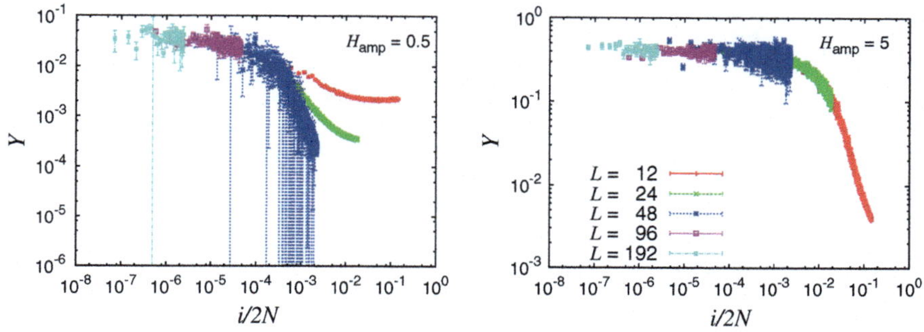

Fig. 6.5 Participation ratio for $H_{\text{amp}} = 0.5$ (*left*) and $H_{\text{amp}} = 5$ (*right*)

Since in a localized state the eigenvectors have a well-defined correlation length, we can use also this criterion to probe the localization. We can define a correlation length from Green's function $\mathcal{G}$, that is defined through the relation $\mathcal{M}\mathcal{G} = \delta_{xy}$, an is commonly used in field theory for two-point correlations. Since $\mathcal{M}^{-1}$ shares eigenvectors ψ_n with $\mathcal{M}$ and has inverse eigenvalues $1/\lambda_n$,[5] Green's function is

$$\mathcal{G}(\boldsymbol{x}, \boldsymbol{y}) = \mathcal{M}^{-1}\delta_{\boldsymbol{xy}} = \sum_n \frac{\psi_n(\boldsymbol{x})\psi_n(\boldsymbol{y})}{\lambda_n}, \tag{6.10}$$

and squaring the relation

$$\mathcal{G}^2(\boldsymbol{x}, \boldsymbol{y}) = \sum_{m,n} \frac{\psi_m(\boldsymbol{x})\psi_m(\boldsymbol{y})\psi_n(\boldsymbol{x})\psi_n(\boldsymbol{y})}{\lambda_m \lambda_n}. \tag{6.11}$$

By averaging over the disorder we gain translational invariance and $\overline{\mathcal{G}^2}$ can be written as a function of the distance $\boldsymbol{r} = \boldsymbol{x} - \boldsymbol{y}$,

$$\overline{\mathcal{G}^2(\boldsymbol{r})} = \overline{\sum_{m,n} \frac{1}{\lambda_m \lambda_n} \sum_{\boldsymbol{x}} \left(\frac{[\psi_m(\boldsymbol{x})\psi_n(\boldsymbol{x})][\psi_m(\boldsymbol{x}+\boldsymbol{r})\psi_n(\boldsymbol{x}+\boldsymbol{r})]}{V} \right)}. \tag{6.12}$$

Making the reasonable assumption that different eigenvectors do not interfere with each other, and exploiting the orthogonality condition $\sum_x \psi_m(\boldsymbol{x})\psi_n(\boldsymbol{x}) = \delta_{mn}$, we obtain the desired correlation function

$$\mathcal{C}(\boldsymbol{r}) = \overline{\mathcal{G}^2(\boldsymbol{r})} = \overline{\sum_n \frac{1}{\lambda_n^2} \psi_n^2(\boldsymbol{x})\psi_n^2(\boldsymbol{x}+\boldsymbol{r})}. \tag{6.13}$$

[5] For simplicity we use N-component eigenvectors $\psi_n(\boldsymbol{x})$ instead of the $2N$-component ones $|\tilde{\pi}\rangle$. The relationship between the two can be recovered through $\psi_n^2(\boldsymbol{x}) = \tilde{\pi}^2 = \vec{\pi}^2$.

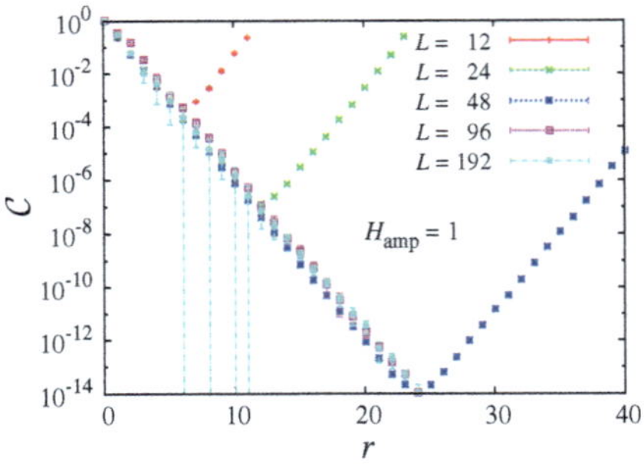

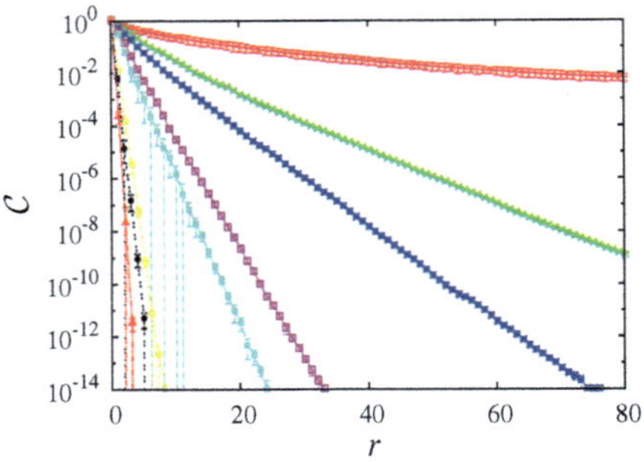

Fig. 6.6 *In the left* set we show the correlation function $C(r)$ for different lattice sizes and $H_{\text{amp}} = 1$. Due to the periodic boundary conditions, when r reaches $L/2$ the correlation function increases again. *On the right* we fix the size to $L = 192$ and show that the correlation length decreases with H_{amp}. The *top curve*, in *red*, is $H_{\text{amp}} = 0.01$, immediately under, in *green*, we have $H_{\text{amp}} = 0.05$, and so on with $H_{\text{amp}} = 0.1, 0.5, 1, 5, 10, 50$

This correlation function favors the softest modes by a factor $1/\lambda_n^2$. This is an advantage, because the bulk modes do not exhibit a finite correlation length, so it is useful to have them suppressed.

We calculated the correlations by inverting $\mathcal{M}$ with a conjugate gradient. A nice exponential decay is visible (Fig. 6.6) to which we can associate a finite correlation length that grows as H_{amp} decreases.

6.5 Anharmonicity

The Hessian matrix $\mathcal{M}$ is a harmonic approximation of the bottom of the valleys that carries plenty of information. Still, we can go beyond and take in account the effects due to the anharmonicity of the potential, and the relationship between different ISs.

The jamming point is characterized by diverging anharmonic effects (the softest modes have the smallest barriers) [Xu10]. We are not able to define an equivalent of the jamming point, but we can seek for a dependency on H_{amp} of the anharmonic effects, and see for example if they diverge in null field.

6.5.1 *Forcings*

Perturbing the Hamiltonian We study the reaction of the system to an additional force along a direction $|\vec{\pi}\rangle$ (identified with the $2N$-dimensional vector $|\tilde{\pi}\rangle$). We are interested in the softest mode, that is localized, and we want to compare it with the behavior of the eigenvectors in bulk of the $\rho(\lambda)$, that are delocalized. Therefore we choose $|\tilde{\pi}\rangle = |\tilde{\pi}_0\rangle$ (softest mode) and $|\tilde{\pi}\rangle = |\tilde{\pi}_{\text{RAND}}\rangle$, a vector whose components are chosen at random, with the condition $\langle\tilde{\pi}_{\text{RAND}}|\tilde{\pi}_{\text{RAND}}\rangle = 1$. The vector $|\tilde{\pi}_{\text{RAND}}\rangle$ is not an eigenvector of $\mathcal{M}$, but it is generally a combination of all the eigenvectors of the system. Since the bulk eigenvectors overwhelm the soft modes by number $|\tilde{\pi}_{\text{RAND}}\rangle$ will be representative of the bulk behavior. The reason why we use $|\tilde{\pi}_{\text{RAND}}\rangle$ instead of an actual bulk eigenvector is that with the Arnoldi algorithm we were able to compute only the lowest eigenvectors, so for the large lattices it was practically impossible to go beyond the localization threshold (recall Fig. 6.5).

With the application of a forcing along $|\vec{\pi}\rangle$, the Hamiltonian is modified in

$$\mathcal{H}_{\text{F}} = -\sum_{\|x-y\|=1} J_{xy}\vec{s}_x\cdot\vec{s}_y - \sum_x^N \left(\vec{h}_x + A_{\text{F}}\vec{\pi}_x\right)\cdot\vec{s}_x, \tag{6.14}$$

where A_{F} is the amplitude of the forcing along $|\vec{\pi}\rangle$, that will be tuned appropriately.

We stimulate the system with forcings of increasing amplitude, and study when this kicks the system out of the original inherent structure. To this scope $A_{\text{F}} = A_{\text{F}}(i_h)$, where $i_h \in \mathbb{N}$ tunes the forcing.[6]

The procedure is conceptually simple. Being N_{F} the number of forcings one wants to impose, for i_h in $\{1,\ldots,N_{\text{F}}\}$

1. Start from the IS $\left|\vec{s}^{(\text{IS})}\right\rangle$ of the unperturbed Hamiltonian $\mathcal{H}_{\text{RF}} \equiv \mathcal{H}_{\text{F}}(i_h = 0)$.
2. From $\left|\vec{s}^{(\text{IS})}\right\rangle$ minimize the energy using $\mathcal{H}_{\text{F}}(i_h)$, and find a new IS for the perturbed system, $|\text{IS}(i_h)\rangle$.
3. From $|\text{IS}(i_h)\rangle$ minimize the energy again, using $\mathcal{H}_{\text{F}}(0) = \mathcal{H}_{\text{RF}}$, and find the IS $|\text{IS}^*\rangle$ (with elements $\vec{s}_x^{(\text{IS})*}$).
4. If $|\text{IS}^*\rangle = \left|\vec{s}^{(\text{IS})}\right\rangle$, the second minimization lead the system back to its original configuration, so the forcing was too weak to break through an energy barrier. On the contrary, if $|\text{IS}^*\rangle \neq \left|\vec{s}^{(\text{IS})}\right\rangle$ the forcing was large enough for a hop to another valley.

Since this is an anharmonicity test, the same procedure for negative i_h yields different results, therefore in our simulations $i_h \in \{-N_{\text{F}},\ldots,0,\ldots,N_{\text{F}}\}$.

To ensure well-defined forcings along $|\vec{\pi}_{\text{RAND}}\rangle$, we normalized A_{F} with $\|\,|\vec{\pi}\rangle\|_1$, since $\left|\sum_x \vec{\pi}_x\cdot\vec{s}_x\right| \le \left|\sum_x \vec{\pi}_x\right| \le \sum_x |\vec{\pi}_x| = \|\,|\vec{\pi}\rangle\|_1$. Because $\|\,|\vec{\pi}\rangle\|_1$ scales nonlinearly with N, we multiplied back by a factor N, obtaining and extensive correction to the energy. For the softest mode we analyzed the effect of intensive forcings of order $O(1)$ because larger forcings lead the system out of the linear response regime. The amplitudes we used can be resumed as

$$A_{\text{F}}(i_h) = \frac{NAi_h}{\|\,|\pi\rangle\|_1} \quad \text{for} \quad |\pi_{\text{RAND}}\rangle\,, \tag{6.15}$$

$$A_{\text{F}}(i_h) = \frac{Ai_h}{\|\,|\pi\rangle\|_1} \quad \text{for} \quad |\pi_0\rangle\,. \tag{6.16}$$

[6] $\mathbb{N}$ is the set of the natural numbers.

The amplitudes A are an external parameter (of order 1), that we tried to tune in order to be in the linear response regime for small i_h, and out of it for i_h approaching N_F. The dependency of the optimal A on L and H_{amp} was highly nonlinear. We list our choices in Table 6.1.

Probing the Linear Regime To make sure that our forcings are not too strong, we monitor the direct reaction of the system to the forcing. We define a "polarized magnetization" $\hat{m} = \langle \mathrm{IS}(i_h) | \vec{\pi} \rangle = \sum_x \vec{s}_x \cdot \pi_x$, that indicates how much the forcing pushed the alignment of the spins along the pion. The amplitude of the forcing is tuned well if $\hat{m}(i_h)$ is close to the linear regime. In Table 6.1 we show the amplitudes A we used in order to be in the linear regime. Figure 6.7 confirms that this was the working condition for the forcings along $|\vec{\pi}_0\rangle$. Figure 6.8 is analogous, but along $|\vec{\pi}_{\mathrm{RAND}}\rangle$. In the latter figure we rescale $\hat{m}$ by a factor $1/\sqrt{N}$ to obtain a collapse. In fact the normalization $\langle \vec{\pi}_{\mathrm{RAND}} | \vec{\pi}_{\mathrm{RAND}} \rangle = 1$ implies that the components of $|\vec{\pi}_{\mathrm{RAND}}\rangle$ are of order $1/\sqrt{N}$, so the polarized magnetization is bounded by $|\hat{m}| = |\langle IS(i_h) | \vec{\pi}_{\mathrm{RAND}} \rangle| \leq \sum_x |\vec{\pi}_x| \sim \sqrt{N}$.

The reader will notice that to be in the regime of quasi-linear response, forcings along $|\pi_{\mathrm{RAND}}\rangle$ can be extensive, whereas the localized forcings along $|\pi_0\rangle$ need to be of order 1.

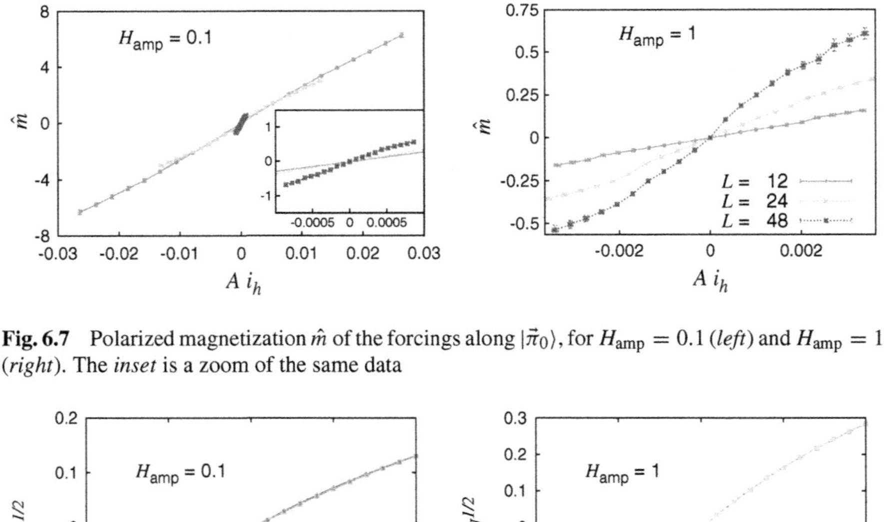

Fig. 6.7 Polarized magnetization $\hat{m}$ of the forcings along $|\vec{\pi}_0\rangle$, for $H_{amp} = 0.1$ (*left*) and $H_{amp} = 1$ (*right*). The *inset* is a zoom of the same data

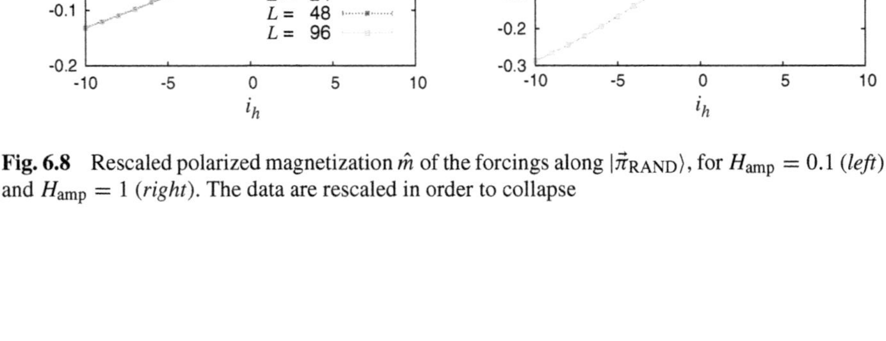

Fig. 6.8 Rescaled polarized magnetization $\hat{m}$ of the forcings along $|\vec{\pi}_{\mathrm{RAND}}\rangle$, for $H_{amp} = 0.1$ (*left*) and $H_{amp} = 1$ (*right*). The data are rescaled in order to collapse

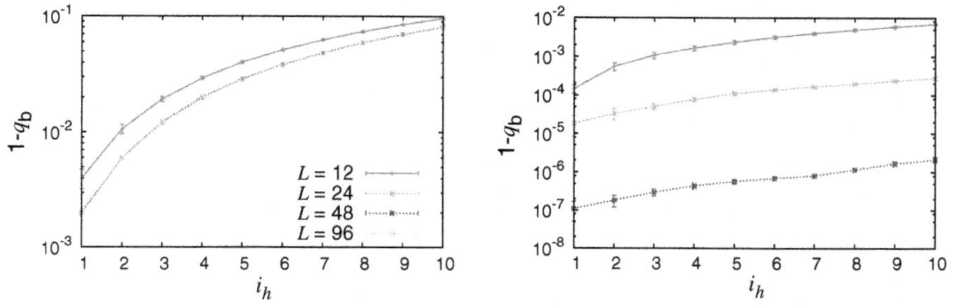

Fig. 6.9 Measurement of how different the configuration is from the initial IS is after the forcing along $|\vec{\pi}_{\mathrm{RAND}}\rangle$ *(left)*, and along $|\vec{\pi}_0\rangle$ *(right)*, for $H_{\mathrm{amp}} = 0.1$. We plot $1 - q_{\mathrm{b}}$ to make the figure clearer. Finite-size effects are neglectable for forcings along $|\vec{\pi}_{\mathrm{RAND}}\rangle$ and sizable along $|\vec{\pi}_0\rangle$

The Perturbed Configuration After the first minimization (with the perturbed Hamiltonian) but before the second, we measure the overlap q_{b} between $\left|\vec{s}^{(\mathrm{IS})}\right\rangle$ and $|\mathrm{IS}(i_h)\rangle$, $q_{\mathrm{b}} = \langle\vec{s}^{(\mathrm{IS})}\,|\mathrm{IS}(i_h)\rangle\,/N$ (Fig. 6.9), and the energy difference ΔE, in terms of $\mathcal{H}_{\mathrm{RF}}$, between $|\mathrm{IS}(i_h)\rangle$ and $\left|\vec{s}^{(\mathrm{IS})}\right\rangle$, $\Delta E = E_{\mathrm{RF}}\big(|\mathrm{IS}(i_h)\rangle\big) - E_{\mathrm{RF}}\big(\left|\vec{s}^{(\mathrm{IS})}\right\rangle\big)$. The maximum value of ΔE before a hop to another valley should give an estimate of height of the barrier. Still, it may happen that the minimum of the energy with Hamiltonian (6.14) have an energy lower than $E_{\mathrm{RF}}\big(\left|\vec{s}^{(\mathrm{IS})}\right\rangle\big)$, so in a strict sense ΔE is not positive definite. To overcome this issue, we resort to the energy difference ΔE^*, in terms of $\mathcal{H}_{\mathrm{RF}}$, between $|\mathrm{IS}(i_h)\rangle$ and $|(\mathrm{IS}^*\rangle$, $\Delta E = E_{\mathrm{RF}}\big(|\mathrm{IS}(i_h)\rangle\big) - E_{\mathrm{RF}}\big(\left|\vec{s}^{(\mathrm{IS})}\right\rangle\big)$, that measures the barrier from the arriving IS instead of the starting one. It has the advantage of being positive definite, and even though the eigenvector $|\vec{\pi}_0\rangle$ of the forcing is not associated to that IS, we will see that the two ISs are so similar that it is a reasonable descriptor.

Ending in a New Valley For each $A_{\mathrm{F}}(i_h)$ we measure the overlap q_{if} between the two minimas of $\mathcal{H}_{\mathrm{RF}}$, the initial IS, $\left|\vec{s}^{(\mathrm{IS})}\right\rangle$, and the final one, $|\mathrm{IS}^*\rangle$. Naïvely, checking that $q_{\mathrm{if}} < 1$ in principle is a good criterion to establish whether the system escaped to another valley. We proceeded similarly, in terms of the spin variations between initial and final configuration, through the quantities

$$w_x = 1 - \left(\vec{s}_x^{\,(\mathrm{IS})} \cdot \vec{s}_x^{\,(\mathrm{IS})\,*}\right), \tag{6.17}$$

$$W = \sum_x^N w_x = N - \langle\vec{s}^{\,(\mathrm{IS})}\,\big|\mathrm{IS}^*\big\rangle = N(1 - q_{\mathrm{if}}), \tag{6.18}$$

$$\mathcal{W} = \frac{\sum_x^N w_x^2}{\left(\sum_x^N w_x\right)^2}. \tag{6.19}$$

The local variation w_x measures the change between the beginning and the end of the process. If the spin stayed the same then $w_x = 0$, while if it became uncorrelated with the initial position $w_x = 1$ in average. If one and only one spin becomes uncorrelated with its initial configuration, the variation of W is $\Delta W = 1/N$. Similar variations ΔW do not mean that one spin has decorrelated and the others have stayed the same, this is impossible because $|\vec{s}^{(\mathrm{IS})}\rangle$ and $|\mathrm{IS}^*\rangle$ are ISs and collective rearrangements are needed. A $\Delta W = 1/N$ means instead that the overall change is equivalent to a single spin becoming independent of its initial state. This is, for a rearrangement, the minimal change in the W that we can define. Since the spins in our model are continuous variables, we impose $\Delta W = 1/N$ as a threshold to state whether there was or not a change of valley.

The cumulant $\mathcal{W}$ is an indicator of the type of rearrangement that took place. If the rearrangement is completely localized (only one spin changes), $\mathcal{W} = 1$, whereas if it is maximally delocalized (all the spins have the same variation), then $\mathcal{W} = 1/N$.

Falling Back in the Same Valley Even though the forcing is along a definite direction, since the energy landscape is very irregular, it may happen that stronger forcings lead the system to the originary valley. For example it may happen that $i_h = 2$ lead the system to a new valley, and $i_h = 3$ lead it once again to the same valley of $i_h = 1$. To exclude these extra apparent valleys we label each visited valley with its W, and assume that two valleys with the same label are the same valley. These events are not probable, and even less likely it is that this happen with two different but equally-labelled valleys, so we neglect the bias due to this unlucky possibility.

6.5.2 Rearrangements

To delineate the effect of the forcings, we want to study, for every couple (H_{amp}, L), the probability that a forcing of amplitude A_{F} lead the system to a new valley, to distinguish the behavior of soft from bulk modes.

Furthermore, once the system made its first jump to a new valley, it is not excluded that a bigger forcing lead it to a third minimum of the energy. One can ask himself what is the probability $P_L^H(A_{\mathrm{F}}, n)$ that n new valleys are reached by forcing the system with an amplitude up to $A_{\mathrm{F}}(i_h)$, and to try to evince a dependency on system L and random field amplitude H_{amp}. Even though n is bounded by i_h, this does not necessarily mean that if we made smaller and more numerous forcings n could not be larger. On another side, if for a certain parameter choice rearrangements are measured only for large i_h, it is reasonable to think that these represent the smallest possible forcings to fall off the IS.

To construct $P_L^H(A_{\mathrm{F}}, n)$, for every replica and sample we start from $i_h = 0$ and increase $|i_h|$ either in the positive or negative direction (the two are accounted for independently). The value we assign to $P_L^H(A_{\mathrm{F}}, n)$ is the number of systems that had n rearrangements after i_h steps, divided by the total number of forcings, that is $2N_{\mathrm{replicas}}N_{\mathrm{samples}}$.

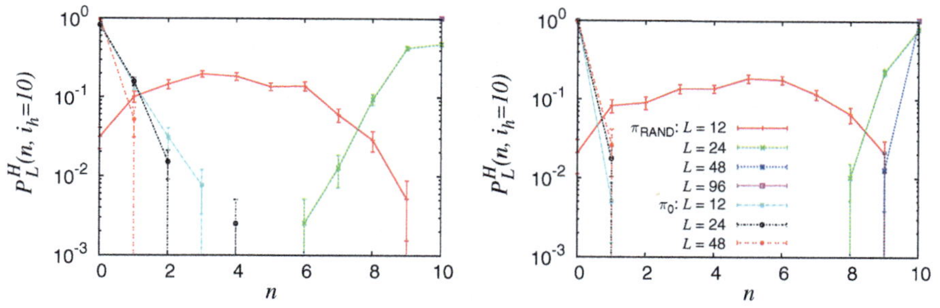

Fig. 6.10 Probability of there being exactly n changes of valley after $i_h = N_F = 10$ forcing steps. The data come from $H_{amp} = 0.1$ (*left*) and $H_{amp} = 1$ (*right*). If $P_L^H(A_F, n) = 1$ for $n = 0$ it means that the forcings were not strong enough to ever get out of the initial IS. On the contrary, $P_L^H(A_F, n) = 1$ for $n = 10$ means that every single step lead the system to a new IS. The latter scenario is realized in the case of forcings along $|\vec{\pi}_{RAND}\rangle$, especially when the system size is large. On the other side, forcings along $|\vec{\pi}_0\rangle$ display a small but finite amount of rearrangements

First rearrangement In Fig. 6.10 we show the probability of measuring exactly n rearrangements after $i_h = N_F = 10$ forcing steps.[7] Even though both for $|\vec{\pi}_{RAND}\rangle$ and $|\vec{\pi}_0\rangle$ we are in the linear response regime, the behavior is very different between the two types of forcing. In the first case every single forcing step we impose leads the system to a new valley. In the second rearrangements are so uncommon that even though the probability of having exactly one rearrangement is finite, that of having more than one becomes negligible for large samples. It is then reasonable to think that any rearrangement we measure for $|\vec{\pi}_0\rangle$, it occurs for the smallest possible forcing, and even when more than one occurs, these jumps are between *neighboring valleys*, where by neighboring we mean that no smaller forcing would lead the system to a different IS. To convince ourselves of this we can give a look at the average number of rearrangements after i_h forcing steps, $n(i_h)$ (Fig. 6.11).[8]

When i_h is small no new ISs are visited and $\langle n\rangle = 0$, while for larger i_h, $\langle n\rangle$ is positive but small, so we can call these changes of valley "first rearrangements", i.e. rearrangement between *neighboring valleys*.

6.5.3 Two-Level Systems

In the spectrum of $\mathcal{M}$, $\rho(\lambda)$, we remarked an extensive number of very soft modes, with a localized eigenstate (Sect. 6.4). The eigenstates can connect different ISs through the forcing procedure described in this section. The connection caused by

---

[7] We do not show data regarding forcings for $H_{amp} = 10, 50$, because no arrangement takes place. Most likely the energy landscape is too trivial.

[8] Because $P_L^H(A_F, n)$ is not defined over all the samples (it is hard to reach many different valleys and it may not happen in all the simulations), the errors on $P_L^H(A_F, n)$ were calculated by resampling over the reduced data sets with the bootstrap method.

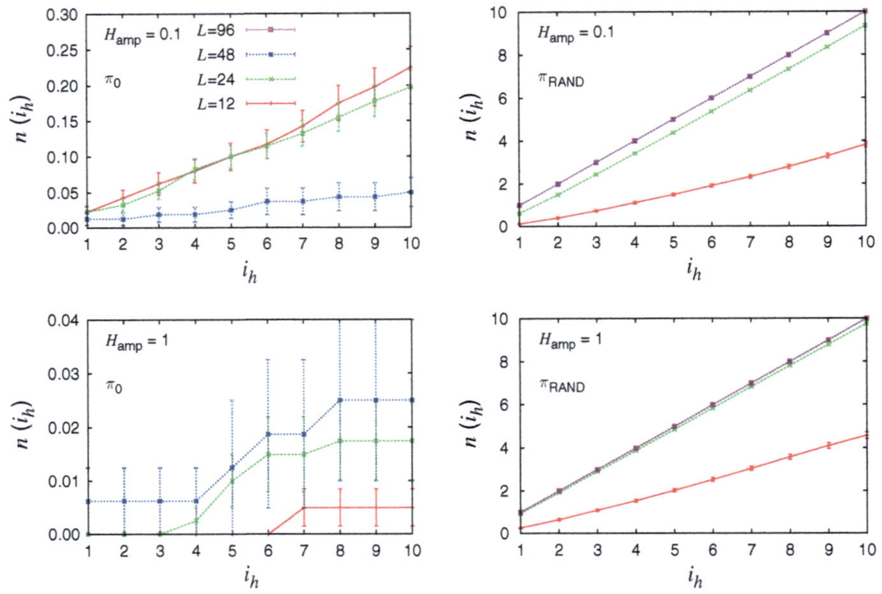

Fig. 6.11 Average number of rearrangements $n(i_h)$ for forcings along $|\vec{\pi}_0\rangle$ (*left*) and along $|\vec{\pi}_{\rm RAND}\rangle$ (*right*). The data come from $H_{\rm amp} = 0.1$ (*top*) and $H_{\rm amp} = 1$ (*bottom*). When the lattice becomes large enough, the forcings along $|\vec{\pi}_{\rm RAND}\rangle$ lead to a new IS every time i_h is increased. The data from the $|\vec{\pi}_0\rangle$ and $H_{\rm amp} = 1$ can be said to be in the regime of first rearrangement

such states is privileged, because the couples of ISs are innaturally near to each other. In Fig. 6.12 we show the mean overlap between initial and final IS, $q_{\rm if} = \langle \vec{s}^{\,(\rm IS)} \,|IS^*\rangle /N$.

As expectable, the rearrangements are localized when we stimulate the system along the softest mode, and delocalized when it is along a random direction (Fig. 6.12, inset). The overlaps $q_{\rm if}$ are much closer to 1 than the overlaps of independent ISs shown in Fig. 6.1, meaning that the ISs are somewhat clustered in tiny groups that are represented by a single IS. This could be an operational definition of classical two-level system, i.e. a system in which there are two very close states connected by a soft mode, where the transitions from one state to the other can be treated as independent of the rest of the system [And72, Phi72, Phi87, Lis15].

To reinforce the idea of two-level system, we see that while the energy barriers from random forcings increase with the system size (the growth is $O(N)$), while those within the two-level system (along the softest mode) do not (Fig. 6.13).

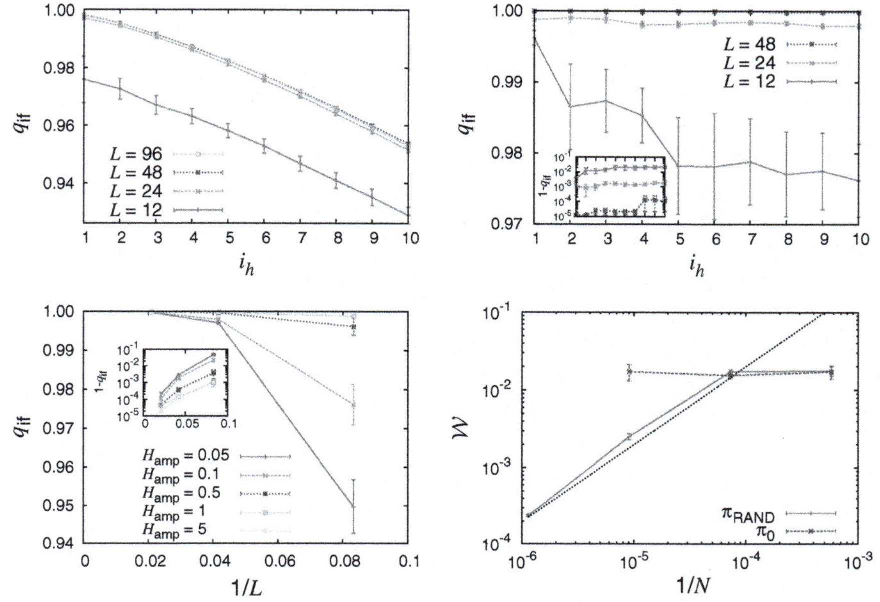

Fig. 6.12 The *two top plots* show the overlap $q_{if} = \langle \vec{s}^{\,(IS)} | IS^* \rangle / N$ between the starting and the final IS, for $H_{amp} = 0.1$. *Top left* q_{if} in forcings along $|\vec{\pi}_{RAND}\rangle$. *Top right* q_{if} in forcings along $|\vec{\pi}_0\rangle$. The overlaps q_{if} are consistently larger than the typical overlap between two inherent structures (Fig. 6.1, *left*), peaked around $P(q_{IS}) \simeq 0.4$. The *bottom left* set shows data for forcings along $|\vec{\pi}_0\rangle$ for all the interesting H_{amp}. A point is missing, for $H_{amp} = 5$, $L = 12$, because we only registered a single rearrangement for this data set. Both *insets* display $1 - q_{if}$ from the same data of the corresponding larger plot, to stress that the overlaps q_{if} never reach 1 (this is redundant, because $q_{if} < 1$ by definition, since it is the overlap between two different configurations). The *bottom right figure* depicts the type of rearrangement that takes place between the initial and final IS. The cumulant $\mathcal{W}$ is defined in (6.19); $\mathcal{W} = 1$ indicates a completely localized rearrangement, where only a single spin changed position, while $\mathcal{W} = 1/N$ indicates a fully delocalized change of the spins. It is visible that a random forcing leads to a completely delocalized rearrangement (*the dotted line* is $\sim 1/N$), whereas a localized forcing implies a localized rearrangement with no appreciable dependency on the system size

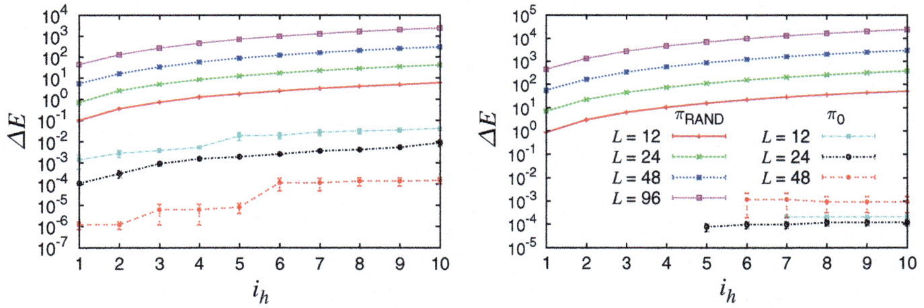

Fig. 6.13 Average energy barrier ΔE^* for forcings along $|\vec{\pi}_{RAND}\rangle$ and $|\vec{\pi}_0\rangle$, for random fields of amplitude $H_{amp} = 0.1$ (*left*) and $H_{amp} = 1$ (*right*). In the right set, some i_h are not represented because for weak forcings along $|\vec{\pi}_0\rangle$ there were no rearrangements

6.6 Overview

The introduction of a random field, besides extinguishing the rotational symmetry, changes the response of the Heisenberg spin glass to soft excitations. In the absence of field the density of states is expected to go as $g(\omega) \propto \omega^2$ [Gri11, Fra15b]. Very strong random fields suppress the soft modes, and a gap appears in the density of states $g(\omega)$. Still, soft modes do resist the application of a random field when it is not too large. The data are compatible with the absence of a gap, where for small ω the density of states grows as $g(\omega) \propto \omega^4$.

It appears that a finite fraction of the modes is localized, suggesting a localization transition when the system size becomes large.

Besides the density of states, that consists in a harmonic approximation of the metastable states, we make an anharmonic analysis by imposing an external force on the system. The reaction of the spin glass has a strong dependency on the direction of application of the force. Extensive corrections to the Hamiltonian are needed to be able to move the spins in the direction of a forcing along a random direction, while order 1 forcings are enough to obtain the same result pushing along the softest mode, suggesting that it is the softest mode that drives the change.

Even though the response appears in both cases concentrated along the softest modes, seldom the softest mode leads the system to a new inherent structure, whereas a delocalized forcing drives it to explore many new valleys of the energy landscape. Forcings of order one along the softest mode are the smallest we can impose in order to have a jump toward another inherent structure. The rearrangement in the change of inherent structure is localized, and the energy barrier does not grow with the system size.

The most attractive feature of the valleys reached with a forcing along the softest mode is that their overlap with the initial inherent structure is very high, much higher than the typical overlap expected for independent inherent structures. This means that there are couples of metastable states with a fundamental relation between them, connected by a soft mode, with a small energy barrier. This could be used as an operational definition of classical two-level system.

References

[Alo01] J.L. Alonso, L.A. Fernández, F. Guinea, V. Laliena, V. Martín-Mayor, Phys. Rev. B **63**, 054411 (Jan 2001). doi:10.1103/PhysRevB.63.054411. http://link.aps.org/doi/10.1103/PhysRevB.63.054411

[And72] P.W. Anderson, B.I. Halperin, C.M. Varma, Phil. Mag. **25**, 1–9 (1972). doi:10.1080/14786437208229210. http://dx.doi.org/10.1080/14786437208229210

[BJ11] M. Baity-Jesi: *Energy landscape in three-dimensional Heisenberg spin glasses*. Master's thesis, Sapienza, Universitá di Roma, Rome, Italy (January 2011). arXiv:1503.08409

[BJ16] M. Baity-Jesi, V. Martín-Mayor, G. Parisi, S. Pérez-Gaviro (2016). In preparation

[Buc84] U. Buchenau, N. Nücker, A.J. Dianoux, Phys. Rev. Lett. **53**, 2316–2319 (Dec 1984). doi:10.1103/PhysRevLett.53.2316. http://link.aps.org/doi/10.1103/PhysRevLett.53.2316

[Cam13] C. Cammarota, G. Biroli, J. Chem. Phys. **138**, 12A547 (2013). doi:10.1063/1.4790400. http://scitation.aip.org/content/aip/journal/jcp/138/12/10.1063/1.4790400

[Cav09] A. Cavagna, Phys. Rep. **476**, 51–124 (2009). arXiv:0903.4264

[Cha15] P. Charbonneau, E.I. Corwin, G. Parisi, F. Zamponi, Phys. Rev. Lett. **114**, 125504 (Mar 2015). doi:10.1103/PhysRevLett.114.125504. http://link.aps.org/doi/10.1103/PhysRevLett.114.125504

[Cha14] P. Charbonneau, J. Kurchan, G. Parisi, P. Urbani, F. Zamponi, Nat. Commun. **5**, 3725 (2014). doi:10.1038/ncomms4725. arXiv:1404.6809

[Chi78] T. Chihara, *An Introduction to Orthogonal Polynomials* (Gordon & Breach, New York, 1978)

[DeG14] E. DeGiuli, E. Lerner, C. Brito, M. Wyart, Proc. Natl. Acad. Sci. **111**, 17054–17059 (2014). doi:10.1073/pnas.1415298111. http://www.pnas.org/content/111/48/17054.full.pdf, http://www.pnas.org/content/111/48/17054.abstract

[Fra15a] S. Franz, G. Parisi, P. Urbani, F. Zamponi, The simplest model of jamming (2015). arXiv:1501.03397

[Fra15b] S. Franz, G. Parisi, P. Urbani, F. Zamponi, Universal spectrum of normal modes in low-temperature glasses: an exact solution (2015). Submitted to PNAS, arXiv:1506.01997

[GM60] M. Gell-Mann, M. Lévy, Il Nuovo Cimento **16**, 705–726 (1960). ISSN 0029-6341. doi:10.1007/BF02859738. http://dx.doi.org/10.1007/BF02859738

[Gri11] T.S. Grigera, V. Martin-Mayor, G. Parisi, P. Urbani, P. Verrocchio, J. Stat. Mech.: Theory Experim. **2011**, P02015 (2011). doi:10.1088/1742-5468/2011/02/P02015. http://stacks.iop.org/1742-5468/2011/i=02/a=P02015

[Gri03] T.S. Grigera, V. Martin-Mayor, G. Parisi, P. Verrocchio, Nature **422**, 289 (2003). doi:10.1038/nature01475

[Hua87] K. Huang, *Statistical Mechanics*, 2nd edn. (John Wiley and Sons, Hoboken, NJ, 1987). ISBN 0471815187

[Key94] T. Keyes, J. Chem. Phys. **101**, 5081–5092 (1994). doi:10.1063/1.468407. http://scitation.aip.org/content/aip/journal/jcp/101/6/10.1063/1.468407

[Kob12] W. Kob, S. Roldán-Vargas, L. Berthier, Nat. Phys. **8**, 164 (2012). doi:10.1038/nphys2133

[Lis15] J. Lisenfeld, G. Grabovskij, C. Müller, J. Cole, G. Weiss, A. Ustinov, Nat. Comm. **6**, 6182 (2015). doi:10.1038/ncomms7182

[Lup15] C. Lupo, G. Parisi, F. Ricci-Tersenghi (2015)

[Mal91] V. Malinovsky, V. Novikov, A. Sokolov, Phys. Lett. A **153**, 63– 66 (1991). ISSN 0375-9601. doi:10.1016/0375-9601(91)90363-D. http://www.sciencedirect.com/science/article/pii/037596019190363D

[Méz84] M. Mézard, G. Parisi, N. Sourlas, G. Toulouse, M. Virasoro, Phys. Rev. Lett. **52**, 1156 (1984). doi:10.1103/PhysRevLett.52.1156

[Méz87] M. Mézard, G. Parisi, M. Virasoro, *Spin-Glass Theory and Beyond* (World Scientific, Singapore, 1987)

[Mon09a] G. Monaco, V.M. Giordano, Proc. Nat. Ac. Sci. **106**, 3659–3663 (2009). doi:10.1073/pnas.0808965106. http://www.pnas.org/content/106/10/3659.full.pdf, http://www.pnas.org/content/106/10/3659.abstract

[Mon09b] G. Monaco, S. Mossa, Proc. Natl. Acad. Sci. **106**, 16907–16912 (2009). doi:10.1073/pnas.0903922106. http://www.pnas.org/content/106/40/16907.full.pdf, http://www.pnas.org/content/106/40/16907.abstract

[O'H03] C.S. O'Hern, L.E. Silbert, A.J. Liu, S.R. Nagel, Phys. Rev. E **68**, 011306 (Jul 2003). doi:10.1103/PhysRevE.68.011306. http://link.aps.org/doi/10.1103/PhysRevE.68.011306

[Phi81] W.A. Phillips, *Amorphous Solids: Low-Temperature Properties*. Topics in Current Physics 24 (Springer-Verlag, Berlin Heidelberg, 1981). doi:10.1007/978-3-642-81534-8

[Phi87] W.A. Phillips, Rep. Prog. Phys. **50**, 1657 (1987). doi:10.1088/0034-4885/50/12/003. http://stacks.iop.org/0034-4885/50/i=12/a=003

[Phi72] W. Phillips, J. Low Temp. Phys. **7**, 351–360 (1972). ISSN 0022-2291. doi:10.1007/BF00660072. http://dx.doi.org/10.1007/BF00660072

[Set98] F. Sette, M. Krisch, C. Masciovecchio, G. Ruocco, G. Monaco, Science **280**, 1550–1555 (1998). doi:10.1126/science.280.5369.1550

[Sor08] D. Sorensen, R. Lehoucq, C. Yang, K. Maschhoff (1996–2008). Arpack, ARnoldi PACKage. www.caam.rice.edu/software/ARPACK/

[Sti95] F.H. Stillinger, Science **267**, 1935–1939 (1995). doi:10.1126/science.267.5206.1935. http://www.sciencemag.org/content/267/5206/1935.full.pdf, http://www.sciencemag.org/content/267/5206/1935.abstract

[Tur82] P. Turchi, F. Ducastelle, G. Treglia, J. Phys. C: Solid State Phys. **15**, 2891 (1982). doi:10.1088/0022-3719/15/13/017. http://stacks.iop.org/0022-3719/15/i=13/a=017

[Wan94] Y. Wan, R.M. Stratt, J. Chem. Phys. **100**, 5123–5138 (1994). doi:10.1063/1.467178. http://scitation.aip.org/content/aip/journal/jcp/100/7/10.1063/1.467178

[Wu92] T.-M. Wu, R.F. Loring, J. Chem. Phys. **97**, 8568–8575 (1992). doi:10.1063/1.463375. http://scitation.aip.org/content/aip/journal/jcp/97/11/10.1063/1.463375

[Wya12] M. Wyart, Phys. Rev. Lett. **109**, 125502 (Sep 2012). doi:10.1103/PhysRevLett.109.125502. http://link.aps.org/doi/10.1103/PhysRevLett.109.125502

[Xu10] N. Xu, V. Vitelli, A.J. Liu, S.R. Nagel, Europhys. Lett. **90**, 56001 (2010). doi:10.1209/0295-5075/90/56001. http://stacks.iop.org/0295-5075/90/i=5/a=56001

Part IV
Conclusions and Outlook

Chapter 7
Conclusions

7.1 General Considerations

It is almost one century that scientists from several domains, going from physics, to chemistry, engineering, mathematics and computer science, gathered to understand the nature of the glass transition. In 1995, Anderson stated: "The deepest and most interesting unsolved problem in solid state theory is probably the theory of the nature of glass and the glass transition" [And95]. Twenty years later, in 2015, despite great steps forward, the main answers on the glass transition are still unanswered.

It would be pretentious to think to make a revolutionary advance in a single Ph.D. thesis, as the scientific advance is usually the sum of a very large series of small contributions. It is like removing all the corns from a huge cob. Every single corn is important, even though from the point of view of the full cob it might seem extremely small.

Removing a corn consists in advancing under any known point of view, from conceiving new theories to developing new methodologies and instruments, to finding some new non-trivial behavior. It is up to the researcher to decide which perspective is more suited to his profile and the problem he tackles, but he should always keep in mind the multidisciplinarity of the problem, and possibly include it in his approach.

In this thesis we dealt with the glassy phase under several points of view, focusing on spin glasses. Our approach was mainly numerical, with a strong imprint due to a theoretical physics background. We worked on simplified systems that carry only few essential features, enough to yield the phenomenology we wish to understand. On one hand we studied the critical behavior of canonical spin glasses, trying to understand how the spin glass phase and transition change under perturbations, focusing on concepts like universality and critical dimensions. On the other hand we tried to get a better view on the energy landscape, a feature with a diverging number of degrees of freedom that is usually described only through a single number, the energy.

M. Baity Jesi, *Spin Glasses*, Springer Theses, DOI 10.1007/978-3-319-41231-3_7

We contributed with a finite amount of small corns, to what one day will hopefully be the full unraveling of the whole cob.

In the following sections we outline shortly the results we achieved in this thesis. More extended conclusions are given at the end of each chapter.

7.2 State of the Art Computing

In this thesis we showed the usefulness of special hardware to achieve meaningful results. The resources we used were never used before for the problems we attacked, so our work is a proof of concept for these approaches.

The data in Chap. 2 were obtained with the dedicated computer JANUS, an FPGA-based computer designed specifically for Ising spin glass simulations. With this machine it was possible to thermalize on unprecedentedly large lattices, at uniquely low temperatures. Even though this machine has been operating since 2008, before the beginning of my research career, each of the results achieved with JANUS represents a proof of the suitability of a dedicated FPGA-based computer.

In Chap. 3 we simulated on Heisenberg spins, for which JANUS is not optimized. We resorted then to GPUs. At the moment of the publication of [BJ14d], despite their popularity, no physical result had been obtained through GPU simulations on a Heisenberg SG (and to our knowledge neither on Ising spins). Even at present date, we are only aware of benchmarks [Yav12, Ber14] performed on Heisenberg spin glasses with GPU. Besides our work, only on Ising systems GPU have been used to obtain new insight on spin glasses [Man15a, Man15b, Lul15].

Moreover, our work can provide even further guiding because we used a large GPU cluster and ran simulations with tens of GPU in parallel.

7.3 The Ising Spin Glass in a Magnetic Field

We studied the three-dimensional Ising Edwards-Anderson spin glass in an external uniform magnetic field. We showed that the finite-size fluctuations are so marked, that searching signs of criticality becomes highly challenging. Taking the averages of the observables hides the behavior of the majority of the measurements, so we needed to develop more sophisticated statistical analysis tools. We classified our measurements through a *conditioning variate*, a function of the observables that helps to distinguish different types of behavior, and proposed a new finite-size scaling ansatz based on the quantiles of the conditioning variate's distribution. In some cases the model appeared critical, and in others it did not. We were not able to extrapolate which of the two dominates in the thermodynamic limit, but we could identify the region where the would-be phase transition should be searched.

7.4 Heisenberg Spin Glass with a Strong Random Exchange Anisotropy

We made equilibrium simulations on the three-dimensional Heisenberg model with the addition of a random exchange anisotropy. We found both the chiral and the spin glass phase transitions. Through a careful finite-size scaling analysis we conclude that the two critical channels are coupled, so the phase transition is unique. The exponents that we calculate are compatible with those of the Ising Edwards-Anderson spin glass, so in the RG sense the exchange anisotropy is a relevant perturbation on the Heisenberg Hamiltonian.

7.5 Energy Landscape of m-Component Spin Glasses

We studied vector spin glasses in three dimensions, focusing on the role of the number of spin components m. We performed an extensive study of the energy landscape and of the zero-temperature dynamics from an excited state. An increase of m implies a decrease of the amount of minima of the free energy, down to the trivial presence of a unique minimum. For little m correlations are small and the dynamics are quickly arrested, while for larger m low-temperature correlations crop up and the convergence is slower, to a limit that appears to be related with the system size.

7.6 Zero-Temperature Dynamics

We analyzed the hysteresis properties of the SK model at zero temperature. The states along the hysteresis loop are marginal, meaning that the density of stability goes to zero as $\rho(\lambda) \propto \lambda^{\theta}$, and exhibit SOC. We analyzed the stability of these configurations, and found that previous scaling arguments on the averages, granting $\theta = 1$, were not exact due to the presence of correlations $C(\lambda)$ between soft spins. This correlation diverges as $1/\lambda$, and implies that they are mutually frustrated. The value $\theta = 1$ was still recovered by analyzing the fluctuations of the mean values.

Also, we stated through scaling arguments that SOC requires that each site have an infinite number of neighbors, so the SK model is critical, and the EA model is not. By mixing both short- and long-range interactions in a duplex network, we argued, giving predictions that we confirmed numerically, that the long-range couplings are a relevant perturbation to the short-range Hamiltonian. That is, as long as there are long-range interactions, a sufficiently large system will display crackling over the whole system.

Finally, we studied the dynamics of the avalanches. We found that the type of spin update influences the power laws of the crackling, but maintains the rest of the features. Furthermore, an avalanche can be represented as a random walk in the number of unstable spins, and this random walk has a bias that depends on how unstable the system is in each moment. Lastly, we described the avalanche dynamics through a random walk in the space of the local stabilities. We found that the correlations between soft spins arise spontaneously during the avalanche, and we saw that the same exponents found statically arise also dynamically.

7.7 Soft Modes and Localization in Spin Glasses

We examined the soft plastic modes in the three-dimensional Heisenberg spin glass under a random field that broke the rotational symmetry, in order to suppress the modes due to symmetries. We studied small fluctuations around the inherent structures, both at at the harmonic and anharmonic levels.

This analysis was motivated by the phenomenology of supercooled liquids, where an excess of low-frequency modes (the boson peak) is found over the Debye behavior of the density of states, $g(\omega) \propto \omega^2$. These excitations are important because they dominate the plastic response of the sample.

We chose Heisenberg spin glasses as a platform to study those behaviors to (i) show that the boson peak appears in diverse types of disordered system, (ii) spin glasses are better understood than structural glasses, and easier to simulate, letting us analyze much larger systems, and (iii) Heisenberg spin glasses, differently from Ising ones, have continuous symmetries and degrees of freedom, so the analysis of the soft modes is analogous to that of sphere systems.

We found that for large random fields a gap appears in the density of states, while when the fields are small the density of states departs from the Debye behavior, with $g(\omega) \propto \omega^4$. These modes are localized, and they connect similar states connected by small energy barriers, that we identify as a classical version of two-level systems.

7.8 Future Challenges and Opportunities

On the long term, the advances exposed in this thesis will represent a small step towards a full comprehension of the glass transition, a few corns of a giant cob.

On a shorter term, a whole set of new research opportunities has been brought to light thanks to this thesis.

From a computational point of view, we opened the door to GPU computing in spin glasses, showing the feasibility and the effectiveness of GPU simulations. Future numerical works on this type of processing units is no more pioneering work aiming to understand whether this possibility is effective, but a full exploitation of the new possibilities of parallel computing. We also proved the usefulness of the construction of machines highly optimized for the resolution of one single type of hard problem, in our case spin glass simulations. The JANUS computer gave us the access to regimes that were unthinkable with any other type of resource, and its success paved the way for its next generation, *Janus II*.

The statistical analyses of the large fluctuations in Chap. 2 open the way for a new approach to this types of problems. Studying large deviation problems as a function of the quantile can reveal interesting results in topics such as temperature chaos [Fer13], and our new finite-size scaling ansatz might reduce drastically the size of the systems necessary to deduce confidently the behavior in the thermodynamic limit with the due further comprehension [Bil14]. Moreover, we showed two promising ways to unravel the question of the existence of a dat line in three dimensions. On one hand we showed in Chaps. 2 and 4 that the link overlap has far less fluctuations than the normal overlap, meaning that a traditional type of analysis on the link overlap could reveal whether or not the spin glass phase persists under an applied magnetic field. On the other hand, we also found the region of temperature in which the phase transition would lie, if it were present. Having a tangible temperature range where to verify the existence of a dat line defines definitively the effort needed to answer this question, that might be to the reach of *Janus II*.

The finding of a coupling between the chiral and spin glass channel in Chap. 3 confirms a part of the Kawamura scenario. The main remaining question is now on the coupling between the two in zero field [Fer09b, Vie09]. It would be also interesting to verify the competition between chiral and spin glass sector in the two-dimensional XY spin glass, where the stiffness of chiral and spin glass sectors appear to be different [Wei08]. We also emphasized that the crossover regime is in practical means more important than the thermodynamic behavior, since in this problem both simulations and experiments are completely immersed in the crossover region. Therefore, quantities such as critical exponents might be more useful if accounted for as a function of the size of the system or of the simulation time, using tools such as the time-length dictionary developed in [AB10a].

A similar type of approach could be used with the crossover of the spin glasses' behavior as a function of m shown in Chap. 4. The fact that the frustration decreases the number of spin components can also be a starting point for models (physical and sociological) where m is a function of the site, meaning that some sites are more susceptible to frustration than others.

The analysis of the dynamics of spin glass avalanches presented in Chap. 5 still has many open points. Random walks in the space of local stabilities could be used to recover more relations between the exponents, and analytical computations on RW in the number of unstable spins can explain what the values of τ and ρ depend on. Furthermore, we expect these results obtained on the correlations in the SK model to extend to other marginally stable systems such as sphere packings: opening a soft contact should imply that contacts carrying small forces should see their force increase.

Finally, the study of the soft modes in Chap. 6 opens a whole new field of studies of the soft modes in spin glasses, and might sign a new trait of unification between sphere packings and spin glasses. To this objective, a search of a boson peak in zero field becomes necessary, possibly as a function of the initial temperature of the relaxations. It would be mostly desirable to be able to search inherent structures from configurations thermalized in the spin glass phase. Also, a study of the correlations lengths in the inherent structures would help to unify the understanding on spin and structural glasses [BJ16].

References

[AB10a] R. Alvarez Baños, A. Cruz, L.A. Fernandez, J.M. Gil-Narvion, A. Gordillo-Guerrero, M. Guidetti, A. Maiorano, F. Mantovani, E. Marinari, V. Martin-Mayor, J. Monforte-Garcia, A. Muñoz Sudupe, D. Navarro, G. Parisi, S. Perez-Gaviro, J.J. Ruiz-Lorenzo, S.F. Schifano, B. Seoane, A. Tarancon, R. Tripiccione, D. Yllanes (Janus Collaboration), J. Stat. Mech. **2010**, P06026 (2010). doi:10.1088/1742-5468/2010/06/P06026. arXiv:1003.2569

[And95] P.W. Anderson, Science **267**, 1615–1616 (1995). doi:10.1126/science.267.5204.1615-e. http://www.sciencemag.org/content/267/5204/1615.6.full.pdf, http://www.sciencemag.org/content/267/5204/1615.6.short

[BJ14d] M. Baity-Jesi, L.A. Fernandez, V. Martin-Mayor, J.M. Sanz, Phys. Rev. **89**, 014202 (2014). doi:10.1103/PhysRevB.89.014202. arXiv:1309.1599

[BJ16] M. Baity-Jesi, V. Martín-Mayor, G. Parisi, S. Pérez-Gaviro, (2016). In preparation

[Ber14] M. Bernaschi, M. Bisson, F. Salvadore, Comput. Phys. Comm. **185**, 2495–2503 (2014). ISSN 0010-4655. doi:10.1016/j.cpc.2014.05.026. http://www.sciencedirect.com/science/article/pii/S0010465514002008

[Bil14] A. Billoire, A. Maiorano, E. Marinari, V. Martin-Mayor, D. Yllanes, Phys. Rev. B **90**, 094201 (2014). doi:10.1103/PhysRevB.90.094201. arXiv:1406.1639

[Fer13] L.A. Fernandez, V. Martin-Mayor, G. Parisi, B. Seoane, *EPL***103**, 67003 (2013). doi:10.1209/0295-5075/103/67003. arXiv:1307.2361

[Fer09b] L.A. Fernandez, V. Martin-Mayor, S. Perez-Gaviro, A. Tarancon, A.P. Young, Phys. Rev. B **80**, 024422 (2009). doi:10.1103/PhysRevB.80.024422

[Lul15] M. Lulli, G. Parisi, A. Pelissetto: Out-of-equilibrium measure of critical parameters for second-order phase transitions (2015). In preparation

[Man15a] M. Manssen, A.K. Hartmann, Phys. Rev. B 91, 174433 (2015). doi:10.1103/PhysRevB.91.174433. http://link.aps.org/doi/10.1103/PhysRevB.91.174433, arXiv:1411.5512

[Man15b] M. Manssen, A.K. Hartmann, A.P. Young, Phys. Rev. B **91**, 104430 (2015). doi:10.1103/PhysRevB.91.104430. arXiv:1501.06760, http://link.aps.org/doi/10.1103/PhysRevB.91.104430

[Vie09] D. Viet, H. Kawamura, Phys. Rev. Lett. **102**, 027202 (2009)

[Wei08] M. Weigel, M.J.P. Gingras, Phys. Rev. B **77**, 104437 (2008). doi:10.1103/PhysRevB.77.104437. http://link.aps.org/doi/10.1103/PhysRevB.77.104437

[Yav12] T. Yavors'kii, M. Weigel, Eur. Phys. J.: Special Topics **210**, 159 (2012)

Appendix A
Monte Carlo on Heisenberg Spin Glasses

The appendix is structured as follows. Sect. A.1 is general about all the MC simulations presented in Chap. 3, though it treats the specific algorithms that we have used with no reference to their implementation, so it is referenced also in Chap. 2. However, the implementation is often crucial. The simulations of Chap. 3 were so demanding that we have used special hardware described in Sect. A.2.1. This special hardware speeds up the simulations thanks to a massive parallelization of the calculations, so in Sect. A.2.2 we give some brief details about it. Finally, we address in Sect. A.2.4 some issues regarding the generation of pseudo-random numbers.

A.1 Simulation Algorithms

For the thermalization of our vector SG we used a blend of several MC dynamics. Specifically, our EMCS consisted of (in sequential order):

- 1 full lattice sweep with the heat bath (HB) algorithm [Ami05, Kra06],
- L lattice sweeps of microcanonical overrelaxation (OR) algorithm [Bro87, Ami05],
- 1 parallel tempering (PT) sweep [Huk96, Mar98, Yll11].

Heatbath by itself would provide correct (but inefficient) dynamics. It actually mimics the natural evolution followed by real SGs (that never reach equilibrium near or below the critical temperature). For this reason we enhance it with two more algorithms. However, HB does play a crucial role, since it is irreducible (i.e. the full configuration space is reachable, at least in principle), at variance with OR, which keeps the total energy constant, and PT, which changes the temperature but not the spin configuration.

Crucial to perform the HB and OR dynamics is the factorization property of the Boltzmann weight for the Hamiltonians (2.1) and (3.9). The conditional probability-density for spin $\vec{s}_x$, given the rest of the spins of the lattice is

M. Baity Jesi, *Spin Glasses*, Springer Theses, DOI 10.1007/978-3-319-41231-3

$$P(\vec{s}_x \mid \{\vec{s}_y\}_{y \neq x}) \propto \mathrm{e}^{(\vec{s}_x \cdot \vec{h}_x)/T}, \tag{A.1}$$

where $\vec{h}_x$ is the *local field* produced by the lattice nearest-neighbors of spin $\vec{s}_x$.[1]

In the HB update, a new orientation for spin $\vec{s}_x$ is drawn from the conditional probability (A.1), see [Ami05] for instance.

The OR update is deterministic. Given a spin $\vec{s}_x$ and its local field, we change the spin as much as possible while keeping the energy constant:

$$\vec{s}_x^{\,\mathrm{new}} = 2\vec{h}_x \frac{\vec{h}_x \cdot \vec{s}_x^{\,\mathrm{old}}}{h_x^2} - \vec{s}_x^{\,\mathrm{old}}. \tag{A.2}$$

Contrarily to HB, the order in which the spins are updated is important in OR. Accessing the lattice randomly increases the autocorrelation time in a substantial way. On the other hand, a sequential update generates a microcanonical wave that sweeps the lattice. The resulting change in the configuration space is significantly larger. A similar microcanonical wave is generated with other types of deterministic lattice sweeps. For instance, one could partition the lattice in a checker-board way and first update all spins in the black sublattice, updating the white spins only afterwards.

The combination of HB and OR has been shown to be effective in the case of isotropic SGs [Pix08] and other models with frustration [Alo96, Mar00a]. However, if one is interested on very low temperatures or large systems, PT is often crucial. For each sample we simulate N_T different copies of the system, each of them at one of the temperatures $T_1 < T_2 < \ldots < T_{N_T}$. A PT update consists in proposing, as configuration change, a swap between configurations at neighboring temperatures. The exchange is accepted with the Metropolis probability

$$P = \min\left[1, \mathrm{e}^{-\beta \Delta E}\right], \tag{A.3}$$

where ΔE is the energy difference between the two configurations and β is the inverse temperature. One of the two systems involved in the swap will decrease its energy, so that change will be automatically accepted. In order to accept the swap both the configuration changes need to be accepted, so the swap is generally accepted with probability $\mathrm{e}^{-\beta|\Delta E|}$. Evidently, the acceptance is higher if the temperatures T_i are closer to each other, since the energy of the configurations will be similar. Notice that exchanging configurations is equivalent to exchange temperatures, so instead of swapping configurations one can swap temperatures, reducing the data transfer to a single number.

---

[1] In the IEA model in a magnetic field of Chap. 2 $h_x = \sum_{y:|x-y|=1}^{d} J_{xy} s_y + h$, in the Heisenberg model with random anisotropic exchange of Chap. 3, $\vec{h}_x = \sum_{y:|x-y|=1}^{d} [J_{xy}\vec{s}_y + D_{xy}\vec{s}_y]$.

A.2 Parallel Computing

We discuss now part of the implementation of our codes on the specific hardware that we disposed of.

A.2.1 Hardware Features

The GPUs we used were of the Tesla generation, produced by NVIDIA, with a SIMD architecture (Single Instruction, Multiple Data), optimized for the parallel processing of large amounts of double precision data.

We had access to Tesla M2050 GPUs in the *Tianhe-1A* supercomputer in Tianjin, China, and Tesla M2090 GPUs on the *Minotauro* cluster in Barcelona, Spain. Despite the extremely high performances claimed by NVIDIA (e.g. 665 Gflops in double precision in the case of the M2090 GPUs), it is practically impossible to reach that limit, because the major bottleneck does not reside in the computing speed, but in the memory access. Yet GPUs keep being a valid tool to simulate on SGs, as they typically allow the same function to be launched concurrently on thousands of threads. This is exactly what we need, since we can update simultaneously different replicas, and also non-neighboring spins within the same replica, because the interactions are only between nearest neighbors.

A.2.2 Effective GPU Coding

The optimization of the GPU code required a great effort. In fact, between the first and the last version of the program, we gained a speed-up factor of 100.

The complexity of the Monte Carlo algorithms, that require the definition of a very large number of variables, is what finally limits the speed of the program, since they exceed the number of registers in the GPU (this effect is called *register spilling* [NVI15]: some of the variables have to be stored in the global memory, slowing down their access).

To limitate the memory access, we opted to simulate the model with binary couplings $D_{xy}^{\alpha\beta} = \pm D$, and $J_{xy} = \pm 1$, in order to be able to store in a single byte the coupling between two sites. Since D_{xy} is symmetric there are 6 independent entries $D_{xy}^{\alpha\beta}$, plus one for J_{xy}. The extra bit stayed unused. Also, we limited the size of the lattice to powers of 2, in order to get be able to evaluate the lattice positions with bitwise operations and to achieve a *coalesced* memory access, as explained in Sect. A.2.3. We also maximized the use of the level 1 cache memory and tiled the system in columns, updating independently two groups of non-neighboring tiles. The black tiles are updated first, and the white are updated in a second kernel call, in order to avoid synchronization conflicts.

Issues of this type with single-GPU coding on spin systems are extensively treated in works such as [Ber11, Yav12, Lul14], so let us focus on the complications related to the use of multiple GPUs. We describe now in practical means the procedure of simulations for $L = 64$ that mixed CUDA and message passing interface (MPI).

For each sample we simulate $2N_T$ replicas, because we need two replicas per temperature to be able to calculate overlaps. We use N_{GPU} GPUs, and each hosts two replicas, not necessarily at the same temperature, hence $N_{\text{GPU}} = N_T$. Since the interactions are only between nearest neighbors, we can update simultaneously up to half of the spins with two independent kernel calls (one for the black tiles and one for the white). Yet, there are only 65,535 threads per GPU [NVI15], and $2L^3 = 524{,}288$ sites, so each thread has to update at least 4 spins. Since the major bottleneck is the memory access, we work with $N_{\text{threads}} = 2^{15} = 32{,}768$ threads, assigning a row of 8 spins to each, along the x axis. This way we can minimize the number of reads from global memory, and we give a direction to the OR spin wave. Adjacent rows are updated in different kernel calls.

A.2.3 Coalescent Memory Reading

Changing the way we read from memory gives GPU programs a dramatic speedup, and the only effort necessary to obtain this is to change the indexing of the memory locations.

When a single multiprocessor is given some thread blocks to deal with, the scheduler executes them in groups of 32 threads, called warps. A warp executes one instruction at a time, and the maximum performance is achieved when all the threads in the warp have a similar execution path. To get coalesced reading, the consecutive threads have to read from consecutive memory positions, in order to maximize bandwidth of the memory bus [NVI15]. So, for example, if thread 1 reads from the memory position 612, thread 2 would make an effective read from position 613. In order to obtain this we have to reorganize the memory indexing in order to have thread 2 pointing to position 613. This is often automatically realized in simple arrays, but not when the spatial geometry comes to play with tiling or with the indexing of the J_{xy}s.

In the specific case of our spin indexing, we want neighboring rows to be called by neighboring threads. Yet, when we say neighboring rows, we mean neighboring rows within the same kernel call, not in the actual lattice. It is like if we compress together all the white tiles and only then we worry about proximity. The first site of the white row i (i runs only over the white tiles) has to be stored besides the first site of row $i + 1$, and so on. This means that their address in memory has to differ only in the least significant bit. The z coordinate is the same both for i and $i + 1$. The same happens for the x coordinate, since both threads sweep the row in the same way. On the y axis, since we update one row of every two, the least significant bit y_0 also is the same. Hence the least significant bit of the coalesced reading has to be y_1, the second least y_2, and so on. On Table A.1, line 4, we give an example of coalesced

Table A.1 A step-by-step example of how to obtain coalescent reading for an $L = 64$ lattice

$i\text{th}$					t_{14}	t_{13}	t_{12}	t_{11}	t_{10}	t_9	t_8	t_7	t_6	t_5	t_4	t_3	t_2	t_1	t_0
i_{row}					r_0	x_5	x_4	x_3	z_5	z_4	z_3	z_2	z_1	z_0	y_5	y_4	y_3	y_2	y_1
i_{site}	r_0	z_5	z_4	z_3	z_2	z_1	z_0	y_5	y_4	y_3	y_2	y_1	y_0	x_5	x_4	x_3	x_2	x_1	x_0
$i_{\text{site}}^{\text{coalesced}}$	r_0	y_0	x_5	x_4	x_3	x_2	x_1	x_0	z_5	z_4	z_3	z_2	z_1	z_0	y_5	y_4	y_3	y_2	y_1

On the **first line** we show the *thread index*. It has 15 significant bits, since we use 2^{15} concurrent threads. We have to use them to identify each tile with the starting point of the row (**second line**). We use the most significant bit to identify the replica. Since L contains 8 rows, we need only 3 digits to identify their starting point on the x axis, but we need all the information on the z axis, and only 5 bits for the y axis, since there is the constraint of having to simulate non-neighboring rows. On the **third line** we show an easy way to organize the bits to identify a site once we started moving along the row, in case of non-coalescent reading. It is straightforwardly deducible from i_{row}. The **last row** shows how to organize the bits to get coalescence. The replica index stays where it is, the eleven following bits are shifted 7 positions to the right, and the final seven are shifted 11 positions to the left. This way consecutive threads access consecutive memory positions. More details in the main text

memory access. Since there are 2^{15} threads, $i\text{th}$ has 15 significant bits. To get the index of the starting site i_{row} of each row we need information on:

- Which replica were updating. There are two replicas, so 1 bit is enough.
- The z coordinate. It can assume $L = 64$ different values, so it requires 6 bits.
- The x coordinate is not constant. We just need the one of the first site of the row. Rows are 8 sites long, so we can only fit 8 along a side. That makes 3 bits.
- The y coordinate. Since adjacent rows are updated in different kernel calls, y has to change of 2 lattice spacings each time we change row, and half of the y choices are forbidden. We need 5 bits for y.

The mapping from $i\text{th}$, associated with the thread to the index i_{row} that indicated the initial site of the tile, is shown on the second line of Table A.1. The index i_{row} needs only 3 bits to store its x position, because since the rows are of 8 sites along the x axis there are only 8 tiles. By adding the three bits (Table A.1, line 3) we obtain an uncoalesced memory read of site i_{site}. From this one we obtain the coalesced read by moving the bits around in order to force the changes of indexing to the least significant bit. Practically, it is obtained by shifting seven positions to the right the 6 z-bits plus the 5 y-bits except y_0, and with an 11 position shift towards left of the remaining y_0 plus the 6 x-bits. Notice that this type of reading is very convenient since it only implies unsophisticated bit-to-bit operations, and it is valid for any L power of 2. This is why almost all our simulations were with $L = 8, 16, 32, 64$.

The remaining information on the actual position on the lattice is given by a binary parity parameter that the kernel gets from the input. The parity tells us whether y is even or odd (if $y_0 = 1$ or $y_0 = -1$), or in other words, if the kernel call regards black

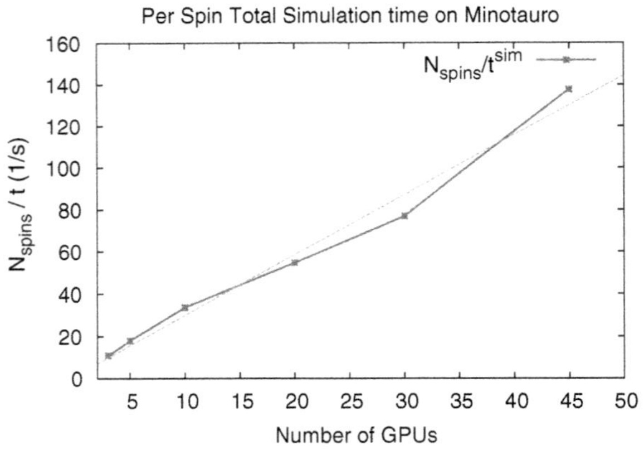

Fig. A.1 Scaling of the computing time with the number of GPUs N_{GPU}. Benchmark performed on the *Minotauro* GPU cluster (Barcelona Supercomputing Center, Barcelona, Spain)

or white cells.[2] The index i_{site} indicates the position of the single site once one took in account the parity and the position along the row.

A.2.3.1 MPI Parallelization

To simulate N_T temperatures with MPI we used $N_T + 1$ cores. N_T of them, called slaves, were in charge of measurements and updates on two lattices, using the resources of a GPU each. The remaining one, called master, did not use any GPU and was dedicated to the PT and to the management of the relationships between slaves. The expedients for the simulations that we described in the previous sections are valid at the level of the slave.

Each 1 HB + L OR sweeps, we do PT. We measure on the device (the GPU) the energy of each replica, and we pass this information to the master. The master makes the PT iterations, that require a negligible amount of time, and assigns a new temperature to each replica. The memory transfer overhead is minimum in this case. It becomes an issue when we have to

1. Perform 2-replica measurements (e.g. overlaps)
2. Write on disk (measurements and backup)

since we are forced to pass the entire configuration via MPI. The nature of the system we are simulating is of help, since we can dilute measures (and writes) almost as much as we desire, as long as we have enough measurements to perform decent averages. The MPI extension turned out to be very effective, since not only the multi-GPU version of the algorithms was as fast as the single-GPU, but also the speed had a linear scaling with the number of GPUs (see Fig. A.1).

[2]For $x = 0$ it tells us if y is even or odd, but for $x = 8$, it tells us if y is odd or even, and so on, because in each layer of rows the parity has to change in order to not update simultaneously neighboring rows.

A.2.4 Parallel Pseudo-Random Number Generator

Pseudo-random number generators (PRNGs) are a critical issue in the implementation of stochastic algorithms [Knu81], but even more in cases like ours, where each of the N_{threads} threads had to carry its own PRNG, and we had a large number of them acting in parallel on the same lattice. This became a major problem especially in the simulations with MPI, where a huge number of PRNGs was concentrated on only two lattices. It was crucial to guarantee the statistical independence of the N_{threads} pseudo-random sequences. We consider three different aspects: (*a*) the PRNG that each thread uses, (*b*) the initialization of the generators and (*c*) our tests on the generators.

A.2.4.1 The Generator

We resorted to a linear combination of Parisi-Rapuano with congruential generators [Fer09c].

With the Parisi-Rapuano sequence [Par85], the nth pseudo-random number P_n is generated through the following relations:

$$\begin{aligned} y_n &= (y_{n-24} + y_{n-55}) \bmod 2^{64} \\ P_n &= y_n \text{ XOR } y_{n-61}, \end{aligned} \tag{A.4}$$

where XOR is the exclusive OR logic operator, and y_i are 64-bit unsigned integers. Although some pathologies have been found in the 32-bit Parisi-Rapuano PRNG [Bal98c], it looks like its 64-bit version is solid [Fer05].

On the other side, we used a 64-bit congruential generator, where the nth element of the sequence, C_n, was given by [Knu81, LÉ99]:

$$C_n = (C_{n-1} \times 3202034522624059733 + 1) \bmod 2^{64}. \tag{A.5}$$

Also this generator is not reliable when used alone [Fer09c, Oss04].

The final pseudo-random number R_n was obtained by summing P_n and C_n:

$$R_n = (P_n + C_n) \bmod 2^{64}. \tag{A.6}$$

A.2.4.2 Initializing the Generators

We have found that problems arise if special care is not devoted to the initialization of the random numbers. This is particularly important in the case of multiple GPUs where $N_{\text{threads}} = 32768$ threads concurrently update the spins in only two lattices.

We need one PRNG for the master, that performs PT, and N_{threads} independent generators for each slave. It is not trivial to avoid periodicities when not only one wants $N_{\text{threads}}N_{\text{GPU}}+1 \sim 1.5\times 10^6$ PRNG, but it is crucial for them to be reproducible, monitorizable and backupable. Starting each simulation with over a million seeds is not a realistic option, but any simplification can be crucial for the simulation.

We decide to use one seed per slave, plus one for the master, and refresh the PRNG every time a backup is done.[3] That makes 46 unsigned long long integer seeds (64 bits each). Passing the PRNG to the kernels is a major bottleneck in our simulations. A combination of a congruential generator with the Parisi-Rapuano wheel is a fair solution in terms of speed and memory passage to the kernel, but the Parisi-Rapuano wheel contains 256 elements (passing them back to the master takes forever), and it is not trivial to initialize properly a very large amount of wheels starting from a single seed.

The starting point for each node is a single seed. From that we have to initialize a whole set of N_{threads} PRNGs, so it is clear that special care is needed to obtain independent initializations.

Implementation For the initialization of the N_{threads} generators through a single seed we resorted to the Luescher generator, we employed the *full luxury* version, which is fireproof but slow [Lue94]. This is how we proceeded to obtain a large set of pseudo-independent PRNG out of a single seed.

1. Use the initial seed to initialize a 64-bit congruential PRNG (A.5).
2. Generate $\sim$1000 random numbers with the congruential PRNG, in case the initial seed was not chosen properly (e.g. it was too small).
3. Use the congruential generator to initialize a Luescher wheel, that requires 256 24-bit elements (although we only need 24 for the initialization, plus an auxiliary variable). Each 24-bit entry for the Luescher wheel is obtained through 3 subsequent call of the congruential. From each call we pick the 8 most significant bits, and append the three together construct the 24-bit number.
4. Generate $\sim$1000 random numbers with Luescher's wheel.
5. Use the Luescher wheel to fill up the state vector of the 64-bit PRNGs in Eq. (A.6). Each entry is obtained through 8 Luescher calls, and taking the 8 most significant bits from each.

In addition to the PRNGs, also the couplings are formed by using Luescher's algorithm. We were probably excessively cautious, given the high quality of the full-luxury generator, but initialization takes only a small fraction of the total computing time, and we wanted to grant the threads sufficiently independent PRNGs.

[3] In order to gain in speed and space in disk, we decided not to save the random wheel when we had to make backups. We limited ourselves to a refresh of the random wheels with new seeds read from the *urandom* device. In this manner, we only had to save $N_{\text{backups}}(N_{\text{GPU}} + 1)$ long long integers per simulation.

A.2.4.3 Tests

We tested with success our random sequences through the whole battery of tests proposed in [Mar95]. To be sure the sequences were reliable also with concurrent threads, we also generated N_{threads} sequences and tested them *horizontally*, i.e. taking first the first number of each sequence, then the second, and so on.

Also, we made simulations with ferromagnetic couplings demanding the energies to be equal, up to the 7th significant digit, to those obtained with an independent CPU program, that had been already used to produce publications such as [Fer09b].

Finally, it has been pointed out that local Schwinger-Dyson relations (see e.g. [Riv90]) can be useful to assess the quality of PRNGs [Bal98c]. The relevant identity here is

$$2T\left\langle \vec{s}_x \cdot \vec{h}_x \right\rangle - \left\langle (\vec{h}_x)^2 - (\vec{s}_x \cdot \vec{h}_x)^2 \right\rangle = 0. \tag{A.7}$$

We averaged it over all the sites in the lattice, in order to obtain a more stringent test for the simulations.

Appendix B
Four-Replica Correlators

In this appendix we give details on the 4-replica correlators used in Chap. 2. In Sect. B.1 we motivate the need of four different replicas, in Sect. B.2 we explain how to find the replicon and longitudinal connected correlation functions G_{R} and G_{L}, we show that the signal carried by G_{L} is much smaller than that of G_{R}, and we give an estimation of the value that the effective anomalous exponent η_{eff} defined in Sect. 2.10 should acquire in the spin glass phase (Sect. B.2.1). Section B.3 is dedicated to an implementation of the MSC technique in our analyses.

In the presence of an external field the overlap is non-zero even in the paramagnetic phase, so the correlation functions $C(\boldsymbol{r})$ [Eq. (1.47)] do not go to zero for large distances. We need therefore to explicitly construct correlators that go to zero. Two natural constructions that can be measured directly are

$$\Gamma_1(\boldsymbol{x},\boldsymbol{y}) = \overline{\left[\langle s_x s_y\rangle - \langle s_x\rangle\langle s_y\rangle\right]^2}, \tag{B.1}$$

$$\Gamma_2(\boldsymbol{x},\boldsymbol{y}) = \overline{\left[\langle s_x s_y\rangle^2 - \langle s_x\rangle^2\langle s_y\rangle^2\right]}. \tag{B.2}$$

In Sect. B.2 we will show how Γ_1 and Γ_2 relate to the correlators of the replicated field theory.

B.1 The Need for Four Replicas

If we use only two replicas to calculate $\Gamma_1(\boldsymbol{x},\boldsymbol{y})$ and $\Gamma_2(\boldsymbol{x},\boldsymbol{y})$, we will introduce an annoying systematic error in our measurements. Let us examine, for example, Γ_2, reexpressing it as a function of the overlaps using Eq. (1.26), $\Gamma_2(\boldsymbol{x},\boldsymbol{y}) = \overline{\langle q_x q_y\rangle - \langle q_x\rangle\langle q_y\rangle}$.

During a single run of N_{MC} EMCS and samplings $q_{x,t}(t = 1,\ldots,N_{\mathrm{MC}})$, we measure an estimator $[q_x]$ of the overlap's thermal average $\langle q_x\rangle$,

M. Baity Jesi, *Spin Glasses*, Springer Theses, DOI 10.1007/978-3-319-41231-3

$$[q_x] = \frac{1}{N_{\mathrm{MC}}} \sum_{t=1}^{N_{\mathrm{MC}}} q_{x,t}. \tag{B.3}$$

The expected value and its estimator are related by

$$[q_x] = \langle q_x \rangle + \eta_x \frac{\sigma_x}{\sqrt{N_{\mathrm{MC}}/(2\tau)}} \tag{B.4}$$

where τ is the integrated time related to q_x[4], η_x is gaussian with $\overline{\eta_x} = 0$ and $\overline{\eta_x^2} = 1$ that stands for the fluctuations around the mean, and σ_x is the amplitude of these fluctuations.

The estimated correlation function is then

$$\begin{aligned}[\Gamma_2(\boldsymbol{x}, \boldsymbol{y})] &= \overline{[q_x q_y] - [q_x]\,[q_y]} = \\ &= \overline{\langle q_x q_y \rangle - \langle q_x \rangle \langle q_y \rangle} \\ &+ \overline{\eta_{xy} \frac{\sigma_{xy}}{\sqrt{N_{\mathrm{MC}}/(2\tau)}}} + \overline{\eta_x \frac{\sigma_x}{\sqrt{N_{\mathrm{MC}}/(2\tau)}}} + \overline{\eta_y \frac{\sigma_y}{\sqrt{N_{\mathrm{MC}}/(2\tau)}}} + \overline{\eta_x \eta_y \frac{\sigma_x \sigma_y}{(N_{\mathrm{MC}}/(2\tau))}}.\end{aligned} \tag{B.5}$$

When averaging over the disorder the three terms that are linear in η are linear in η disappear because $\overline{\eta} = 0$. On the contrary, since η_x and η_y are correlated $\overline{\eta_x \eta_y} \neq 0$, therefore the last term represents a bias of order $o(N_{\mathrm{MC}}^{-1})$ that does not disappear with an average over the disorder.

Since the disorder fluctuations are $o(N_{\mathrm{samples}}^{-1/2})$, as long as $N_{\mathrm{MC}} \gg N_{\mathrm{samples}}$ we can neglect this bias. As this is not necessarily true, so we recur to four-replica measurements to have uncorrelated fluctuations. With an analogous procedure to the one we just presented, the reader will notice that there is no bias in the four-replica estimators we present in the next sections.

B.2 Computing the Replicon and Longitudinal Correlation Functions

With 4 replicas we can construct 3 different correlators

$$\begin{aligned}G_1(\boldsymbol{x}, \boldsymbol{y}) &= \overline{\langle s_x s_y \rangle^2} = \\ &= \overline{\langle s_x^{(a)} s_y^{(a)} s_x^{(b)} s_y^{(b)} \rangle},\end{aligned} \tag{B.6}$$

$$\begin{aligned}G_2(\boldsymbol{x}, \boldsymbol{y}) &= \overline{\langle s_x s_y \rangle \langle s_x \rangle \langle s_y \rangle} = \\ &= \overline{\langle s_x^{(a)} s_y^{(a)} s_x^{(b)} s_y^{(c)} \rangle},\end{aligned} \tag{B.7}$$

[4] See e.g. [Ami05] for informations on the relation between integrated time and number of independent measurements.

$$G_3(\boldsymbol{x},\boldsymbol{y}) = \overline{\langle s_{\boldsymbol{x}}\rangle^2 \langle s_{\boldsymbol{y}}\rangle^2} = \\ = \overline{\langle s_{\boldsymbol{x}}^{(a)} s_{\boldsymbol{y}}^{(b)} s_{\boldsymbol{x}}^{(c)} s_{\boldsymbol{y}}^{(d)}\rangle}. \tag{B.8}$$

None of those goes to zero for large distances $\|\boldsymbol{x} - \boldsymbol{y}\|$, but, in the paramagnetic phase they all tend to the same value, q_{EA}, when $\|\boldsymbol{x} - \boldsymbol{y}\| \longrightarrow \infty$. So, to create connected correlators, we can make two linearly independent combinations of them, and obtain the basic connected propagators of the replicated field theory [Dom98, Dom06][5]

$$G_{\mathrm{R}} = G_1 - 2G_2 + G_3, \tag{B.9}$$

$$G_{\mathrm{L}} = G_1 - 4G_2 + 3G_3. \tag{B.10}$$

G_{R} the G_{L} are easily related to Γ_1 and Γ_2 by expanding their expressions in Eqs. (B.1, B.2). The first relation is direct,

$$\Gamma_1(\boldsymbol{x},\boldsymbol{y}) = \overline{\left[\langle s_{\boldsymbol{x}} s_{\boldsymbol{y}}\rangle - \langle s_{\boldsymbol{x}}\rangle^2 \langle s_{\boldsymbol{y}}\rangle\right]^2} = \\ = \overline{\left\langle s_{\boldsymbol{x}} s_{\boldsymbol{y}}\right\rangle^2 - 2\left\langle s_{\boldsymbol{x}} s_{\boldsymbol{y}}\right\rangle \langle s_{\boldsymbol{x}}\rangle \left\langle s_{\boldsymbol{y}}\right\rangle + \langle s_{\boldsymbol{x}}\rangle \left\langle s_{\boldsymbol{y}}\right\rangle} = G_{\mathrm{R}}(\boldsymbol{x},\boldsymbol{y}). \tag{B.11}$$

To expand Γ_2 we complete a square

$$\begin{aligned}\Gamma_2(\boldsymbol{x},\boldsymbol{y}) &= \overline{\left[\langle s_{\boldsymbol{x}} s_{\boldsymbol{y}}\rangle^2 - \langle s_{\boldsymbol{x}}\rangle^2 \langle s_{\boldsymbol{y}}\rangle^2\right]} = \\ &= \overline{\left(\left\langle s_{\boldsymbol{x}} s_{\boldsymbol{y}}\right\rangle^2 - 2\left\langle s_{\boldsymbol{x}} s_{\boldsymbol{y}}\right\rangle \langle s_{\boldsymbol{x}}\rangle \left\langle s_{\boldsymbol{y}}\right\rangle + \langle s_{\boldsymbol{x}}\rangle^2 \left\langle s_{\boldsymbol{y}}\right\rangle^2\right)} + \\ &+ 2\left(\overline{\left\langle s_{\boldsymbol{x}} s_{\boldsymbol{y}}\right\rangle \langle s_{\boldsymbol{x}}\rangle \left\langle s_{\boldsymbol{y}}\right\rangle - \langle s_{\boldsymbol{x}}\rangle^2 \left\langle s_{\boldsymbol{y}}\right\rangle^2}\right) = \\ &= G_{\mathrm{R}}(\boldsymbol{x},\boldsymbol{y}) + 2\left[G_2(\boldsymbol{x},\boldsymbol{y}) - G_3(\boldsymbol{x},\boldsymbol{y})\right].\end{aligned} \tag{B.12}$$

We can rewrite Eq. (B.12) in the more convenient form $\Gamma_2 - \Gamma_1 = 2(G_2 - G_3)$. Notice finally from Eqs. (B.9, B.10) and Eq. (B.12) that $G_{\mathrm{L}} = G_{\mathrm{R}} - 2(G_2 - G_3) = 2\Gamma_1 - \Gamma 2$.

The relations between Gs and Γs can be resumed as

$$\begin{aligned} G_{\mathrm{R}} &= \Gamma_1, \\ G_{\mathrm{L}} &= 2\Gamma_1 - \Gamma_2, \\ 2\,(G_2 - G_3) &= \Gamma_2 - \Gamma_1 = G_{\mathrm{R}} - G_{\mathrm{L}}. \end{aligned} \tag{B.13}$$

The definitions (B.9, B.10), valid at equilibrium, were used in [BJ14b] in an out-of-equilibrium context, for lattices of size $L = 80$. In that work it had been noticed that the replicon is the only correlator that carries a significant signal.

Also in the present work we measured both signals, and we can confirm that the same phenomenology is observed in completely thermalised systems. In Fig. B.1

[5] In the effective field theory the longitudinal (G_{L}) and anomalous (G_{A}) propagators are degenerated. G_{R} is the replicon propagator.

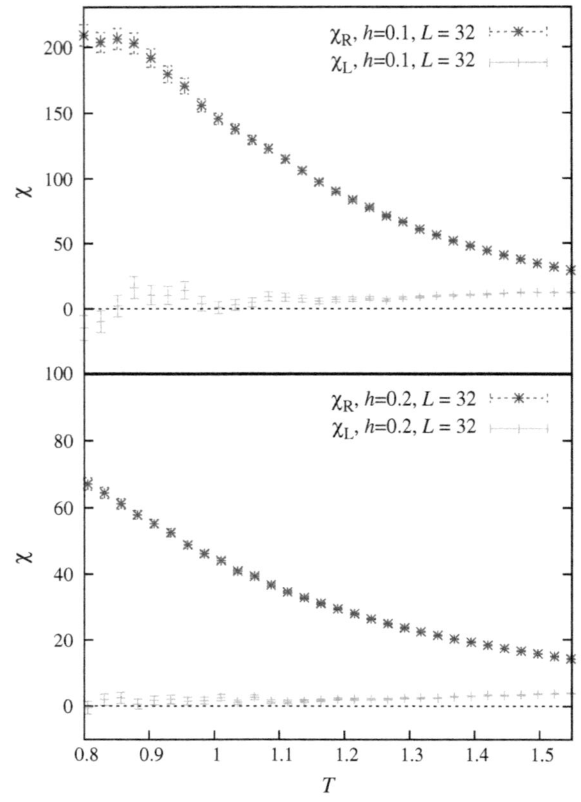

Fig. B.1 Replicon and longitudinal susceptibilities as a function of T in our equilibrium simulations, for the fields $h = 0.1, 0.2$ in our largest lattice sizes ($L = 32$). Just as in [BJ14b] the signal carried by the longitudinal propagator is much smaller than that of the replicon

we plot both the replicon susceptibility χ_R and the longitudinal susceptibility χ_L, at $h = 0.1, 0.2$. The figure is qualitatively very similar to Fig. 13 of [BJ14b], where it is shown that χ_R carries a significant signal, while χ_L is very close to zero.

B.2.1 The Effective Anomalous Dimension in the Spin-Glass Phase

The value of the effective anomalous exponent η_{eff} (Sect. 2.10) in the deep spin-glass phase can be predicted by using the fact that G_R is dominant with respect to G_L.

In fact, in a RSB situation the overlap q is defined over a finite range, so the overlap's variance $\sigma_q^2 = E(q^2) - E(q)^2$ is of order one:

$$\text{RSB} \Rightarrow \sigma_q^2 \sim 1. \tag{B.14}$$

Now, on general grounds (see for instance [Fis91]) we can expect

$$\left[E(q^2) - E(q)^2\right] \sim \overline{\langle q^2 \rangle - \langle q \rangle^2}, \tag{B.15}$$

and remark that the r.h.s. is $\hat{\Gamma}_2(\mathbf{0})/N$, the zero-moment Fourier transform of Γ_2 [defined in (B.2)].[6] We have then that in RSB conditions

$$\hat{\Gamma}_2(\mathbf{0}) \sim N\sigma_q^2 \overset{\text{RSB}}{\sim} N. \tag{B.16}$$

$\hat{\Gamma}_2$ can be related to the replicon and longitudinal susceptibilities through (B.13), that imply that $\hat{\Gamma}_2(\mathbf{0}) = 2\chi_{\mathrm{R}} + \chi_{\mathrm{L}}$. Now, in the beginning of this section we found out empirically that the longitudinal susceptibility is subdominant with respect to the replicon channel (Fig. B.1), so in the large-volume limit, in the presence of RSB, the replicon susceptibility scales like the volume:

$$\text{RSB} \Rightarrow \chi_R \sim N. \tag{B.17}$$

Let us recall (2.19) and impose the just-found implication. We have then

$$2^D \overset{\text{RSB}}{=} \frac{\chi_{\mathrm{R},2L}}{\chi_{\mathrm{R},L}} \equiv 2^{2-\eta_{\text{eff}}}, \tag{B.18}$$

therefore in the spin-glass phase we would have $\eta_{\text{eff}} = -1$.

B.3 Measuring the Propagators with Multi-spin Coding

We now write the correlators in a way that is useful for multi-spin coding (MSC), and then we show explicitly how MSC coding was done on these quantities.

B.3.1 Correlators as Simple Functions of Simple Fields

A simple way to construct unbiased quantities is to define them as functions of fields of differences. With four replicas we can define

$$\begin{aligned} X_1(\boldsymbol{x}) &= (s_{\boldsymbol{x}}^{\mathrm{a}} - s_{\boldsymbol{x}}^{\mathrm{b}})(s_{\boldsymbol{x}}^{\mathrm{c}} - s_{\boldsymbol{x}}^{\mathrm{d}}), \\ X_2(\boldsymbol{x}) &= s_{\boldsymbol{x}}^{\mathrm{a}} s_{\boldsymbol{x}}^{\mathrm{b}} - s_{\boldsymbol{x}}^{\mathrm{c}} s_{\boldsymbol{x}}^{\mathrm{d}}. \end{aligned} \tag{B.19}$$

[6] The correlation functions $G(\boldsymbol{x}, \boldsymbol{y})$ and $\Gamma(\boldsymbol{x}, \boldsymbol{y})$ are averaged over the disorder. Once this average is performed we can integrate out one of the two spatial dependencies and write them as $G(\boldsymbol{r})$ and $\Gamma(\boldsymbol{r})$. There is no ambiguity in this notation: when these function are written as depending on two parameters, it is the two positions $\boldsymbol{x}$ and $\boldsymbol{y}$, when there is only one parameter it is $\boldsymbol{r} = \boldsymbol{x} - \boldsymbol{y}$.

These are the quantities we actually measure, we want to relate them with the correlation functions G_R and G_L (Eqs. B.9, B.10).

Expanding the X_1 field correlator we get

$$\langle X_1(\boldsymbol{x})X_1(\boldsymbol{y})\rangle = 4\left\langle s_x^a s_x^c s_y^a s_y^c\right\rangle - 8\left\langle s_x^a s_x^c s_y^a s_y^d\right\rangle + 4\left\langle s_x^a s_x^c s_y^b s_y^d\right\rangle . \tag{B.20}$$

On the other side rewriting the replicon propagator G_R as a function of four replicas yields

$$G_R(\boldsymbol{x},\boldsymbol{y}) = \overline{\left\langle s_x^a s_y^a s_x^b s_y^b\right\rangle - 2\left\langle s_x^a s_y^a s_x^b s_y^c\right\rangle + \left\langle s_x^a s_y^a s_x^c s_y^d\right\rangle}, \tag{B.21}$$

so

$$G_R(\boldsymbol{x},\boldsymbol{y}) = \frac{1}{4}\overline{\langle X_1(\boldsymbol{x})X_1(\boldsymbol{y})\rangle}. \tag{B.22}$$

Equivalently, an expansion of the X_2 field correlator returns

$$\begin{aligned}\langle X_2(\boldsymbol{x})X_2(\boldsymbol{y})\rangle &= \left\langle s_x^a s_x^b s_y^a s_y^b\right\rangle - \left\langle s_x^a s_x^b s_y^c s_y^d\right\rangle - \left\langle s_x^c s_x^d s_y^a s_y^b\right\rangle + \left\langle s_x^c s_x^d s_y^c s_y^d\right\rangle = \\ &= 2\left(\left\langle s_x s_y\right\rangle^2 - \langle s_x\rangle^2\left\langle s_y\right\rangle^2\right).\end{aligned} \tag{B.23}$$

By averaging it over the disorder we can relate it to the non-connected correlators of Eqs. (B.6, B.7, B.8),

$$\frac{1}{2}\overline{\langle X_2(\boldsymbol{x})X_2(\boldsymbol{y})\rangle} = G_1(\boldsymbol{x},\boldsymbol{y}) - G_3(\boldsymbol{x},\boldsymbol{y}) = \tag{B.24}$$

$$= 2G_R(\boldsymbol{x},\boldsymbol{y}) - G_L(\boldsymbol{x},\boldsymbol{y}), \tag{B.25}$$

where for the second relation we used Eq. (B.13). The expression of G_L in terms of the fields X_i becomes

$$G_L(\boldsymbol{x},\boldsymbol{y}) = \frac{1}{2}\overline{\langle X_1(\boldsymbol{x})X_1(\boldsymbol{y})\rangle} - \frac{1}{2}\overline{\langle X_2(\boldsymbol{x})X_2(\boldsymbol{y})\rangle}. \tag{B.26}$$

Since it is possible to construct the fields X_i with three independent permutations of the replicas (X_i^{abcd}, X_i^{acbd} and X_i^{adbc}), we compute correlators starting from each of those permutations and then average to reduce the fluctuations.

B.3.2 Plane Correlators

Since we average over the disorder, the replicon and longitudinal correlation functions can be rewritten as a function of the distance vector $\boldsymbol{r}$. We concentrate on the $G_R(\boldsymbol{r})$ because it carries the most signal and it is the one we used in our analyses. It is expressed as

$$G_{\mathrm{R}}(\boldsymbol{r}) = \overline{\frac{1}{4}\sum_{\boldsymbol{x}} \langle X_1(\boldsymbol{x}) X_1(\boldsymbol{x}+\boldsymbol{r})\rangle}. \tag{B.27}$$

For the convolution theorem, analogously as we did in Eq. (1.42), we can write its Fourier transform as

$$\hat{G}_{\mathrm{R}}(\boldsymbol{k}) = \overline{\frac{N}{4}\left\langle |\hat{X}_1(\boldsymbol{k})\hat{X}_1(-\boldsymbol{k})|^2\right\rangle}, \tag{B.28}$$

where, for $\boldsymbol{k} = (k, 0, 0)$,

$$\hat{X}_1(\boldsymbol{k}) = \frac{1}{N}\sum_{\ell}^{L} \mathrm{e}^{\mathrm{i}k\ell} P(\ell), \tag{B.29}$$

and $P(\ell)$ is the field averaged over a plane with $x_1 = \ell$

$$P(\ell) = \sum_{y,z} X_1(\ell, y, z). \tag{B.30}$$

Clearly, one can choose any plane orientation, though some are easier to code than others. In our analyses we chose planes orthogonal to the vectors of the euclidean basis and to the diagonals of the lattice [vectors of the type (1, 1, 0) and (1, 1, 1)].

The computationally demanding part of the computation of $G_{\mathrm{R}}(\boldsymbol{r})$ consists in creating the plane fields P for all the samples and replica choices. Once we have those, the remaining operations are of order L and are quickly performed. In the next section we show how it was possible to speed up this problematic part of the analysis.

B.3.3 Multi-spin Coding

We present now MSC [Jac81] as an extremely fast technique to be able to calculate the elementary bricks through which we can construct our correlators. We will show how to use MSC to extract the plane sums $\sum_{\boldsymbol{x}\in\text{plane}} X_1(\boldsymbol{x})$ from the configurations. Once they are calculated for all the planes of each direction (for example the directions can be x, y, z and the single planes are the L possible plains one can construct along each direction), the core of the arithmetic operations is done, and correlation functions are constructed quickly.

In a MC simulation on Ising spins, the naïve approach is to store the information of each spin with an integer variable. This results in a large waste of memory, since an integer number of n_b bits could store information for n_b spins at a time. Since a bit assumes the values $b = 0, 1$, the spin's value is $s = 1 - 2b$. If instead of using an integer for a single spin we use it for n_b spins, not only do we gain in memory, but also in speed. In fact, operations on the spins are highly parallelizable, so if one performs bitwise operations on the integers storing the spins, he can ideally gain a performance factor of n_b. This is the idea of multispin coding.

Clearly this technique presents a long series of caveats and complications, since *only* bitwise operations are allowed. Storing binary magnitudes such as spins is easy, but updating them in a MC simulations is non trivial, since the energy barriers can assume several values, and also it is not possible to use the same random number to update spins of the same lattice.

Storing the lattices The easiest way to parallelize is to treat groups of n_b samples at a time, assigning to an n_b-bit integer, that we call a *word*, the value of the spin $s_{\boldsymbol{x}}$ (or $b_{\boldsymbol{x}}$, if we want to talk in terms of bits) for each of the samples. The bits of the word $u_{\boldsymbol{x}}^{\mathrm{a}}$, indicating site $\boldsymbol{x}$ and replica 'a' will be

$$u_{\boldsymbol{x}}^{\mathrm{a}} = [b_{\boldsymbol{x}}^{\mathrm{a},1}, b_{\boldsymbol{x}}^{\mathrm{a},2}, \ldots, b_{\boldsymbol{x}}^{\mathrm{a},n_b}], \tag{B.31}$$

where we labelled with an extra superscript the different samples (i.e. bits). To store the full configuration of the n_b samples we need $4N$ words: a word per site per replica.

The words $u_{\boldsymbol{x}}^{\mathrm{a}}$ are stored in variables of type `MYWORD`, where `MYWORD` is usually an n_b-bit integer. In this work we used $n_b = 128$. In our C code we use triple arrays to store the configurations, so the full configurations are stored in arrays of the type `MYWORD u[NT] [NR] [N];` where `NT` is the number of measurements N_{m} we use (recall Sect. 2.2.2) and `NR` is the number of replicas, and `N` is the number of spins N in a single lattice.

If N_{samples} is a multiple of n_b the method is then fully optimized, otherwise it is enough to discard a number of bits from the last group of samples.

B.3.4 Replicon correlator with MSC

We will not face the task of explaining how to perform a MSC simulation, that is already done in literature, for example in [Jac81, Seo13]. We will instead focus on how we multi-spin coded the analysis of the correlation function G_{R} (G_{L} is similar).

We already described in Sect. B.3.1 how it is possible to obtain G_{R} and G_{L} from the fields X_1 and X_2 [Eq. (B.19)], that are simple enough to allow for a MSC computation: The field X_1 takes only the values $-4, 0, 4$, while X_2 takes $-2, 0, 2$, so they can be stored with two bytes each (per site per sample). We want to use MSC to construct the plane average $P(\ell)$ of X_1 and X_2, that is the most computationally demanding part of the analyses.

At the beginning of the MSC computation we have 4 replicas $u_{\boldsymbol{x}}^{\mathrm{a}}, u_{\boldsymbol{x}}^{\mathrm{b}}, u_{\boldsymbol{x}}^{\mathrm{c}}, u_{\boldsymbol{x}}^{\mathrm{d}}$ with which to construct $X_1(\boldsymbol{x})$ (for G_{R} we do not need X_2).[7]

The MSC operations have to be iterated over all the sites. Once the loop over the sites is finished the per-site analysis is over, global quantities are created and the

[7]We do it with the three independent permutations of the replicas $X_1^{\mathrm{abcd}}(\boldsymbol{x})$, $X_1^{\mathrm{acbd}}(\boldsymbol{x})$ and $X_1^{\mathrm{adbc}}(\boldsymbol{x})$.

MSC part is finished. The loop over the sites is the bulk of MSC, where we compute n_b per-site observables at a time through bitwise operations. In C the loops appears as (along with some variable declaration)

Listing B.1: C code for MSC: Variable declarations. The // symbols indicate that the rest of the line is commented.

```
//Where we store the final overlaps - 6 permutations, n_b samples
int q        [ 6][NUMBITS];
int temporal[12][NUMBITS];

//Temporary variables to store the four spins
MYWORD spinA,spinB,spinC,spinD;
MYWORD temp;

//Temporary variables to store the six overlaps
MYWORD spinAB,spinAC,spinAD,spinBC,spinBD,spinCD;

//space_N[12] is a set of buffers, used to store large vectors
//They are defined as global variables
//MYWORD space_N[12][N];

//Buffers for large vectors
aguja1_AB_plus =space_N[0]; //Store positive values of X1_ABCD
aguja1_AB_minus=space_N[1]; //Negative values of X1_ABCD
aguja1_AC_plus =space_N[2]; //Positive values of X1_ACBD
aguja1_AC_minus=space_N[3]; //Negative values of X1_ACBD
aguja1_AD_plus =space_N[4]; //Positive values of X1_ADBC
aguja1_AD_minus=space_N[5]; //Negative values of X1_ADBC
agujaQ_AB=space_N[6];       //Overlaps qAB
agujaQ_AC=space_N[7];       //Overlaps qAC
agujaQ_AD=space_N[8];       //Overlaps qAD
agujaQ_BC=space_N[9];       //Overlaps qBC
agujaQ_BD=space_N[10];      //Overlaps qBD
agujaQ_CD=space_N[11];      //Overlaps qCD

for(site=0; site<N; site++)
{
    spinA=u[i0][0][site];
    spinB=u[i1][1][site];
    spinC=u[i2][2][site];
    spinD=u[i3][3][site];
       .
       .
       .
```

The first step is calculating the overlaps between couples of replicas. The XOR logic gate ($\wedge$ in C) between two bits returns 1 if they are different, and 0 if they are the same. It can be used to represent the overlap between two spins. Calling $b_x^{\mathrm{ab}} = b_x^{\mathrm{a}} \wedge b_x^{\mathrm{b}}$ the value of the bit representing the overlap $q_x^{(\mathrm{ab})}$, will be

$$q_x^{(\mathrm{ab})} = \begin{cases} +1, & \text{if } b_x^{\mathrm{ab}} = 0 \\ -1, & \text{if } b_x^{\mathrm{ab}} = 1 \end{cases}. \tag{B.32}$$

Calling `_my_xor(out,in1,in2)` a function (or macro) that returns as `out` the bitwise XOR between `in1` and `in2`, the code continues as

Listing B.2: C code for MSC: Computing overlaps with MSC

```
    .
    .
    .
//Overlaps computed with XOR gates
    _my_xor(spinAB,spinA,spinB); // AB=A^B
    _my_xor(spinAC,spinA,spinC); // AC=A^C
    _my_xor(spinAD,spinA,spinD); // AD=A^D
    _my_xor(spinBC,spinB,spinC); // BC=B^C
    _my_xor(spinBD,spinB,spinD); // BD=B^D
    _my_xor(spinCD,spinC,spinD); // CD=C^D

//Store the local overlaps
    agujaQ_AB[site]=spinAB;
    agujaQ_AC[site]=spinAC;
    agujaQ_AD[site]=spinAD;
    agujaQ_BC[site]=spinBC;
    agujaQ_BD[site]=spinBD;
    agujaQ_CD[site]=spinCD;
    .
    .
    .
```

For the fields X_1 the calculation is more involved, because we need to use 2 bits. Among the several possibilities, we decide to use the two necessary bits independently. One bit stores the positive values, and the other stores the negative values. So, if the two are the same, the value of the variable is zero, otherwise it is $+1$ or -1 depending on which of the two is non-zero.

The difference between two spins $s_x^{\mathrm{a}} - s_x^{\mathrm{b}}$ can assume the values $-2, 0, 2$. It is zero if the are the same, i.e. if their overlap is equal to $q_x^{\mathrm{ab}} = 1$ [and $b_x^{\mathrm{ab}} = 0$, for Eq. (B.32)]. Taking the example of the field X_1^{abcd}, if either $b_x^{\mathrm{ab}} = 0$ or $b_x^{\mathrm{cd}} = 0$, then the whole product is zero. For the field X_1^{abcd} to be non zero we need $q_x^{\mathrm{ab}} = q_x^{\mathrm{cd}} = -1$ [$b_x^{\mathrm{ab}} = b_x^{\mathrm{cd}} = 1$].

The AND gate (& in C), returns a 0 unless both input bits are 1, so $X_1^{\mathrm{abcd}} \neq 0$ if and only if $b_x^{\mathrm{ab}} \,\&\, b_x^{\mathrm{cd}} = 1$. In that case we have to understand what sign it assumes.

Given $s_x^{\mathrm{a}} - s_x^{\mathrm{b}} \neq 0$, if $s_x^{\mathrm{a}} = 1$ then $s_x^{\mathrm{a}} - s_x^{\mathrm{b}} = 2$, and if $s_x^{\mathrm{a}} = -1$ then $s_x^{\mathrm{a}} - s_x^{\mathrm{b}} = -2$. The same holds for $s_x^{\mathrm{c}} - s_x^{\mathrm{d}}$. So, the product between the aforementioned differences, X_1^{abcd}, is inferable by comparing s_x^{a} with s_x^{c}

$$\mathrm{sign}(X_1^{\mathrm{abcd}}) = \begin{cases} +, & \text{if } q_x^{\mathrm{ac}} = +1 \ (b_x^{\mathrm{ac}} = 0) \\ -, & \text{if } q_x^{\mathrm{ac}} = -1 \ (b_x^{\mathrm{ac}} = 1). \end{cases} \tag{B.33}$$

To represent this with bitwise operations first we calculate the auxiliary value `temp`. Having `temp=1` is a necessary condition for a positive X_1^{abcd}, so (`temp` AND b^{ac}) is 1 if and only if $X_1^{\mathrm{abcd}} = 1$. This means that we can store the bit (`temp` AND b^{ac}) for the negative values of X_1. Equivalently, for the positive values we can use a NAND

[NOT AND, ∼& in C (the simple not is ∼)] gate. The following commented code clarifies the procedure[8]

Listing B.3: C code for MSC: Creating the fields with MSC

```
    .
    .
    .
//////////////////////////////////////////////////
// We want to create the following fields       //
//                                              //
// X1[0][i]=(u[a][i]-u[b][i])*(u[c][i]-u[d][i]) //
// X1[1][i]=(u[a][i]-u[c][i])*(u[b][i]-u[d][i]) //
// X1[2][i]=(u[a][i]-u[d][i])*(u[b][i]-u[c][i]) //
//////////////////////////////////////////////////


//First field
// X1AB=(sA-sB)*(sC-sD) //
    _my_and(temp,spinAB,spinCD);                  // temp=AB&CD :
                                                  // temp=0 ==> X1=0
                                                  // temp=1 ==> X1=-2,+2
// (~AC)&(AB&CD) :
//~AC gives positive values ==> store in aguja1_AB_plus
    _my_andnot(aguja1_AB_plus[0],spinAC,temp);
//   AC &(AB&CD) :
// AC gives negative values ==> store in aguja1_AB_minus
    _my_and(aguja1_AB_minus[0],spinAC,temp);
    aguja1_AB_plus++;                             //Pass to the next site
    aguja1_AB_minus++;                            //Pass to the next site



//Second field
// X1AC     [B<>C]
    _my_and(temp,spinAC,spinBD);                  // temp=AC&BD :
                                  // temp=0 ==> X1=0
                                  // temp=1 ==> X1=-2,+2
// (~AB)&(AC&BD) :
//~AB gives positive values ==> store in aguja1_AC_plus
    _my_andnot(aguja1_AC_plus[0],spinAB,temp);
//   AB &(AC&BD) :
// AB gives negative values ==> store in aguja1_AC_minus
    _my_and(aguja1_AC_minus[0],spinAB,temp);
```

[8]The code contains the logic-gate macros for the AND gate, `_my_and(out,in1,in2)`, and for the NAND, `_my_andnot(out,in1,in2)`. In both cases the two words `in1` and `in2` are the input, and `out` is the output.

```
        aguja1_AC_plus++;                        //Pass to the next site
        aguja1_AC_minus++;                       //Pass to the next site



    //Third field
    // X1AD    [B<>D]
        _my_and(temp,spinAD,spinBC);             // temp=AD&BC :
                                    // temp=0 ==> X1=0
                                    // temp=1 ==> X1=-2,+2
    // (~AD)&(AD&BC) :
    //~AD gives positive values ==> store in aguja1_AD_plus
        _my_andnot(aguja1_AD_plus[0],spinAC,temp); // (~AD)&(AD&BC) :~AD
           gives positive values ==> store in aguja1_AD_plus
    //   AD &(AD&BC) :
    // AD gives negative values ==> store in aguja1_AD_minus
        _my_and(aguja1_AD_minus[0],spinAC,temp);   //   AD &(AD&BC) : AD
           gives negative values ==> store in aguja1_AD_minus
        aguja1_AD_plus++;                        //Pass to the next site
        aguja1_AD_minus++;                       //Pass to the next site

    }// close the loop for(site=0; site<N; site++)
       .
       .
       .
```

Once the MSC loop is finished we have $3N$ words (one per site per permutation) each containing the site-dependent field $X_1^{\mathrm{abcd}}(\boldsymbol{x})$ for the set of n_b samples. The final step is to transform this in a sample-dependent quantity over which it is possible to perform normal arithmetic operations. Practically, we want to transform the bits in numbers.

To this objective we call a generic function `suma_booleana(buffer, size, n_bits, obs)` that takes the `buffer` where the N n_b-words are stored, and yields an array of n_b elements—one per sample—each containing information on the variable over the whole system. In other words we pass from N words each describing a site, to n_b values, each describing a sample. This can be done through $O(\log(N))$ operations.

In the following listing we show how this was done with the overlap, with the array `q[6][NUMBITS]`, defined in listing B.1, that contains the count of how many overlaps $q_{\boldsymbol{x}} = -1$ there are in each system, for the six combinations of the replicas and the n_b samples. In general this function will need as extra input also the size of the lattice `size=`N, and the number of bits `n_bits` that are necessary to construct that number (usually `n_bits=` $\log_2 N$).

Listing B.4: C code for MSC: From the multi-spin to the traditional formalism

```
    .
    .
    .
for (k=0;k<6;k++) //Loop over the 6 overlaps (AB,AC,AD,BC,BD,CD)
{
    //q[i][ibit] counts how many local overlaps q_x=-1 there are
    suma_booleana(space_N[6+k],N,bits_of_N, q[k] );
}
    .
    .
    .
```

Regarding the correlation functions the situation is slightly more complicated. We want to average the field X_1 not over the whole lattice, but over specific planes, in order to be able to compute the correlation at distance r. We define NPLANES planes, along the directions we want to average over (privileged directions are easier to code), and loop over them. For each direction we make a loop over the distances, and for each distance we perform the following operations:

(A) The first step to average X_1 over the plane is to create a buffer with only the sites regarding that plane. This is done for the 3 permutations of the replica indices. For each permutation we have the positive- and the negative-value buffer, that makes 6 buffers in total.

(B) We expand each of the six buffers with sum_booleana, this time over an $L * L$-dimensional space. We store those data, regarding a single r of a single direction, in 6 temporal variables temporal (declared in listing B.1).

(C) We store each plane with an array sumplane (declared in listing B.1) that depends on the parameters of all the nested loops: plane direction o, plane position r, replica permutation k, and sample ibit. The storage has to be performed through the operation temporal[2*k] [ibit]-temporal[2*k+1] [ibit], because temporal[2*k] [ibit] stores the number of sites with $X_1(\boldsymbol{x}) = 1$, and temporal[2*k+1] [ibit] has information on the number of sites with $X_1(\boldsymbol{x}) = -1$, so the full sum $\sum_{\boldsymbol{x} \in \text{plane}} X_1(\boldsymbol{x})$ is obtained by subtracting one from the other.

The C code is as follows

Listing B.5: C code for MSC: Storing the X_1 regarding each plane

```
    .
    .
    .
//space_N[12] & space_S[6] are sets of buffers to store large vectors
//They are defined as global variables
//MYWORD space_N[12][N],
//MYWORD space_S[ 6][L*L];

for (o=0;o<NPLANES;o++) // Loop in plane orientations
{
    for (r=0;r<L;r++)
    {
        /////////
        // (A) //
        /////////
        for (j=0;j<L*L;j++)
        {
            i=plane[o][r][j];
            for (k=0;k<6;k++)
                space_S[k][j]=space_V[k][i];
        }
        /////////
        // (B) //
        /////////
        for (k=0;k<6;k++)
        {
            suma_booleana(space_S[k],L*L,bits_de_S,temporal[k]);

            //In temporal[k] we have, for each sample,
            //the sum of the X1 of type k of a plane:
            // 0 <= temporal[k][ibit] <= L*L
            //
            //k even: counts how many positive X1
            //k  odd: counts how many negative X1
            }
        /////////
        // (C) //
        /////////
        //Loop over the n_b samples
        for (ibit=0;ibit<NUMBITS;ibit++)
        {
            //Loop over the 3 permutations of the replica indices
            for (k=0;k<3;k++)
                sumplane[k][o][r][ibit]=temporal[2*k][ibit]-temporal
                   [2*k+1][ibit];

            //sumplane is declared as a global variable
            //int sumplane[6][NPLANES][L][NUMBITS];
            //
            //temporal[2*k][ibit] counts the number of times X1(x)=1
            //temporal[2*k+1][ibit] the number of times X1(x)=-1
            //temporal[2*k][ibit]-temporal[2*k+1][ibit]: sum_x X1(x)
        }//ibit
    }//r
}//o
```

At this point the analysis can procede in the traditional way, by computing the plane correlators with `sumplane`.

At the end of the full procedure we will have to proceed with the correct normalization of the correlators, taking in account for example that the X_1 we calculated is a factor 4 smaller than its actual value.

Appendix C
Technical Details on the Creation of Quantiles

To grant the reproducibility of our results in Chap. 2, we give details on how we proceeded in the labelling of the observables with the conditioning variate (CV), and over the definition of the quantiles. Section C.1 is dedicated to the construction of the pdf of the CV and Sect. C.2 to that of the quantiles. In Sect. C.3 we show that by using two-replica instead of four-replica correlation functions the quantile description give a similar result, with the first quantiles do not show signs of scale invariance, but the ξ/L and R_{12} related to the median do suggest a phase transition.

C.1 Creating the $P(\hat{q})$

As already explained in Sect. 2.2 the analysis we conduct uses instantaneous realisations of the observables, instead of the average over the equilibrium regime. This is because computing $P(\hat{q})$ properly requires as many instances of the overlap as possible.

Operatively, we discard the first half of each simulation from the measurements because out of equilibrium. We divide the second half of the simulation time-series in 16 blocks, and for the 4 replicas we save the final configuration of each block. This gives us 16^4 configurations over which we can potentially compute overlaps for a single sample. Since it is not feasible to make measurements over the 16^4 times per sample, for N_t times we pick 4 random numbers between 1 and 16 to create an instant measure. This way we increase our statistics of a factor N_t, obtaining $\mathcal{N}_m = N_{samples}(L, T, h) \times N_t$ measures for each triplet (L, T, h). We used $N_t = 1000$.

With the 4 replicas it is possible to compute 6 different overlaps q_i $(i = 1, \ldots, 6)$, and one instance of most observables, for example the replicon susceptibility χ_R. Our ansatz is that χ_R and the overlaps have some type of correlation, so we label χ_R with some function of the overlaps $\hat{q}(q_1, \ldots, q_6)$, that we called conditioning variate.

The random variable $\hat{q}$ will have a probability distribution function $P(\hat{q})$ that we want to calculate numerically, in order to be able to work on the quantiles. Since

M. Baity Jesi, *Spin Glasses*, Springer Theses, DOI 10.1007/978-3-319-41231-3

our objective is not to individuate exactly the quantiles, but to compute observables related to a particular quantile, we coarse grain the range of definition of the $P(\hat{q})$. This is done by making a binning of the $P(\hat{q})$ [Eqs. (C.1, C.2) here below]. This way, each conditioned expectation value of a generic observable, $E(\mathcal{O}|\hat{q})$, can be calculated over a reasonable amount of measurements, and we have exactly one conditioned expectation value for each bin of the $P(\hat{q})$. Integrals such as those in (2.4) and (2.7) are computed as sums over the histogram bins. Furthermore, the described histogramming procedure has the advantage that errors can be calculated in a very natural way with the JK method.

In order to have, as L increases, both a growing number of bins, and of points per bin, we choose bins of width $\Delta\hat{q} = 1/\sqrt{aV}$. We add the restriction of having at least 150 bins, in order to be able to define the quantiles properly (with large bins it could happen that a single bin contain more than 10 % of the pdf, and we want to avoid the eventuality of two quantiles in the same bin). We verified that there is no appreciable difference in the results between $a = 1, 2, 4$. Larger a implies a too large error, because the bins are too small, while with smaller a the bins are too few. The results we show throughout this thesis have $a = 2$.

To compute the conditional expectation values defined in Sect. 2.4 we use the following estimators:

$$E(\mathcal{O}|\hat{q} = c) \approx \frac{\frac{1}{\mathcal{N}_{\mathrm{m}}} \sum_i^{\mathcal{N}_{\mathrm{m}}} \mathcal{O}_i \mathcal{X}_c(\hat{q}_i)}{\frac{1}{\mathcal{N}_{\mathrm{m}}} \sum_i^{\mathcal{N}_{\mathrm{m}}} \mathcal{X}_c(\hat{q}_i)}, \tag{C.1}$$

$$P(\hat{q}) \approx \frac{1}{\mathcal{N}_{\mathrm{m}}} \sum_i^{\mathcal{N}_{\mathrm{m}}} \mathcal{X}_{c=\hat{q}}(\hat{q}_i), \tag{C.2}$$

where with the symbol "$\approx$" we stress that the quantity is an estimator that converges to the exact value only in the limit of an infinite number of measurements $\mathcal{N}_{\mathrm{m}}$. $\mathcal{X}_c$ is the characteristic function defined in Eq. (2.3).

C.2 Defining the Quantiles

As stated in Sect. 2.5, the quantiles are the points that separate definite areas under $P(\hat{q})$. Therefore, the ith quantile $\tilde{q}_i$ is defined by means of the cumulative distribution $X(\hat{q})$ of $P(\hat{q})$, via the implicit relation

$$X(\tilde{q}_i) = \int_{-1}^{\tilde{q}_i} \mathrm{d}\hat{q}\; P(\hat{q}) = \frac{i}{10}. \tag{C.3}$$

Since this is a continuous relation, and our binning is discrete, it is most probable that the quantile fall between two neighbouring bins. To evaluate the observables right at the position of the quantile, we make linear interpolations between the two bins.

Let us call $i_{\text{bin}}^{-}(i_{\text{bin}}^{+})$ the bin just under (over) quantile i. Observable $\mathcal{O}_i$ at quantile i will be a linear combination of the values it assumes at i_{bin}^{-} and i_{bin}^{+}:

$$\mathcal{O}_i = p\,\mathcal{O}_{i_{\text{bin}}^{-}} + (1-p)\,\mathcal{O}_{i_{\text{bin}}^{+}}, \tag{C.4}$$

where the interpretation of the indices is straightforward, and $0 \le p \le 1$ is the interpolation weight

$$p = \frac{X(\tilde{q}_i) - X(\hat{q}_{i_{\text{bin}}^{+}})}{X(\hat{q}_{i_{\text{bin}}^{-}}) - X(\hat{q}_{i_{\text{bin}}^{+}})}. \tag{C.5}$$

C.3 Quantiles with 2-Replica Correlators

To have well behaving (connected) correlators in the presence of a magnetic field we needed to use 4 replicas for each instance of them. As explained in Sects. 2.4 and 2.5, since the overlap is a 2-replica observable, we had to choose a function of the 6 overlaps in order to have a one-to-one correspondence between conditioning variates and the correlators. The functions we tried out were the minimum, the maximum, the median and the average of the 6 overlaps.

Now, it is legitimate to ask oneself if the fluctuations we observed would also be visible having q as conditioning variate. Although this is not possible with the replicon correlation function G_{R}, we can renounce to have a connected correlation function, and study the fluctuations of the 2-replica point-to-plane correlator

$$G_2^{\text{nc}}(r) = \sum_{y,z} E(q_{(0,0,0)}\, q_{(r,y,z)}), \tag{C.6}$$

which allows us to have q as a conditioning variate. $G_2^{\text{nc}}(r)$ is the total correlation between the origin, (0, 0, 0), and the plane $x = r$. Of course, one could equivalently consider the planes $y = r$ or $z = r$. One can displace freely the origin, as well. We average over all these $3V$ choices.

At this point, it is possible to compare with previous work that studied fluctuations with 2-replica correlators [Par12a]. Furthermore, we can construct the pseudoconnected correlation function

$$G_2^{\text{c}}(r) = \frac{G_2^{\text{nc}}(r) - G_2^{\text{nc}}(L/2)}{G_2^{\text{nc}}(0) - G_2^{\text{nc}}(L/2)}, \tag{C.7}$$

which forcedly is one for $r = 0$, and goes to zero for $r = L/2$. In Fig. C.1 we show that the same dramatic fluctuations encountered with G_{R} (Fig. 2.3) are also present here.

The overall results, Fig. C.2, are consistent with the picture we draw in Sect. 2.9. On the one hand, the standard data average hides all signs of a phase transition. On the other hand, the fifth quantile displays signs of scale invariance.

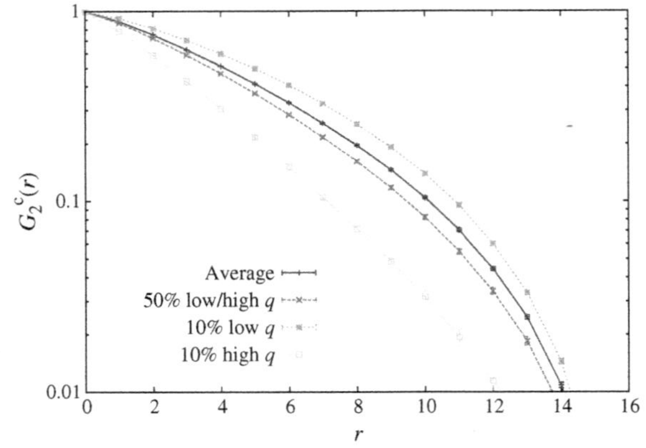

Fig. C.1 Same as Fig. 2.3, but for the 2-replica connected correlation function $G_2^c(r)$ (C.7). We show $L = 32$ data from $h = 0.2$, $T = 0.805128$. Note that $G_2^c(r)$ is bound to be 1 at $r = 0$, and 0 at $r = L/2$, so the fluctuations between different quantiles are even stronger than they may appear

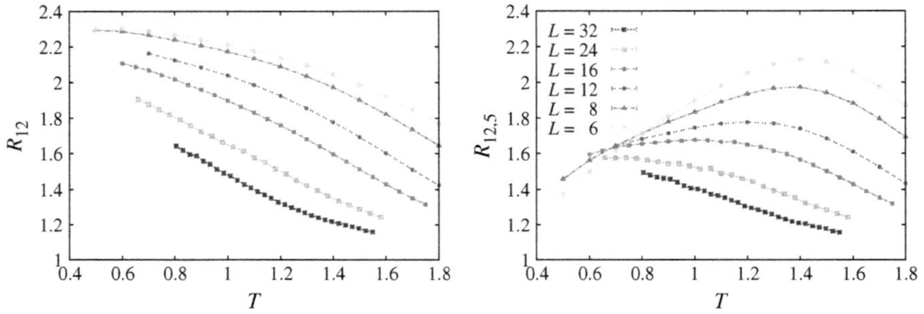

Fig. C.2 The R_{12} cumulant computed from the two-replica correlation function (C.6) rather than from four replicas. The field is $h = 0.2$. On the *left side* we show the average behavior, and on the *right*, the 5th quantile, with the plain overlap q (1.27) as conditioning variate

Appendix D
Decomposing Conditional Expectations

We want to derive here some useful relations pertinent to the conditioned expectations of Chap. 2 that can be used to have a quantitative criterion for the conditioning variate (Sect. 2.4.3) and to check that the statistical analysis code is reliable.

D.1 Variance

In Sect. 2.4.3 we used the integral rule

$$\mathrm{var}(\mathcal{O}) = E\left(\left[\mathcal{O} - E(\mathcal{O})\right]^2\right) = \int_{-1}^{1} d\hat{q}\, P(\hat{q}) \left\{\mathrm{var}(\mathcal{O}|\hat{q}) + \left[E(\mathcal{O}) - E(\mathcal{O}|\hat{q})\right]^2\right\}, \tag{D.1}$$

$$\mathrm{var}(\mathcal{O}|\,\hat{q}) = E\left(\left[\mathcal{O} - E(\mathcal{O}|\hat{q})\right]^2 \mid \hat{q}\right), \tag{D.2}$$

to choose the best CV. The $P(\hat{q}) = E[\mathcal{X}_{\hat{q}}]$, when computed numerically, is actually an empirical probability over the whole set of $\mathcal{N}_{\mathrm{m}}$ measurements,

$$P(\hat{q}) \approx h(\hat{q}) = \frac{\sum_i^{\mathcal{N}_{\mathrm{m}}} \mathcal{X}_{\hat{q}_i}(\hat{q})}{\int_{-1}^{1} d\hat{q} \sum_i^{\mathcal{N}_{\mathrm{m}}} \mathcal{X}_{\hat{q}_i}(\hat{q})}, \tag{D.3}$$

where the i labels the measurements, and $\hat{q}_i$ the value of the CV for measurement i.

Relation (D.1) is easily shown to be true by applying Eq. (2.4) to the variance of $\mathcal{O}$, $\mathrm{var}(\mathcal{O})$ and summing zero to it:

$$E\left(\left[\mathcal{O} - E(\mathcal{O})\right]^2\right) =$$

$$= \int_{-1}^{1} d\hat{q}\, P(\hat{q}) E\left(\left[\mathcal{O} - E(\mathcal{O})\right]^2 \middle| \hat{q}\right) =$$

M. Baity Jesi, *Spin Glasses*, Springer Theses, DOI 10.1007/978-3-319-41231-3

$$= \int_{-1}^{1} d\hat{q}\, P(\hat{q}) \left\{ E\left(\mathcal{O}^2 | \hat{q}\right) + E(\mathcal{O})^2 - 2E(\mathcal{O})E(\mathcal{O}|\,\hat{q}) + \left[E(\mathcal{O}|\,\hat{q})^2 - E(\mathcal{O}|\,\hat{q})^2\right]\right\} =$$

$$= \int_{-1}^{1} d\hat{q}\, P(\hat{q}) \left\{ \mathrm{var}(\mathcal{O}|\,\hat{q}) + \left[E(\mathcal{O}) - E(\mathcal{O}|\hat{q})\right]^2 \right\}. \tag{D.4}$$

D.2 Higher Moments

The same procedure can be used to find a relation for higher moments. The skewness of observable $\mathcal{O}$ is

$$\mathrm{S}(\mathcal{O}) = E\left[\left[\mathcal{O} - E(\mathcal{O})\right]^3\right] = \int_{-1}^{1} d\hat{q}\, P(\hat{q}) E\left(\left[\mathcal{O} - E(\mathcal{O})\right]^3 |\, \hat{q}\right). \tag{D.5}$$

To simplify the notation let us write

$$\begin{aligned} \hat{E}(\mathcal{O}) &= E(\mathcal{O}|\,\hat{q}), \\ \hat{\mathrm{S}}(\mathcal{O}) &= \mathrm{S}(\mathcal{O}|\,\hat{q}) = E\left(\left[\mathcal{O} - E(\mathcal{O}|\,\hat{q})\right]^3 |\, \hat{q}\right), \end{aligned} \tag{D.6}$$

so, opening the cube,

$$\mathrm{S}(\mathcal{O}) =$$

$$= \int_{-1}^{1} d\hat{q}\, P(\hat{q}) \left\{ \hat{E}\left(\mathcal{O}^3 - E(\mathcal{O})^3 - 3\mathcal{O}^2 E(\mathcal{O}) + 3\mathcal{O} E(\mathcal{O})^2\right)\right\} =$$

$$= \int_{-1}^{1} d\hat{q}\, P(\hat{q}) \left\{ \left[\hat{E}\left(\mathcal{O}^3\right) + 2\hat{E}(\mathcal{O})^3 - 3\hat{E}\left(\mathcal{O}^2\right)\hat{E}(\mathcal{O})\right] + \right.$$

$$\left. - E(\mathcal{O})^3 + 3\hat{E}\left(\mathcal{O}^2 \hat{E}(\mathcal{O}) + 3\hat{E}\left(\mathcal{O}^2\right) E(\mathcal{O}) - 3\hat{E}\left(\mathcal{O}^2\right) E(\mathcal{O}) - 2\hat{E}(\mathcal{O})^3\right)\right\} =,$$

the term in square brackets is equal to $\hat{\mathrm{S}}(\mathcal{O})$

$$= \int_{-1}^{1} d\hat{q}\, P(\hat{q}) \Big\{ \hat{\mathrm{S}}(\mathcal{O}) + 3\hat{E}\left(\mathcal{O}^2\right)\left[\hat{E}(\mathcal{O}) - E(\mathcal{O})\right] + \hat{E}(\mathcal{O})^3 - \hat{E}(\mathcal{O})^3 - E(\mathcal{O})^3 + $$
$$-3\hat{E}(\mathcal{O})^2 E(\mathcal{O}) + 3\hat{E}(\mathcal{O})E(\mathcal{O})^2 - 2\hat{E}(\mathcal{O})^3 + 3\hat{E}(\mathcal{O})^2 E(\mathcal{O}) \Big\} =$$
$$= \int_{-1}^{1} d\hat{q}\, P(\hat{q}) \Big\{ \hat{\mathrm{S}}(\mathcal{O}) + 3\hat{E}\left(\mathcal{O}^2\right)\left[\hat{E}(\mathcal{O}) - E(\mathcal{O})\right] + \left[\hat{E}(\mathcal{O}) - E(\mathcal{O})\right]^3 + $$
$$-3\hat{E}(\mathcal{O})^2\left[\hat{E}(\mathcal{O}) - E(\mathcal{O})\right]\Big\} =$$
$$= \int_{-1}^{1} d\hat{q}\, P(\hat{q}) \left\{ \hat{\mathrm{S}}(\mathcal{O}) + \left[\hat{E}(\mathcal{O}) - E(\mathcal{O})\right]\left(3\,\mathrm{var}(\mathcal{O}|\,\hat{q}) + \left[\hat{E}(\mathcal{O}) - E(\mathcal{O})\right]^2\right)\right\},$$

that can also be rewritten as

$$\mathrm{S}(\mathcal{O}) = \int_{-1}^{1} d\hat{q}\, P(\hat{q}) \left\{ \hat{\mathrm{S}}(\mathcal{O}) + 3\,\mathrm{var}(\mathcal{O}|\,\hat{q})\left[\hat{E}(\mathcal{O}) - E(\mathcal{O})\right] + \left[\hat{E}(\mathcal{O}) - E(\mathcal{O})\right]^3\right\}. \tag{D.7}$$

Operatively, in our spin systems we define two types of skewness of the overlap, depending on the replicas we use

$$\mathrm{S}_2(q) = E\left[\left(q^{(\mathrm{ab})} - E(q)\right)^3\right], \tag{D.8}$$
$$\mathrm{S}_3(q) = E\left[\left(q^{(\mathrm{ab})} - E(q)\right)\left(q^{(\mathrm{ac})} - E(q)\right)\left(q^{(\mathrm{bc})} - E(q)\right)\right]. \tag{D.9}$$

Applying Eq. (D.7) to $\mathrm{S}_2(q)$ is straightforward, while for $\mathrm{S}_3(q)$ we have to apply some little modification specifying the replica

$$\mathrm{S}_3(q) = \int_{-1}^{1} d\hat{q}\, P(\hat{q}) \Big\{ \hat{E}\left(q^{(\mathrm{ab})}q^{(\mathrm{ac})}q^{(\mathrm{bc})}\right) - E(q)\hat{E}\left(q^{(\mathrm{ab})}q^{(\mathrm{bc})} + q^{(\mathrm{ac})}q^{(\mathrm{bc})} + q^{(\mathrm{ab})}q^{(\mathrm{ac})}\right)$$
$$+E(q)^2\hat{E}\left(q^{(\mathrm{ab})} + q^{(\mathrm{ac})} + q^{(\mathrm{bc})}\right) - E(q)^3\Big\}. \tag{D.10}$$

The terms in Eq. D.10 can be easily computed in our analysis out of the four simulated replicas

$$\hat{E}\left(q^{(\mathrm{ab})}q^{(\mathrm{ac})}q^{(\mathrm{bc})}\right) = \frac{1}{4}\sum_{\alpha\neq\beta\neq\gamma}\hat{E}\left(q^{(\alpha\beta)}q^{(\alpha\gamma)}q^{(\beta\gamma)}\right), \tag{D.11}$$

$$\hat{E}\left(q^{(\mathrm{ab})}q^{(\mathrm{bc})}\right) = \frac{1}{12}\sum_{\alpha\neq\beta\neq\gamma}\hat{E}\left(q^{(\alpha\beta)}q^{(\beta\gamma)} + q^{(\alpha\gamma)}q^{(\beta\gamma)} + q^{(\alpha\beta)}q^{(\alpha\gamma)}\right), \tag{D.12}$$

$$\hat{E}\left(q^{(\mathrm{ab})}\right) = \frac{1}{6}\sum_{\alpha\neq\beta}\hat{E}\left(q^{(\alpha\beta)}\right), \tag{D.13}$$

where the indices α, β, γ in the sums indicate the different replicas.

We give the same expression for the kurtosis $\mathrm{K} = E\left[\left(q^{(\mathrm{ab})} - E(q)\right)^4\right]$

$$\mathrm{K} = \int_{-1}^{1} d\hat{q}\, P(\hat{q})\left\{\hat{\mathrm{K}}(q) + \left[\hat{E}\,(q) - E(q)\right]^4 + 4\,\hat{\mathrm{S}}(q)\left[\hat{E}\,(q) - E(q)\right] + \right.$$
$$\left. + 6\,\mathrm{var}(q|\hat{q})\left[\hat{E}\,(q) - E(q)\right]^2\right\}, \tag{D.14}$$

where we introduced $\hat{\mathrm{K}}(q) = \hat{E}\left(\left[q - \hat{E}\,(q)\,\right]^4\right)$.

More in general, we find that for the nth moment $K_n(\mathcal{O})$

$$K_n(\mathcal{O}) = \int_{-1}^{1} d\hat{q}\; h(\hat{q})\sum_{i=0}^{n}\binom{n}{i} K_i(\mathcal{O}|\hat{q})\left[E(\mathcal{O}) - E(\mathcal{O}|\hat{q})\right]^{n-i}, \tag{D.15}$$

where we have to notice that $K_1(\mathcal{O}|\hat{q}) = 0$.

D.3 Consistency Checks on the Correlation Functions

Since in our analyses we often measure both the correlation function $C(\boldsymbol{r})$ [Eq. 1.47)] and its Fourier transform $\chi(\boldsymbol{k})$ [Eq. 1.48)], it is useful from a programming point of view to have some constraints that tie one to the other. Our programs were quite intricated, and these constraints, despite their easy derivation, revealed crucial to keep the code under control.

Since $C(\boldsymbol{r}) = C(-\boldsymbol{r})$, and because of the periodic boundary conditions, when we calculate correlation functions along an axis, $C(r) = C(L-r)$, and $\hat{C}(k) = \hat{C}(L-k)$. Moreover, the wave numbers restrict to $k = 2\pi n/L$ ($n = 0, \ldots, L-1$), so let us label them with the integer index n, $\hat{C}(k(n)) = \hat{C}(n)$.

These symmetries give us the chance to create simple constraints on the correlators to check their consistency. The correlation function has to be expressable as anti Fourier transform of the $\hat{C}(k)$ through

$$C\left(r\right)=\hat{C}\left(0\right)+2\sum_{n=1}^{L/2-1}\hat{C}(n)\cos\left(\frac{2\pi n}{L}\right)+\hat{C}\left(\frac{L}{2}\right). \tag{D.16}$$

On the reverse way, we easily get basic constraints on the $\hat{C}(n)$ for some specific value of n:

$$\hat{C}\left(0\right)=C\left(0\right)+2\sum_{r=1}^{L/2-1}C\left(r\right)+C\left(\frac{L}{2}\right), \tag{D.17}$$

$$\hat{C}\left(\frac{L}{2}\right)=C\left(0\right)+2\sum_{r=1}^{L/2-1}C\left(r\right)(-1)^{r}+C\left(\frac{L}{2}\right)(-1)^{L/2}. \tag{D.18}$$

We can also get a constraint for $\hat{C}(L/4)$,

$$\hat{C}\left(\frac{L}{4}\right)=C\left(0\right)+2\sum_{r=1}^{L/2-1}C\left(r\right)\cos\left(\frac{\pi r}{2}\right)+C\left(\frac{L}{2}\right)\cos\left(\frac{\pi L}{4}\right),$$

and since r is an integer index and the cosines' arguments are multiples of $\pi/2$, we can reexpress it as

$$\hat{C}\left(\frac{L}{4}\right)=C\left(0\right)+2\sum_{r=1}^{L/2-1}C\left(r\right)\left[1+(-1)^{r}\right](-1)^{r/2}+C\left(\frac{L}{2}\right)\cos\left(\frac{\pi L}{4}\right). \tag{D.19}$$

These tests were performed both on the average and on the per-quantile correlation functions.

Appendix E
Managing the Errors

The observables $\mathcal{O}$ measured in the numerical experiments shown in this dissertation suffer from two noises, one due to thermal fluctuations during a single run, and a second one deriving from the disorder. Since we perform measurements at equilibrium, we can treat these measurements as independent identically distributed (i.i.d.) random variables with two independent noises.

Given a a set of $\mathcal{N}$ measurements $\mathcal{O}_i$, their expected value $E(\mathcal{O})$ can be evaluated through an estimator

$$\tilde{E}(\mathcal{O}) = \frac{1}{\mathcal{N}} \sum_{i=1}^{\mathcal{N}} \mathcal{O}_i \tag{E.1}$$

that for the central limit theorem is at a $o(\mathcal{N}^{-1/2})$ distance from e($\mathcal{O}$).

Nonlinear functions of the observables, $f(\mathcal{O})$,[9] can be estimated by evaluating them over the estimator. This results in an estimator $f(\tilde{E}(\mathcal{O}))$ that reproduces the actual expected value $f\,(E(\mathcal{O}))$ with a bias of order $o(\mathcal{N}^{-1})$ (see Sect. B.1). Since this bias is smaller than the statistical error we can neglect it.

We present in this appendix the jackknife and the bootstrap method, that are the two resampling methods that were used to calculate error bars throughout this dissertation. Since these techniques are treated extendedly in literature (see e.g. [You12]), we will limit ourselves to a description of the methodology, with no pretention of originality.

E.1 The Jackknife Method

Being the central value of the linear functions of the observables $f(\mathcal{O})$ estimated as $f\,(E(\mathcal{O}))$, the jackknife (JK) method provides us a way to compute an appropriate uncertainty on it. The idea is to block the data in a way that suppresses fluctuations

[9]For simplicity of notation we treat functions of a single observable, but our statements are also valid for functions of many observables.

M. Baity Jesi, *Spin Glasses*, Springer Theses, DOI 10.1007/978-3-319-41231-3

and time correlations. Given the full set $\mathcal{B}'$ of measurements $\mathcal{O}_i (i = 1, \ldots, \mathcal{N})$, we group them in n blocks $b_j (j = 0, \ldots, n-1)$ of size ℓ, so $n\ell = \mathcal{N}$, getting n per-block estimators

$$\tilde{E}_j\left(\mathcal{O}\right) = \frac{1}{\ell} \sum_{i \in b_j}^{\ell} \mathcal{O}_i \tag{E.2}$$

of the expectation value $E(\mathcal{O})$. From those we contruct JK estimators by creating new JK bins. Each JK bin $b_j^{(\mathrm{JK})}$ contains the full data except that regarding precisely b_j, so $b_j^{(\mathrm{JK})} = \mathcal{B}' \backslash b_j$. The JK estimators are

$$\tilde{E}_j^{(\mathrm{JK})}\left(\mathcal{O}\right) = \frac{1}{\mathcal{N} - \ell} \sum_{i \notin b_j}^{\mathcal{N} - \ell} \mathcal{O}_i = \frac{1}{\mathcal{N} - \ell} \sum_{i \in b_j^{(\mathrm{JK})}}^{\mathcal{N} - \ell} \mathcal{O}_i, \tag{E.3}$$

and over each of them we evaluate the nonlinear function $f_j^{(\mathrm{JK})} = f\left(\tilde{E}_j^{(\mathrm{JK})}(\mathcal{O})\right)$. The JK error estimate σ_f is then

$$\sigma_f = \sqrt{(n-1)\left[\frac{1}{n}\sum_{j=0}^{n-1} f_j^{(\mathrm{JK})2} - \left(\frac{1}{n}\sum_{j=0}^{n-1} f_j^{(\mathrm{JK})}\right)^2\right]}. \tag{E.4}$$

From a programming point of view, it is often useful to define $n+1$ JK blocks, using the the extra one, block n, to store the average, so in the following section we will use the notation $f_n^{(\mathrm{JK})} = f\left(\tilde{E}_n^{(\mathrm{JK})}(\mathcal{O})\right) = f\left(\tilde{E}(\mathcal{O})\right)$.

E.1.1 Variations on the Jackknife Blocks to Reduce the Numerical Rounding Errors

Reducing the rounding errors often reveals fundamental in numerical analyses, since computers only have a finite number of decimal digits to perform arithmetical operations (we always used double precision).

Had we an infinite precision, we would calculate the variance of an observable $\mathcal{O}$ as

$$\mathrm{var}(\mathcal{O}) = E(\mathcal{O}^2) - E(\mathcal{O})^2. \tag{E.5}$$

Yet, this approach is not always numerically stable. If the relative fluctuations are very small there is a very large amount of significant digits between the most significant digit of the averages and the most significant digit of the deviations. This gap may be larger than the numerical precision, and could imply, for instance, that positive-definite quantities such as (E.5) assume negative values. To suppress these rounding errors we exploit the translational invariance of the variance

$$E(\mathcal{O}^2) - E(\mathcal{O})^2 = E\left((\mathcal{O} - c)^2\right) - E(\mathcal{O} - c)^2, \quad \forall c \in \mathbb{R} \tag{E.6}$$

to enhance numerical stability with the convenient choice $c = \tilde{E}(\mathcal{O})$. By measuring quantities with this offset we contain the gap that causes the rounding errors.

Consequently, when we construct the JK blocks we do it in two steps. First we calculate $\tilde{E}(\mathcal{O})$, and only later the variance (or higher moments). With this election Eq. (E.6) becomes

$$\mathrm{var}(\mathcal{O}) = E\left(\left(\mathcal{O} - \tilde{E}(\mathcal{O})\right)^2\right) - E\left(\mathcal{O} - \tilde{E}(\mathcal{O})\right)^2. \tag{E.7}$$

This translates in a correction that we have to apply to every JK block but the nth one, the one that stores the average, because in that case the second term is zero.

One can extend this reasonment to the rth moment of the observable. We show it for the quantile-dependent moments of q, since they were widely used in our programs. Let us use the contracted notations $\tilde{E}_j \equiv \tilde{E}_j(q|\hat{q})$ when the estimator is not followed by parentheses, and expand the polinomial

$$\tilde{E}_j\left(\left[q - \tilde{E}_j(q|\hat{q})\right]^r\right) = \tag{E.8}$$

$$= \tilde{E}_j\left(\left[(q - \tilde{E}_n) - (\tilde{E}_j - \tilde{E}_n)\right]^r\right) = \tag{E.9}$$

$$= \sum_{s=0}^{r} \binom{r}{s} \tilde{E}_j\left((q - \tilde{E}_n)^{r-s}\right)(\tilde{E}_j - \tilde{E}_n)^s. \tag{E.10}$$

The first moments $r = 2, 3, 4$ are

$$\tilde{E}_j\left(\left[q - \tilde{E}_j(q|\hat{q})\right]^2\right) = \tilde{E}_j\left(\left[q - \tilde{E}_n(q|\hat{q})\right]^2\right) - \left(\tilde{E}_j(q|\hat{q}) - \tilde{E}_n(q|\hat{q})\right)^2, \tag{E.11}$$

$$\begin{aligned} \tilde{E}_j\left(\left[q - \tilde{E}_j(q|\hat{q})\right]^3\right) =& \tilde{E}_j\left(\left[q - \tilde{E}_n(q|\hat{q})\right]^3\right) - 2\left(\tilde{E}_j(q|\hat{q}) - \tilde{E}_n(q|\hat{q})\right)^3 + \\ & - 3\tilde{E}_j\left(\left(\left[q - \tilde{E}_n(q|\hat{q})\right]^2\right)\left(\tilde{E}_j(q|\hat{q}) - \tilde{E}_n(q|\hat{q})\right),\right. \end{aligned} \tag{E.12}$$

$$\begin{aligned} \tilde{E}_j\left(\left[q - \tilde{E}_j(q|\hat{q})\right]^4\right) =& \tilde{E}_j\left(\left[q - \tilde{E}_n(q|\hat{q})\right]^4\right) - 3\left(\tilde{E}_j(q|\hat{q}) - \tilde{E}_n(q|\hat{q})\right)^4 + \\ & - 4\tilde{E}_j\left(\left(\left[q - \tilde{E}_n(q|\hat{q})\right]^3\right)\left(\tilde{E}_j(q|\hat{q}) - \tilde{E}_n(q|\hat{q})\right) +\right. \end{aligned} \tag{E.13}$$

$$+\,6\tilde{E}_j\left(\left(\left[q-\tilde{E}_n(q|\hat{q})\right]^2\right)\left(\tilde{E}_j(q|\hat{q})-\tilde{E}_n(q|\hat{q})\right)^2,\right. \tag{E.14}$$

where it is clear that in the nth block all the terms of the right hand sides disappear except the first.

E.2 The Bootstrap Method

The bootstrap method is a valuable tool to calculate mean and variance of an estimator, as well as other moments (see [Efr94] for a detailed treatise). It comes in a wide variety of variants, and we will give the procedure for a very simple one, that we have used in the work here described. Similarly to the JK method, the estimator of the central value is the one described in Eq. (E.1), and the procedure concerns the determination of its uncertainty.

Given a population $\mathcal{X}_0$ of $\mathcal{N}$ measurements we resample it N_b times. Each resampling consists in recreating a population of $\mathcal{N}$ elements, by picking them at random from the initial population. This means that each element of $\mathcal{X}_0$ can appear several times or not appear at all in the generic resampled population $\mathcal{X}_i\,(i=1,\ldots,N_b)$.

From each of the N_b populations we extract quantities x such as the average or the median, and calculate their simple and quadratic averages

$$x_i^{(1)}=\frac{1}{\mathcal{N}}\sum_{j\in\mathcal{X}_i}^{\mathcal{N}}x_j,$$

$$x_i^{(2)}=\frac{1}{\mathcal{N}}\sum_{j\in\mathcal{X}_i}^{\mathcal{N}}x_j^2.$$

The bootstrap error is then

$$\sigma_b=\sqrt{\left(\frac{N_b}{N_b-1}\right)\left[\frac{1}{N_b}\sum_{i=1}^{N_b}x^{(2)}-\left(\frac{1}{N_b}\sum_{i=1}^{N_b}x^{(1)}\right)^2\right]}. \tag{E.15}$$

The magnitude of σ_b does not depend on the number of resamplings N_b, but to take best advantage out of the method it is good that each data point be represented in the resampling, so as a general rule we adopted $N_b=10\mathcal{N}$ to be able to make a proper resampling of the data set.

Appendix F
The Inherent Structures

This appendix refers mainly to Chap. 6 (Sect. F.1.1 refers to Chap. 4), and it is dedicated to show how we found the inherent structures (ISs) (Sect. F.1), to the comparison between ISs reached with different protocols (Sect. F.2), and to the derivation of the Hessian matrix at the local minimum of the energy.

An IS is the configuration to which the system converges when we decide to relax it. When we talk about relaxing, we mean to give the best satisfaction to all the local constraints, that is moving towards the nearest energy minimum. Although this concept seems well-defined, there is an ambiguity related to what one means by *nearest*.

One could in principle define a distance, find all the minima of the energy, and see which of those minimizes this distance. Yet, different definitions of a distance can give different results, and especially in discrete models degeneracies are not excluded by this definition.[10] Moreover, we do not have a way to measure all the local minima of the energy, and even if we had, it is not granted that the physical evolution converge to a minimum defined this way.

More in general, since when we minimize the energy we are following a non-equilibrium procedure, there is a component of arbitrariety on the protocol we use. The mostly used way to minimize the energy in spin systems is through a quench, i.e. with the Gauss-Seidel algorithm (Sect. F.1.1), that is local and minimizes maximally the energy in each update, and can be seen as a zero-temperature MC. Nonetheless, there is no solid reason to state that ISs found with one algorithm are more representative than others, but there also is none to say that all the inherent structures are equivalent. It has been shown in [BJ11] that the algorithm choice does imply some differences on the average properties of the ISs, but we show in this appendix that they are small enough to be neglected.

[10]Two minima can be equivalent candidates for being the IS of an excited configuration.

M. Baity Jesi, *Spin Glasses*, Springer Theses, DOI 10.1007/978-3-319-41231-3

F.1 Minimizing the Energy

We discuss two very simple algorithms of energy minimization that were used in this thesis.

F.1.1 Gauss-Seidel

The most commonly used way to minimize the energy of a SG is the Gauss-Seidel algorithm, that consists in successive local rearrangements of the spins that decrease maximally the local energy. The spin update with Gauss-Seidel consists in aligning each spin to its local field

$$\vec{s}_x^{\,\mathrm{Q}} = \frac{\vec{h}_x}{|\vec{h}_x|}, \quad \vec{h}_x = \sum_{\|x-y\|} J_{xy}\vec{s}_x \tag{F.1}$$

Energy minimizations with the Gauss-Seidel algorithm are often called quenches, since they consist in lowering abruptly the energy (temperature) of the system. Since sometimes in literature also variants of Gauss-Seidel have also been called quenches, one also refers to Gauss-Seidel as a greedy quench.

The problem with Gauss-Seidel is that despite a very fast initial decrease of the energy, after few steps its convergence to a local minimum becomes so slow that the algorithm is not usable to obtain ISs on large lattices (see e.g. [Sok92], where it is explained that in systems with continuous degrees of freedom convergence problems arise).

F.1.2 Successive Overrelaxation

To overcome the convergence trouble of the quenches, we recur to the successive overrelaxation (SOR), that consists in an interpolation, through a parameter Λ, between a greedy quench with the Gauss-Seidel algorithm, and the microcanonical OR update shown in Appendix A.1.

We propose sequential single-flip updates with the rule

$$\vec{s}_x^{\,\mathrm{SOR}} = \frac{\vec{h}_x + \Lambda\vec{s}_x^{\,\mathrm{OR}}}{||\vec{h}_x + \Lambda\vec{s}_x^{\,\mathrm{OR}}||}. \tag{F.2}$$

The limit $\Lambda = 0$ corresponds to a direct quench that notoriously presents convergence problems. On the other side, with $\Lambda = \infty$ the energy does not decrease.

It is shown in [BJ11] that the optimal value of Λ in terms of convergence speed is $\Lambda \approx 300$.

F.2 Testing the Dependency on T and Λ

In Chap. 6 we used SOR with $\Lambda = 300$ because the Gauss-Seidel algorithm, that is recovered by setting $\Lambda = 0$, has strong convergence problems and it was not possible to reach the ISs for the system sizes we needed. To validate the generality of our results we compared the ISs reached with $\Lambda = 300$ and $\Lambda = 1$, at $H_{\text{amp}} = 0$ over a wide range of temperatures. We took advantage, for this comparison, of the $L = 48$ configurations that were thermalized in [Fer09b], that go from T_{SG} to $\frac{5}{3}T_{\text{SG}}$.

In Fig. F.1 we plot the energy e_{IS} of the reached ISs, as a function of the temperature T. We show ten random samples, each minimized with $\Lambda = 1{,}300$. Increasing Λ the energy of the inherent structures decreases but this variation is smaller than the dispersion between different samples. The energy of the ISs also decreases with T, but this decrease too is smaller than the fluctuation between samples. Since the dispersion on the energy is dominated by the disorder, rather than by Λ or T, we can think of putting ourselves in the most convenient situation: $T = \infty$, that does not require thermalization and $\Lambda = 300$, that yields the fastest minimization.

Also the spectrum of the dynamical matrix, to which a great attention is dedicated in the whole Chap. 6, does not show relevant signs of dependency on either T of Λ, as shown in Fig. F.2.

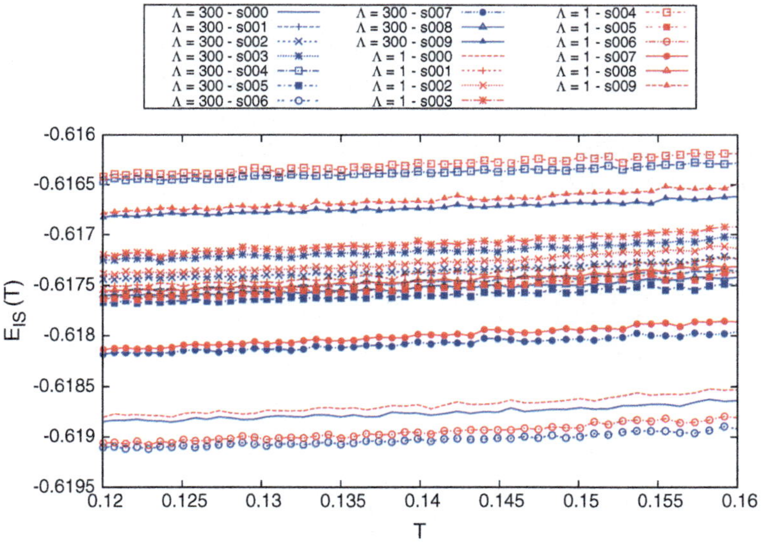

Fig. F.1 Energy of the inherent structure as a function of temperature for 10 samples chosen at random, for $H_{\text{amp}} = 0$. We use the same symbol for the same sample. ISs obtained with $\Lambda = 300$ are in *blue*. *Red* represents $\Lambda = 1$. Sample-to-sample fluctuations are the largest source of dispersion, compared with Λ and T

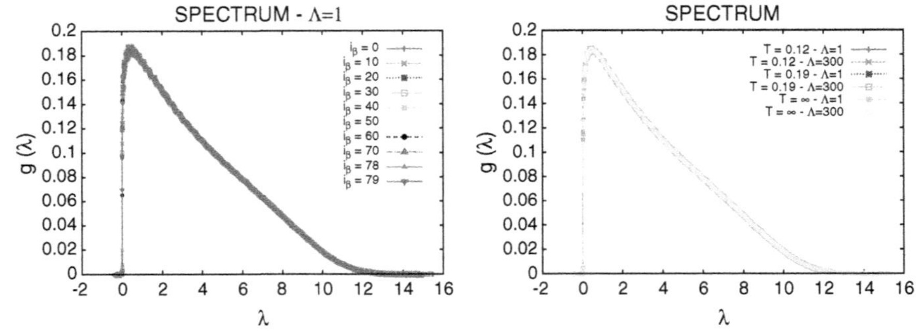

Fig. F.2 Spectrum $\rho(\lambda)$ of the Hessian matrix calculated at the inherent structure for $H_{\text{amp}} = 0$. *Left* $\rho(\lambda)$ for different temperatures from $T = 0.12$ to $T = \infty$. *Right* Comparison of the spectrum between $\Lambda = 1$ and $\lambda = 300$ at $T = 0.12, 0.19, \infty$ vary Λ

F.3 Derivation of $\mathcal{M}$

In this section we derive the expression of the Hessian matrix $\mathcal{M}$ of the Hamiltonian $\mathcal{H}_{\text{RF}}$ (6.1) that we implemented in our programs. In terms of pionic perturbations, recall (6.2), $\mathcal{M}$ would be defined as $\mathcal{M}_{xy}^{\alpha\beta} = \frac{\partial^2 \mathcal{H}_{\text{RF}}}{\partial \pi_{x,\alpha} \pi_{y,\beta}}$. An easy way to extract the Hessian is to write $\mathcal{H}_{\text{RF}}$ as perturbations around the IS and to pick only the second-order terms.

To rewrite $\mathcal{H}_{\text{RF}}$ as a function of the pionic perturbations, it is simpler to compute separately the dot products $\left(\vec{s}_x \cdot \vec{s}_y\right)$ and $\vec{h}_x \cdot \vec{s}_x$. Including the ϵ factors into the perturbation π_x, the generic spin near the IS is expressed as $\vec{s}_x = \vec{s}_x^{\,(\text{IS})}\sqrt{1-\vec{\pi}_x^2} + \vec{\pi}_x$. We can make a second-order expansion of the non-diagonal part of the Hamiltonian by taking the first-order expansion of the square root $\sqrt{1-\vec{\pi}_x^2} \simeq 1 - \vec{\pi}_x^2/2$,

$$
\begin{aligned}
&\left(\vec{s}_x \cdot \vec{s}_y\right) = \qquad\qquad\qquad\qquad\qquad\qquad\qquad\qquad\qquad\qquad\qquad\qquad\qquad \text{(F.3)}\\
&= \left(\vec{s}_x^{\,(\text{IS})}\sqrt{1-\vec{\pi}_x^2} + \vec{\pi}_x\right) \cdot \left(\vec{s}_y^{\,(\text{IS})}\sqrt{1-\vec{\pi}_y^2} + \vec{\pi}_y\right) = \\
&= \sqrt{1-\vec{\pi}_x^2}\sqrt{1-\vec{\pi}_y^2}\left(\vec{s}_x^{\,(\text{IS})} \cdot \vec{s}_y^{\,(\text{IS})}\right) + \sqrt{1-\vec{\pi}_x^2}\left(\vec{\pi}_y \cdot \vec{s}_x^{\,(\text{IS})}\right) + \\
&\qquad\qquad + \sqrt{1-\vec{\pi}_y^2}\left(\vec{\pi}_x \cdot \vec{s}_y^{\,(\text{IS})}\right) + \left(\vec{\pi}_x \cdot \vec{\pi}_y\right) = \\
&= \left(1 - \frac{\vec{\pi}_x^2}{2}\right)\left(1 - \frac{\vec{\pi}_y^2}{2}\right)\left(\vec{s}_x^{\,(\text{IS})} \cdot \vec{s}_y^{\,(\text{IS})}\right) + \left(1 - \frac{\vec{\pi}_x^2}{2}\right)\left(\vec{\pi}_y \cdot \vec{s}_y^{\,(\text{IS})}\right) + \\
&\qquad\qquad + \left(1 - \frac{\vec{\pi}_y^2}{2}\right)\left(\vec{\pi}_y \cdot \vec{s}_x^{\,(\text{IS})}\right) + \left(\vec{\pi}_x \cdot \vec{\pi}_y\right) + o(|\vec{\pi}|^3) \simeq \\
&\simeq \left(\vec{s}_x^{\,(\text{IS})} \cdot \vec{s}_y^{\,(\text{IS})}\right) + \left(\vec{s}_x^{\,(\text{IS})} \cdot \vec{\pi}_y\right) + \left(\vec{s}_y^{\,(\text{IS})} \cdot \vec{\pi}_x\right) + \qquad\qquad\qquad\qquad \text{(F.4)}\\
&\qquad\qquad + \frac{1}{2}\left[\left(-\vec{\pi}_x^2 - \vec{\pi}_y^2\right)\left(\vec{s}_x^{\,(\text{IS})} \cdot \vec{s}_y^{\,(\text{IS})}\right) + 2\vec{\pi}_x \cdot \vec{\pi}_y\right].
\end{aligned}
$$

On the other hand the random-field term is

$$\left(\vec{h}_x \cdot \vec{s}_x\right) = \vec{h}_x \cdot \left(\vec{s}_x^{\,(\mathrm{IS})}\sqrt{1-\vec{\pi}_x^2} + \vec{\pi}_x\right) \quad \simeq \tag{F.5}$$

$$\simeq \vec{h}_x \cdot \left[\vec{s}_x^{\,(\mathrm{IS})}\left(1-\frac{\vec{\pi}_x^2}{2}\right) + \vec{\pi}_x\right] \quad = \left(\vec{h}_x \cdot \vec{s}_x^{\,(\mathrm{IS})}\right) + \left(\vec{h}_x \cdot \vec{\pi}_x\right) - \frac{\vec{\pi}_x^2}{2}\left(\vec{h}_x \cdot \vec{s}_x^{\,(\mathrm{IS})}\right).$$

By inserting Eqs. (F.3, F.5) and taking only the second-order terms we obtain how the Hessian matrix acts on the fields $|\pi\rangle$

$$\frac{1}{2}\left\langle \vec{\pi}_x \right| \mathcal{M} \left| \vec{\pi}_y \right\rangle = \tag{F.6}$$

$$= -\frac{1}{2}\sum_{<x,y>} J_{x,y}\left[\left(-\vec{\pi}_x^2 - \vec{\pi}_y^2\right)\left(\vec{s}_x^{\,(\mathrm{IS})} \cdot \vec{s}_y^{\,(\mathrm{IS})}\right) + 2\vec{\pi}_x \cdot \vec{\pi}_y\right] + \sum_x^N \frac{\vec{\pi}_x^2}{2}\left(\vec{h}_x \cdot \vec{s}_x^{\,(\mathrm{IS})}\right) =$$

$$= \frac{1}{2}\sum_x^N \vec{\pi}_x^2\left[\vec{s}_x^{\,(\mathrm{IS})} \cdot \left(\vec{h}_x^{\,(\mathrm{IS})} + \vec{h}_x\right)\right] + \frac{1}{2}\sum_x \vec{\pi}_x \cdot \sum_{y:|x-y|=1} J_{xy}\vec{\pi}_y,$$

where we called $\vec{h}_x^{\,(\mathrm{IS})}$ the local field of the IS. The just-obtained expression represents a sparse matrix with a matrix element $\mathcal{M}_{xy}$ that comfortably splits as $\mathcal{M}_{xy} = \mathcal{D}_{xy} + \mathcal{N}_{xy}$ into a diagonal term $\mathcal{D}_{xy}$ and a nearest-neighbor one $\mathcal{N}_{xy}$, with

$$\mathcal{D}_{xy} = \delta_{xy}\left[\vec{s}_x^{\,(\mathrm{IS})} \cdot \left(\vec{h}_x^{\,(\mathrm{IS})} + \vec{h}_x\right)\right], \tag{F.7}$$

$$\mathcal{N}_{xy} = -\sum_{\mu=-d}^{d} J_{xy}\delta_{x+\hat{e}_\mu,y}, \tag{F.8}$$

where $\hat{e}_\mu$ is the unit vector towards one of the $2d$ neighbors.

$\mathcal{M}$ in the local reference frame The last step is to get an expression of the Hessian matrix in the local reference frame, that includes the spin normalization constraint.

In the local reference frame the pions are written like $\vec{\pi} = a_1\hat{e}_{1,x} + a_2\hat{e}_{2,x}$ because they are perpendicular to the first vector of the basis, $\vec{s}_x^{\,(\mathrm{IS})}$, and that is why we write them in a two-dimensional representation as $\tilde{\pi} = (a_1, a_2)$ (see Sect. 6.2.2).

In this local basis, the matrix element acting on the pions is written as

$$\vec{\pi}_x\mathcal{M}_{xy}\vec{\pi}_y = (a_{1,x},\ a_{2,x})\begin{pmatrix}\mathcal{M}_{xy}(\hat{e}_{1,x}\cdot\hat{e}_{1,y}) & \mathcal{M}_{xy}(\hat{e}_{2,x}\cdot\hat{e}_{1,y})\\ \mathcal{M}_{xy}(\hat{e}_{1,x}\cdot\hat{e}_{2,y}) & \mathcal{M}_{xy}(\hat{e}_{2,x}\cdot\hat{e}_{2,y})\end{pmatrix}\begin{pmatrix}a_{1,y}\\ a_{2,y}\end{pmatrix}, \tag{F.9}$$

so in the $2N$-dimensional reference $\mathcal{M}$ is expressed as

$$\mathcal{M}_{xy}^{\alpha\beta} = \mathcal{M}_{xy}\left(\hat{e}_{\alpha,x} \cdot \hat{e}_{\beta,y}\right), \tag{F.10}$$

and the elements of the diagonal and nearest-neighbor operators $\mathcal{D}$ and $\mathcal{N}$ become

$$\mathcal{D}^{\alpha\beta}_{xy} = \delta_{xy}\,\delta^{\alpha\beta}\left[\vec{s}_x^{\,(\mathrm{IS})}\cdot\left(\vec{h}_x^{(\mathrm{IS})}+\vec{h}_x\right)\right], \tag{F.11}$$

$$\mathcal{N}^{\alpha\beta}_{xy} = -\sum_{\mu=-d}^{d} J_{xy}\delta_{x+\hat{e}_\mu,y}\left(\hat{e}_{\alpha,x}\cdot\hat{e}_{\beta,y}\right). \tag{F.12}$$

A consistency check A consistency and debugging check we could run with the Hessian matrix is to control that the configurations were actually inherent structures, by verifying that for small perturbations of order ϵ the energy variations were quadratic in ϵ

$$\mathcal{H}-\mathcal{H}(\epsilon) = \frac{\epsilon^2}{2}\langle\pi|\,\mathcal{M}\,|\pi\rangle + o(\epsilon^3). \tag{F.13}$$

References

[Alo96] J.L. Alonso, A. Tarancón, H. Ballesteros, L. Fernández, V. Martín-Mayor, A. Muñoz, Sudupe. Phys. Rev. B **53**, 2537 (1996). doi:10.1103/PhysRevB.53.2537

[Ami05] D.J. Amit, V. Martin-Mayor, *Field Theory, the Renormalization Group and Critical Phenomena*, 3rd edn. (World Scientific, Singapore, 2005). doi:10.1142/9789812775313_bmatter. http://www.worldscientific.com/worldscibooks/10.1142/5715

[BJ11] M. Baity-Jesi, *Energy landscape in three-dimensional Heisenberg spin glasses*. Master's thesis, Sapienza, Universitá di Roma, Rome, Italy (January 2011). arXiv:1503.08409

[BJ14b] M. Baity-Jesi, R.A. Baños, A. Cruz, L.A. Fernandez, J.M. Gil-Narvion, A. Gordillo-Guerrero, D. Iniguez, A. Maiorano, M.F.E. Marinari, V. Martin-Mayor, J. Monforte-Garcia, A. Muñoz, Sudupe, D. Navarro, G. Parisi, S. Perez-Gaviro, M. Pivanti, F. Ricci-Tersenghi, J.J. Ruiz-Lorenzo, S. F. Schifano, B. Seoane, A. Tarancon, R. Tripiccione, D. Yllanes. Phys. Rev. E **89**, 032140 (2014). doi:10.1103/PhysRevE.89.032140. arXiv:1307.4998

[Bal98c] H.G. Ballesteros, V. Martin-Mayor, Phys. Rev. E **58**, 6787 (1998)

[Ber11] M. Bernaschi, G. Parisi, L. Parisi, Comput. Phys. Commun. **182**, 1265–1271 (2011)

[Bro87] F.R. Brown, T.J. Woch, Phys. Rev. Lett. **58**, 2394–2396 (1987). doi:10.1103/PhysRevLett.58.2394

[Dom06] C. de Dominicis, I. Giardina, *Random Fields and Spin Glasses: A Field Theory Approach* (Cambridge University Press, Cambridge, 2006)

[Dom98] C. de Dominicis, I. Kondor, T. Temesvári, *Spin Glasses and Random Fields*, eds. by A.P. Young (World Scientific, Singapore, 1998). arXiv:cond-mat/9705215

[Efr94] B. Efron, R.J. Tibshirani, *An Introduction to Bootstrap* (Chapman & Hall/CRC, London, 1994)

[Fer09b] L.A. Fernandez, V. Martin-Mayor, S. Perez-Gaviro, A. Tarancon, A.P. Young, Phys. Rev. B **80**, 024422 (2009). doi:10.1103/PhysRevB.80.024422

[Fer05] L.A. Fernandez, V. Martin-Mayor, D. Sciretti, A. Tarancón, J. Velasco, Phys. Lett. B **628**, 281 (2005)

[Fer09c] L.A. Fernandez, V. Martin-Mayor, D. Yllanes, Nucl. Phys. B **807**, 424–454 (2009)

[Fis91] K. Fisher, J. Hertz, Spin Glasses (Cambridge University Press. Cambridge (1991). doi:10.1017/CBO9780511628771

[Huk96] K. Hukushima, K. Nemoto, J. Phys. Soc. Japan **65**, 1604 (1996). doi:10.1143/JPSJ.65.1604. arXiv:cond-mat/9512035

[Jac81] L. Jacobs, C. Rebbi, J. Comput. Phys. **41**, 203 (1981). doi:10.1016/0021-9991(81)90089-9

[Knu81] D. Knuth, *The Art of Computer Programming*, vol. 2, 2nd edn. (Addison-Wesley, Reading, Massachussetts, 1981)

[Kra06] W. Krauth, *Statistical Mechanics: Algorithms and Computations* (Oxford University Press, Oxford, 2006)

[LÉ99] P. L' Écuyer, Math. Comp. **68**, 249 (1999)

[Lue94] M. Luescher, P. Weisz, U. Wolff, Comput. Phys. Commun. **79**, 100–110 (1994)

[Lul14] M. Lulli, M. Bernaschi, G. Parisi, (2014). doi:10.1016/j.cpc.2015.06.019. In press, arXiv:1411.0127

[Mar98] E. Marinari, G. Parisi, D. Rossetti: Eur. Phys. J. B **2**, 495 (1998)

[Mar00a] E. Marinari, V. Martin-Mayor, A. Pagnani, Phys. Rev. B **62**, 4999 (2000)

[Mar95] G. Marsaglia, Diehard battery of tests of randomness (1995). http://www.stat.fsu.edu/pub/diehard

[NVI15] NVIDIA Corporation, *Cuda C Programming Guide*, 7.0 edn. (2015). docs.nvidia.com/cuda/cuda-c-programming-guide/index.html

[Oss04] G. Ossola, D. Sokal, Alan. Nucl. Phys. B **691**, 259 (2004)

[Par85] G. Parisi, F. Rapuano, Phys. Lett. B **157**, 301–302 (1985). ISSN:0370-2693. doi:10.1016/0370-2693(85)90670-7. http://www.sciencedirect.com/science/article/pii/0370269385906707

[Par12a] G. Parisi, F. Ricci-Tersenghi, Phil. Mag. **92**, 341 (2012). doi:10.1080/14786435.2011.634843. arXiv:1108.0759v1

[Pix08] J. Pixley, A. Young, Phys. Rev. B **78**, 014419 (2008)

[Riv90] R.J. Rivers, *Path Integral Methods in Quantum Field Theories* (Cambridge University Press, 1990)

[Seo13] B. Seoane, *Spin glasses, the quantum annealing, colloidal glasses and crystals: exploring complex free-energy landscapes*. Ph.D. thesis, Universidad Complutense de Madrid (January 2013)

[Sok92] A. Sokal, *Quantum Fields on the Computer* (Ed. Michael Creutz, World Scientific, Singapore, 1992)

[Yav12] T. Yavors'kii, M. Weigel, Eur. Phys. J. Special Topics **210**, 159 (2012)

[Yll11] D. Yllanes, *Rugged Free-Energy Landscapes in Disordered Spin Systems*. Ph.D. thesis, Universidad Complutense de Madrid (2011). arXiv:1111.0266

[You12] A.P. Young, Everything you wanted to know about data analysis and fitting but were afraid to ask. School on "Efficient Algorithms in Computational Physics", Bad Honnef (September 2012). arXiv:1210.3781v3

Index

M. Baity Jesi, *Spin Glasses*, Springer Theses, DOI 10.1007/978-3-319-41231-3

Printed by Printforce, the Netherlands